VOLUME ONE HUNDRED AND FOURTEEN

INTERNATIONAL REVIEW OF NEUROBIOLOGY

Modern Concepts of Focal Epileptic Networks

INTERNATIONAL REVIEW OF NEUROBIOLOGY

VOLUME 114

SERIES EDITORS

R. ADRON HARRIS

Waggoner Center for Alcohol and Drug Addiction Research
The University of Texas at Austin
Austin, Texas, USA

PETER JENNER

Division of Pharmacology and Therapeutics
GKT School of Biomedical Sciences
King's College, London, UK

VOLUME ONE HUNDRED AND FOURTEEN

INTERNATIONAL REVIEW OF NEUROBIOLOGY

Modern Concepts of Focal Epileptic Networks

Edited by

PREMYSL JIRUSKA

Department of Developmental Epileptology, Institute of Physiology, Academy of Sciences of Czech Republic and Department of Neurology, 2nd Faculty of Medicine, Charles University in Prague, Motol University Hospital, Prague, Czech Republic

MARCO de CURTIS

Department of Epileptology and Experimental Neurophysiology, Fondazione IRCCS, Istituto Neurologico C Besta, Milan, Italy

JOHN G.R. JEFFERYS

Neuronal Networks Group, School of Clinical and Experimental Medicine, University of Birmingham, Birmingham, and Department of Pharmacology, University of Oxford, Oxford, United Kingdom

ELSEVIER

AMSTERDAM • BOSTON • HEIDELBERG • LONDON
NEW YORK • OXFORD • PARIS • SAN DIEGO
SAN FRANCISCO • SINGAPORE • SYDNEY • TOKYO
Academic Press is an imprint of Elsevier

Academic Press is an imprint of Elsevier
225 Wyman Street, Waltham, MA 02451, USA
525 B Street, Suite 1800, San Diego, CA 92101-4495, USA
32 Jamestown Road, London NW1 7BY, UK
The Boulevard, Langford Lane, Kidlington, Oxford OX5 1GB, UK

First edition 2014

Notices
Knowledge and best practice in this field are constantly changing. As new research and experience broaden our understanding, changes in research methods, professional practices, or medical treatment may become necessary.

Practitioners and researchers must always rely on their own experience and knowledge in evaluating and using any information, methods, compounds, or experiments described herein. In using such information or methods they should be mindful of their own safety and the safety of others, including parties for whom they have a professional responsibility.

To the fullest extent of the law, neither the Publisher nor the authors, contributors, or editors, assume any liability for any injury and/or damage to persons or property as a matter of products liability, negligence or otherwise, or from any use or operation of any methods, products, instructions, or ideas contained in the material herein.

ISBN: 978-0-12-418693-4
ISSN: 0074-7742

For information on all Academic Press publications
visit our website at store.elsevier.com

CONTENTS

Contributors *ix*

1. Modern Concepts of Focal Epileptic Networks **1**
Premysl Jiruska, Marco de Curtis, and John G.R. Jefferys

Acknowledgments 5
References 5

2. Neocortical Focus: Experimental View **9**
Igor Timofeev, Sylvain Chauvette, and Sara Soltani

1. Introduction 10
2. Neocortical Neuronal Activities During States of Vigilance 10
3. Neocortical Neuronal Activities During Seizures 15
4. Alterations of Synaptic and Intrinsic Excitability During Seizures 17
5. Unbalance of Excitatory and Inhibitory Influences Leading to Seizure Generation 18
6. Focus of Epilepsy Versus Seizure and Interaction of Epileptic Focus with Intact Cortex 20
7. Neuronal Plasticity Leading to Development of Epilepsy 22
8. Conclusions 25

Acknowledgments 26
References 26

3. Malformations of Cortical Development and Neocortical Focus **35**
Heiko J. Luhmann, Werner Kilb, and Hans Clusmann

1. Introduction 36
2. Normal and Abnormal Development of the Cerebral Cortex 37
3. The (Un)identified Neocortical Focus 45
4. Removal of a Neocortical Focus 50
5. Conclusion 53

Acknowledgments 53
References 53

4. Limbic Networks and Epileptiform Synchronization: The View from the Experimental Side **63**

Charles Behr, Margherita D'Antuono, Shabnam Hamidi, Rochelle Herrington, Maxime Lévesque, Pariya Salami, Zahra Shiri, Rüdiger Köhling, and Massimo Avoli

1. Background 64
2. Mechanisms Underlying Epileptiform Synchronization 66
3. Interactions Between Limbic Areas 73
4. Seizure-Onset Types 75
5. Modulation of Epileptiform Synchronization by Neurosteroids 78
6. Conclusions 81

Acknowledgments 81

References 81

5. Limbic Networks: Clinical Perspective **89**

Aylin Y. Reid and Richard J. Staba

1. Introduction 90
2. Clinical Features of Limbic Epilepsy 91
3. Structural Networks in Limbic Epilepsy 95
4. Functional Networks in Limbic Epilepsy 97
5. Metabolic Networks in Limbic Epilepsy 109
6. Conclusions 111

References 112

6. Modern Concepts of Seizure Modeling **121**

Christophe Bernard, Sebastien Naze, Timothée Proix, and Viktor K. Jirsa

1. Introduction 122
2. Detailed Biophysical Models 122
3. Lumped Models 137
4. Analogies 146
5. Discussion 148

Acknowledgments 149

References 149

7. Mechanisms of Ictogenesis **155**

Thomas Blauwblomme, Premysl Jiruska, and Gilles Huberfeld

1. Introduction 156
2. Recording the Brain and its Networks During the Transition to Seizures 159

3. Cellular Substrates of Ictogenesis: Neurons and Neurotransmission 162
4. The Environment of Neurons: Ions, Astrocytes, and the Extracellular Space 170
5. Conclusion 174
Acknowledgments 175
References 175

8. Seizure Termination 187

Frédéric Zubler, Andreas Steimer, Heidemarie Gast, and Kaspar A. Schindler

1. Introduction 188
2. Metabolic Mechanisms of Seizure Termination 189
3. Network Aspects of Seizure Termination 193
4. Conclusions 202
References 203

9. Epileptic Focus and Alteration of Metabolism 209

Jakub Otáhal, Jaroslava Folbergrová, Richard Kovacs, Wolfram S. Kunz, and Nicola Maggio

1. Introduction 210
2. Acute Changes in Energy Metabolism During Seizures 212
3. Metabolic Dysfunction of Chronic Epileptic Tissue 216
4. Alteration of Mitochondrial Genome in Chronic Epileptic Tissue 220
5. Changes in Blood Flow Regulation in Epileptic Tissue 224
6. Alteration of Blood Brain Barrier 228
7. Conclusions 232
Acknowledgments 233
References 233

10. Modern Techniques of Epileptic Focus Localization 245

Lukas Martinkovic, Hrvoje Hecimovic, Vlastimil Sulc, Radek Marecek, and Petr Marusic

1. Introduction 246
2. Structural Imaging, Volumetrics, and Morphometric Analysis 247
3. Functional Neuroimaging in Epilepsies Using PET 253
4. Ictal Perfusion 260
5. Irritative Zone as Assessed by BOLD fMRI 265
6. Conclusions 271
Acknowledgments 271
References 271

11. From Treatment to Cure: Stopping Seizures, Preventing Seizures, and Reducing Brain Propensity to Seize **279**

Ivan Pavlov and Stephanie Schorge

1. From Anticonvulsant Therapy to Disease-Modifying Treatment 280
2. Seizure Prevention: Toward Personalized Medicine to Stop Seizures from Happening 280
3. Seizure Prevention: Stopping Epileptogenesis 282
4. Reducing Potential Adverse Effects by Treating Seizures in Real Time 283
5. Real-Time Seizure Treatments: What to Deliver? 283
6. Optogenetics 284
7. Chemical Genetics 287
8. What to Hit: Are Interneurons a Good Target for Interventions? 288
9. Gene Therapy: Introducing Genes to Selected Neurons 290
10. Treatments to Restore Excitability: Is there a Possibility of a Cure? 291
11. How Far to Cure(s)? 293

Acknowledgments 294

References 294

Index *301*

Contents of Recent Volumes *307*

CONTRIBUTORS

Massimo Avoli
Department of Neurology, Neurosurgery and Physiology, Montréal Neurological Institute, Montréal, Québec, Canada, and Department of Experimental Medicine, Facoltà di Medicina e Odontoiatria, Sapienza Università di Roma, Roma, Italy

Charles Behr
Department of Neurology, Neurosurgery and Physiology, Montréal Neurological Institute, Montréal, Québec, Canada

Christophe Bernard
Institut de Neurosciences des Systèmes, and Inserm UMR_S 1106, Aix Marseille Université, Marseille, France

Thomas Blauwblomme
Neurosurgery Unit, Hopital Necker—Enfants Malades, APHP; Université Paris Descartes; INSERM U1129—Infantile Epilepsies and Brain Plasticity; University Paris Descartes, PRES Sorbonne Paris Cité, and CEA, Gif sur Yvette, Paris, France

Sylvain Chauvette
Centre de recherche de l'Institut universitaire en santé mentale de Québec (CRIUSMQ), Université Laval, Québec, Canada

Hans Clusmann
Department of Neurosurgery, RWTH Aachen University, Aachen, Germany

Margherita D'Antuono
Department of Neurology, Neurosurgery and Physiology, Montréal Neurological Institute, Montréal, Québec, Canada

Marco de Curtis
Department of Epileptology and Experimental Neurophysiology, Fondazione IRCCS, Istituto Neurologico C Besta, Milan, Italy

Jaroslava Folbergrová
Institute of Physiology, v.v.i., Academy of Sciences of the Czech Republic, Prague, Czech Republic

Heidemarie Gast
Department of Neurology, Inselspital, University Hospital Bern, University of Bern, Bern, Switzerland

Shabnam Hamidi
Department of Neurology, Neurosurgery and Physiology, Montréal Neurological Institute, Montréal, Québec, Canada

Hrvoje Hecimovic
Zagreb Epilepsy Center, Department of Neurology, University Hospital, Zagreb, Croatia

Rochelle Herrington
Department of Neurology, Neurosurgery and Physiology, Montréal Neurological Institute, Montréal, Québec, Canada

Gilles Huberfeld
INSERM U1129—Infantile Epilepsies and Brain Plasticity; University Paris Descartes, PRES Sorbonne Paris Cité; CEA, Gif sur Yvette; Clinical Neurophysiology Department, CHU Pitié-Salpêtrière, APHP, and Université Pierre et Marie Curie, Paris, France

John G.R. Jefferys
Neuronal Networks Group, School of Clinical and Experimental Medicine, University of Birmingham, Birmingham, and Department of Pharmacology, University of Oxford, Oxford, United Kingdom

Viktor K. Jirsa
Institut de Neurosciences des Systèmes, and Inserm UMR_S 1106, Aix Marseille Université, Marseille, France

Premysl Jiruska
Department of Developmental Epileptology, Institute of Physiology, Academy of Sciences of Czech Republic, and Department of Neurology, 2nd Faculty of Medicine, Charles University in Prague, Motol University Hospital, Prague, Czech Republic

Werner Kilb
Institute of Physiology, University Medical Center of the Johannes Gutenberg University, Mainz, Germany

Rüdiger Köhling
Institute of Physiology, University of Rostock, Rostock, Germany

Richard Kovacs
Institute for Neurophysiology, Charité–Medical University Berlin, Berlin, Germany

Wolfram S. Kunz
Department of Epileptology, University of Bonn, Bonn, Germany

Maxime Lévesque
Department of Neurology, Neurosurgery and Physiology, Montréal Neurological Institute, Montréal, Québec, Canada

Heiko J. Luhmann
Institute of Physiology, University Medical Center of the Johannes Gutenberg University, Mainz, Germany

Nicola Maggio
Department of Neurology, The Joseph Sagol Neuroscience Center, and Talpiot Medical Leadership Program, The Chaim Sheba Medical Center, Tel HaShomer, Israel

Radek Marecek
CEITEC—Central European Institute of Technology, and Brno Epilepsy Center, Department of Neurology, St. Anne's University Hospital and Faculty of Medicine, Masaryk University, Brno, Czech Republic

Lukas Martinkovic
Department of Neurology, Charles University in Prague, 2nd Faculty of Medicine, Motol University Hospital, Prague, Czech Republic

Petr Marusic
Department of Neurology, Charles University in Prague, 2nd Faculty of Medicine, Motol University Hospital, Prague, Czech Republic

Sebastien Naze
Institut de Neurosciences des Systèmes, and Inserm UMR_S 1106, Aix Marseille Université, Marseille, France

Jakub Otáhal
Institute of Physiology, v.v.i., Academy of Sciences of the Czech Republic, Prague, Czech Republic

Ivan Pavlov
Department of Clinical and Experimental Epilepsy, UCL Institute of Neurology, London, United Kingdom

Timothée Proix
Institut de Neurosciences des Systèmes, and Inserm UMR_S 1106, Aix Marseille Université, Marseille, France

Aylin Y. Reid
Department of Neurology, David Geffen School of Medicine, University of California Los Angeles, Los Angeles, California, USA

Pariya Salami
Department of Neurology, Neurosurgery and Physiology, Montréal Neurological Institute, Montréal, Québec, Canada

Kaspar A. Schindler
Department of Neurology, Inselspital, University Hospital Bern, University of Bern, Bern, Switzerland

Stephanie Schorge
Department of Clinical and Experimental Epilepsy, UCL Institute of Neurology, London, United Kingdom

Zahra Shiri
Department of Neurology, Neurosurgery and Physiology, Montréal Neurological Institute, Montréal, Québec, Canada

Sara Soltani
Centre de recherche de l'Institut universitaire en santé mentale de Québec (CRIUSMQ), Université Laval, Québec, Canada

Richard J. Staba
Department of Neurology, David Geffen School of Medicine, University of California Los Angeles, Los Angeles, California, USA

Andreas Steimer
Department of Neurology, Inselspital, University Hospital Bern, University of Bern, Bern, Switzerland

Vlastimil Sulc
Department of Neurology, Charles University in Prague, 2nd Faculty of Medicine, Motol University Hospital, Prague, Czech Republic

Igor Timofeev
Centre de recherche de l'Institut universitaire en santé mentale de Québec (CRIUSMQ), and Department of Psychiatry and Neuroscience, Université Laval, Québec, Canada

Frédéric Zubler
Department of Neurology, Inselspital, University Hospital Bern, University of Bern, Bern, Switzerland

CHAPTER ONE

Modern Concepts of Focal Epileptic Networks

Premysl Jiruska*,†,1, Marco de Curtis‡, John G.R. Jefferys§,¶

*Department of Developmental Epileptology, Institute of Physiology, Academy of Sciences of Czech Republic, Prague, Czech Republic

†Department of Neurology, 2nd Faculty of Medicine, Charles University in Prague, Motol University Hospital, Prague, Czech Republic

‡Department of Epileptology and Experimental Neurophysiology, Fondazione IRCCS, Istituto Neurologico C Besta, Milan, Italy

§Neuronal Networks Group, School of Clinical and Experimental Medicine, University of Birmingham, Birmingham, United Kingdom

¶Department of Pharmacology, University of Oxford, Oxford, United Kingdom

[1]Corresponding author: e-mail address: jiruskapremysl@gmail.com

Contents

Acknowledgments 5
References 5

Early experiments with topical application of convulsants and with local cortical lesions suggested that epileptic activity and seizures can be generated within a highly restricted cortical region, defined as the epileptic focus. Such foci were thought to contain populations of abnormally behaving cells (neurons and glia) that sustain a range of hyperexcitable phenomena including seizures, interictal epileptiform discharges, and pathological network oscillations. This concept was further supported by pioneering recordings and surgical resections from human brain during the early days of epilepsy surgery by Penfield and Jasper (1954). Seizures originating from the epileptic focus were termed focal seizures and the corresponding epilepsy was classified as focal epilepsy (Commission on Classification and Terminology of the International League Against Epilepsy, 1981). Focal seizures can propagate outside the focus to secondarily involve regions that are not functionally altered but are recruited by the epileptiform discharge. When discharge propagation is widespread and also involves subcortical areas, secondary generalization occurs. Altered network dynamics at the focus and local/distant effects of epileptiform activity can disrupt ongoing physiological processes

International Review of Neurobiology, Volume 114
ISSN 0074-7742
http://dx.doi.org/10.1016/B978-0-12-418693-4.00001-7

and may result in neurological and cognitive deficits observed in some patients with epilepsy.

Neurobiological research in the field of epilepsy aims to identify specific structural, functional, or genetic abnormalities that can reliably explain how focal epilepsy develops (the study of epileptogenesis) and what are the main mechanisms responsible for seizure initiation (the study of ictogenesis). Advances in experimental and clinical studies brought novel insights into the cellular dynamics and network organization of the epileptic brain that question the concept of the epileptic focus. Specifically, experience from surgical treatment has shown that the concept of a restricted focus is not optimal for planning of epilepsy surgery and could be responsible for failure to achieve seizure freedom in a substantial population of patients who underwent resection. Intracranial recordings demonstrate that the region involved in the generation of seizures and interictal events often includes nonlesional areas and involves spatially distant regions within the same or different lobes. The concept of the epileptic focus was, therefore, redefined and replaced by the identification and the definition of overlapping pathological and pathophysiological zones that generate epileptiform activities: the seizure-onset zone, the irritative zone (the region that generates interictal discharges), the epileptogenic lesion, and the epileptogenic zone, the resection or disconnection of which is necessary and sufficient for seizure freedom (Kahane, Landre, Minotti, Francione, & Ryvlin, 2006; Rosenow & Luders, 2001). The importance of altered network organization in focal epilepsies was stressed by Spencer (2002) and the new classification proposal of epilepsies insists on the concept of the networks when focal epilepsies are discussed (Berg et al., 2010). More recently, the concept of "system epilepsy" was introduced (Avanzini et al., 2012) that suggests that specific networks are prone to generate seizures, possibly only when a part of the network is damaged or functionally altered. The existence of system-specific susceptibility to seizures is also supported by the demonstration that systemic applications of proconvulsive drugs selectively alter specific networks and induce interictal and ictal epileptiform patterns segregated into specific cortical systems (Boido, Jesuthasan, de Curtis, & Uva, 2014; Carriero et al., 2012).

The seeming contradiction between the concepts of epileptic focus and of epileptic networks can be reconciled when time is considered. In hypothetical terms, very localized, focal changes in excitability could in principle develop after an acute injury and the evolution of the acute damage, together with the occurrence of seizure-like discharges, may establish the later development of network changes. Acute recordings in patients with hemorrhagic strokes show that seizure patterns recorded during the acute stage differ from late seizures (Dreier et al., 2012).

The importance of the network concept was substantially bolstered by introduction into epilepsy research of approaches from the fields of complex dynamics of networks and graph theory (Chapter 6; Bullmore & Sporns, 2009; van Diessen, Diederen, Braun, Jansen, & Stam, 2013). This new and rapidly expanding mathematical field has a substantial impact on epilepsy research, and on neuroscience in general. It further demonstrates how important connectivity is for understanding the abnormal behavior generated within epileptic networks and how structural and functional connectivity can shape the epileptic phenomena (Stefan & Lopes da Silva, 2013; Wendling, Chauvel, Biraben, & Bartolomei, 2010). Experimental and clinical findings, together with the advanced use of mathematical and physical approaches in epilepsy research, revealed that epilepsy and seizures are very complex dynamical phenomena and that understanding these processes requires integrating together information from different spatial and temporal domains (Jiruska et al., 2013; Chapter 8). The importance of complexity is well demonstrated by genetic studies. Several genes have been identified within families in which epilepsy occurred across several generations (Lerche et al., 2013). However, people within families sharing the same mutations were identified but some of them did not develop epileptic phenotypes (Lerche et al., 2013). The twentieth century reductionist approach to identify specific genes responsible for diseases proved disappointing in many types of epilepsy, and it is now widely accepted that the epileptic phenotype is the result of complex interactions between genes and cellular networks within the organism, and environmental factors. In addition, development of antiepileptic drugs targeting specific mechanisms implicated in seizure initiation failed to live up to expectations and raised the possibility of much more complex mechanisms being responsible for seizures (Brodie et al., 2011). It is equally necessary to consider multiple factors and complex interactions for understanding how seizures are generated. It is well known that seizure (network) patterns observed *in vivo* are poorly reproduced in slices, suggesting that wider networks and more complex interactions are required to generate specific patterns of seizures.

On the local (cellular) scale, epileptiform phenomena are the result of the complex interaction between multiple neuronal subtypes. It is well known that behavior of isolated cells may substantially change when the cells are mutually connected; the pattern of connection determines the population behavior. One of the best examples comes from interneurons which, if connected into networks, generate spontaneous oscillations in the gamma band (Whittington, Traub, & Jefferys, 1995). In epilepsy, for a long time it was assumed that epileptic seizures are caused mainly by altered dynamics within the network of epileptic pyramidal neurons; this excitatory theory dominated the field of epilepsy research for decades. Traditionally, epilepsy is described as an imbalance between excitation and inhibition (Westbrook, 1991). Molecular reorganization of pyramidal cells, changes in intrinsic properties increasing neuronal excitability and newly developed pathological communication between pyramidal cells were seen as the main causes of this imbalance. The second factor of this imbalance represented weakened inhibition due to loss of specific interneurons, loss of excitatory drive onto interneurons, etc. (Pavlov, Kaila, Kullmann, & Miles, 2013; Sloviter, 1987; Vreugdenhil, Hack, Draguhn, & Jefferys, 2002). For many years, the shift of the balance between excitation and inhibition toward enhanced excitation dominated the theories on the interictal behavior within the focus and on the mechanisms responsible for initiation of seizures. Contemporary research implicates interneurons in playing more causal roles in seizure initiation and, paradoxically, has shown that intense activity of interneurons may induce complex changes that alter potassium and chloride homeostasis resulting in increased excitability, depolarization, and synchronization of principal cells and a shift brain dynamics toward the seizure (Chapter 7; de Curtis & Gnatkovsky, 2009; Gnatkovsky, Librizzi, Trombin, & de Curtis, 2008; Pavlov et al., 2013).

Glial cells and their interactions with the neuronal compartment play a crucial role in normal brain function and dysfunction. Altered glia has been demonstrated in several forms of experimental and human focal epilepsy (Losi, Cammarota, & Carmignoto, 2012). It was also demonstrated that altered glia–neuron interaction is required for the generation, maintenance, and propagation of focal seizures (Carmignoto & Haydon, 2012; Devinsky, Vezzani, Najjar, De Lanerolle, & Rogawski, 2013; Steinhauser, Seifert, & Bedner, 2012; Tian et al., 2005). The complexity of the interactions responsible for focal seizure generation and focal epileptogenesis is further expanded when the contributions of the extracellular compartment (Dityatev & Fellin, 2008; Jefferys, 1995) and of the neuro-glia-vascular unit (Chapter 9; Stanimirovic & Friedman, 2012) to pathological hyperexcitability are considered.

Interactions between the network components may influence transition to seizure. The network concept is equally important for neocortical epilepsies (Chapter 2) and especially in epilepsies which are due to abnormalities of cortical development, where the pathological network organization largely depends on the nature of the malformation (Chapter 3). Experimentally, it has also been demonstrated that temporal lobe epilepsy in particular involves several limbic structures several, or all, of which may initiate seizures in each patient (Chapters 4 and 5). In human mesial temporal lobe epilepsy, the entire lobe is functionally affected (Bartolomei, Chauvel, & Wendling, 2008) and the histopathological damage cannot be considered focal, but involves different hippocampal and parahippocampal structures with variable longitudinal extent (Blumcke, Coras, Miyata, & Ozkara, 2012).

Is the knowledge about the dynamical interactions within epileptic networks critical for development of new and more effective therapy? Contemporary science and technological advances open prospects for developing novel therapeutic approaches particularly in patients with drug-resistant epilepsy. Such novel approaches may include delivery of genes controlling cell excitability (including optogenetics), new techniques of brain stimulation, local drug delivery, and implantation of stem cells to restore the specific cell loss (Chapter 11). A fundamental advantage of some of these therapies is that they are not anticonvulsant (i.e., targeting at the main symptom) but rather they target the epilepsy as a disease process thus having disease-modifying (and potentially curative) effects. However, these therapies must be targeted to appropriate areas of the epileptic network. This can be achieved only by precisely understanding the dynamics and interactions within the epileptic circuits (Chapters 6, 8, and 10). This involves not only understanding the electrophysiology and function of the neurons but also metabolic networks. It is being repeatedly demonstrated that epileptic tissue displays alterations of metabolism and mitochondrial function and is often observed across various models of epilepsy and in human patients (Chapter 9).

ACKNOWLEDGMENTS

This work was supported by grants from the Ministry of Health of the Czech Republic (IGA NT/14489-3), Neuron Fund for Support of Science (Czech Republic, 2012/10), Czech Science Foundation (Czech Republic, P303/14-02634S), Epilepsy Research UK (P1102), Medical Research Council of the UK (G0802162), and Italian Health Ministry grants (Ricerca Corrente 2012-2014 and RF 151; EPICARE of Associazione Paolo Zorzi per le Neuroscience).

REFERENCES

Avanzini, G., Manganotti, P., Meletti, S., Moshe, S. L., Panzica, F., Wolf, P., et al. (2012). The system epilepsies: A pathophysiological hypothesis. *Epilepsia*, *53*, 771–778.

Bartolomei, F., Chauvel, P., & Wendling, F. (2008). Epileptogenicity of brain structures in human temporal lobe epilepsy: A quantified study from intracerebral EEG. *Brain*, *131*, 1818–1830.

Berg, A. T., Berkovic, S. F., Brodie, M. J., Buchhalter, J., Cross, J. H., van Emde Boas, W., et al. (2010). Revised terminology and concepts for organization of seizures and epilepsies: Report of the ILAE Commission on Classification and Terminology, 2005-2009. *Epilepsia, 51*, 676–685.

Blumcke, I., Coras, R., Miyata, H., & Ozkara, C. (2012). Defining clinico-neuropathological subtypes of mesial temporal lobe epilepsy with hippocampal sclerosis. *Brain Pathology, 22*, 402–411.

Boido, D., Jesuthasan, N., de Curtis, M., & Uva, L. (2014). Network dynamics during the progression of seizure-like events in the hippocampal-parahippocampal regions. *Cerebral Cortex, 24*, 163–173.

Brodie, M. J., Covanis, A., Gil-Nagel, A., Lerche, H., Perucca, E., Sills, G. J., et al. (2011). Antiepileptic drug therapy: Does mechanism of action matter? *Epilepsy and Behavior, 21*, 331–341.

Bullmore, E., & Sporns, O. (2009). Complex brain networks: Graph theoretical analysis of structural and functional systems. *Nature Reviews Neuroscience, 10*, 186–198.

Carmignoto, G., & Haydon, P. G. (2012). Astrocyte calcium signaling and epilepsy. *Glia, 60*, 1227–1233.

Carriero, G., Arcieri, S., Cattalini, A., Corsi, L., Gnatkovsky, V., & de Curtis, M. (2012). A guinea pig model of mesial temporal lobe epilepsy following nonconvulsive status epilepticus induced by unilateral intrahippocampal injection of kainic acid. *Epilepsia, 53*, 1917–1927.

Commission on Classification and Terminology of the International League Against Epilepsy (1981). Proposal for revised clinical and electroencephalographic classification of epileptic seizures. From the Commission on Classification and Terminology of the International League Against Epilepsy. *Epilepsia, 22*, 489–501.

de Curtis, M., & Gnatkovsky, V. (2009). Reevaluating the mechanisms of focal ictogenesis: The role of low-voltage fast activity. *Epilepsia, 50*, 2514–2525.

Devinsky, O., Vezzani, A., Najjar, S., De Lanerolle, N. C., & Rogawski, M. A. (2013). Glia and epilepsy: Excitability and inflammation. *Trends in Neurosciences, 36*, 174–184.

Dityatev, A., & Fellin, T. (2008). Extracellular matrix in plasticity and epileptogenesis. *Neuron Glia Biology, 4*, 235–247.

Dreier, J. P., Major, S., Pannek, H. W., Woitzik, J., Scheel, M., Wiesenthal, D., et al. (2012). Spreading convulsions, spreading depolarization and epileptogenesis in human cerebral cortex. *Brain, 135*, 259–275.

Gnatkovsky, V., Librizzi, L., Trombin, F., & de Curtis, M. (2008). Fast activity at seizure onset is mediated by inhibitory circuits in the entorhinal cortex in vitro. *Annals of Neurology, 64*, 674–686.

Jefferys, J. G. (1995). Nonsynaptic modulation of neuronal activity in the brain: Electric currents and extracellular ions. *Physiological Reviews, 75*, 689–723.

Jiruska, P., de Curtis, M., Jefferys, J. G., Schevon, C. A., Schiff, S. J., & Schindler, K. (2013). Synchronization and desynchronization in epilepsy: Controversies and hypotheses. *Journal of Physiology, 591*, 787–797.

Kahane, P., Landre, E., Minotti, L., Francione, S., & Ryvlin, P. (2006). The Bancaud and Talairach view on the epileptogenic zone: A working hypothesis. *Epileptic Disorders, 8*(Suppl. 2), S16–S26.

Lerche, H., Shah, M., Beck, H., Noebels, J., Johnston, D., & Vincent, A. (2013). Ion channels in genetic and acquired forms of epilepsy. *Journal of Physiology, 591*, 753–764.

Losi, G., Cammarota, M., & Carmignoto, G. (2012). The role of astroglia in the epileptic brain. *Frontiers in Pharmacology, 3*, 132.

Pavlov, I., Kaila, K., Kullmann, D. M., & Miles, R. (2013). Cortical inhibition, pH and cell excitability in epilepsy: What are optimal targets for antiepileptic interventions? *Journal of Physiology, 591*, 765–774.

Penfield, W., & Jasper, H. (1954). *Epilepsy and the functional anatomy of the human brain.* Boston: Little Brown.

Rosenow, F., & Luders, H. (2001). Presurgical evaluation of epilepsy. *Brain*, *124*, 1683–1700.

Sloviter, R. S. (1987). Decreased hippocampal inhibition and a selective loss of interneurons in experimental epilepsy. *Science*, *235*, 73–76.

Spencer, S. S. (2002). Neural networks in human epilepsy: Evidence of and implications for treatment. *Epilepsia*, *43*, 219–227.

Stanimirovic, D. B., & Friedman, A. (2012). Pathophysiology of the neurovascular unit: Disease cause or consequence? *Journal of Cerebral Blood Flow and Metabolism*, *32*, 1207–1221.

Stefan, H., & Lopes da Silva, F. H. (2013). Epileptic neuronal networks: Methods of identification and clinical relevance. *Frontiers in Neurology*, *4*, 8.

Steinhauser, C., Seifert, G., & Bedner, P. (2012). Astrocyte dysfunction in temporal lobe epilepsy: $K+$ channels and gap junction coupling. *Glia*, *60*, 1192–1202.

Tian, G. F., Azmi, H., Takano, T., Xu, Q., Peng, W., Lin, J., et al. (2005). An astrocytic basis of epilepsy. *Nature Medicine*, *11*, 973–981.

van Diessen, E., Diederen, S. J., Braun, K. P., Jansen, F. E., & Stam, C. J. (2013). Functional and structural brain networks in epilepsy: What have we learned? *Epilepsia*, *54*, 1855–1865.

Vreugdenhil, M., Hack, S. P., Draguhn, A., & Jefferys, J. G. (2002). Tetanus toxin induces long-term changes in excitation and inhibition in the rat hippocampal CA1 area. *Neuroscience*, *114*, 983–994.

Wendling, F., Chauvel, P., Biraben, A., & Bartolomei, F. (2010). From intracerebral EEG signals to brain connectivity: Identification of epileptogenic networks in partial epilepsy. *Frontiers in Systems Neuroscience*, *4*, 154.

Westbrook, G. L. (1991). Seizures and epilepsy. In E. R. Kandel, T. M. Jessel, & J. H. Schwartz (Eds.), *Principles of neural science* (pp. 910–935). New York: McGraw-Hill.

Whittington, M. A., Traub, R. D., & Jefferys, J. G. (1995). Synchronized oscillations in interneuron networks driven by metabotropic glutamate receptor activation. *Nature*, *373*, 612–615.

CHAPTER TWO

Neocortical Focus: Experimental View

Igor Timofeev*^,†,1, Sylvain Chauvette*, Sara Soltani*
*Centre de recherche de l'Institut universitaire en santé mentale de Québec (CRIUSMQ), Université Laval, Québec, Canada
†Department of Psychiatry and Neuroscience, Université Laval, Québec, Canada
[1]Corresponding author: e-mail address: igor.timofeev@phs.ulaval.ca

Contents

1. Introduction 10
2. Neocortical Neuronal Activities During States of Vigilance 10
3. Neocortical Neuronal Activities During Seizures 15
4. Alterations of Synaptic and Intrinsic Excitability During Seizures 17
5. Unbalance of Excitatory and Inhibitory Influences Leading to Seizure Generation 18
6. Focus of Epilepsy Versus Seizure and Interaction of Epileptic Focus with Intact Cortex 20
7. Neuronal Plasticity Leading to Development of Epilepsy 22
8. Conclusions 25
Acknowledgments 26
References 26

Abstract

All brain normal or pathological activities occur in one of the states of vigilance: wake, slow-wave sleep, or REM sleep. Neocortical seizures preferentially occur during slow-wave sleep. We provide a description of neuronal behavior and mechanisms mediating such a behavior within neocortex taking place in natural states of vigilance as well as during seizures pointing to similarities and differences exhibited during sleep and seizures. A concept of epileptic focus is described using a model of cortical undercut, because in that model, the borders of the focus are well defined. In this model, as in other models of acquired epilepsy, the main factor altering excitability is deafferentation, which upregulates neuronal excitability that promotes generation of seizures. Periods of disfacilitation recorded during slow-wave sleep further upregulate neuronal excitability. It appears that the state of neurons and neuronal network in the epileptic focus produced by deafferentation are such that seizures cannot be generated there. Instead, seizures always start around the perimeter of the undercut cortex. Therefore, we define these areas as the seizure focus. In this zone, neuronal connectivity and excitability are moderately enhanced, lowering the threshold for seizure generation.

International Review of Neurobiology, Volume 114
ISSN 0074-7742
http://dx.doi.org/10.1016/B978-0-12-418693-4.00002-9

1. INTRODUCTION

Epilepsy is a word grouping several neurological disorders. The main common feature of these disorders is the occurrence of unprovoked seizures. A seizure is an episode of abnormal (paroxysmal) brain electrical activity. Usually, abnormal local and then long-range synchronization accompanies seizures. The mechanisms of local versus long-range synchronization are different (Timofeev, Bazhenov, Seigneur, & Sejnowski, 2012). Convulsive seizures are associated with body movements, while nonconvulsive seizures (i.e., absence seizures) are not. There are two main causes of epilepsy: genetically determined and acquired or secondary. In the majority of genetically determined epilepsies, the altered genes are present in every cell of the body (including the brain), increasing the likelihood that these epilepsies will be generalized. Recent progress indicates that genome as a whole and not single genes are responsible for epileptic encephalopathies (Epi K. Consortium Epilepsy Phenome/Genome P, 2013). It suggests that genetic epilepsies might also lead to focal epilepsies. In contrast, acquired epilepsies are usually primarily focal. Brain trauma, stroke, brain tumor, and infections are leading causes of acquired epilepsy. Therefore, the common feature of conditions triggering acquired epilepsy is a local lesion. Two main processes underlay all brain functions: excitation and inhibition. In neocortical system, roughly 80% of neurons are excitatory (Gabbott & Somogyi, 1986) and all long-range connections are excitatory too. Even if some interneurons might be more sensitive to pathological conditions induced by cortical lesions (Avramescu, Nita, & Timofeev, 2009), the major damage is done to the excitatory network. Therefore, an overall reduction of excitation is a major factor accompanying lesion-dependent epilepsy.

2. NEOCORTICAL NEURONAL ACTIVITIES DURING STATES OF VIGILANCE

All brain activities, whether normal or pathological, are executed in one of three states of vigilance: wake, slow-wave sleep, or REM sleep. Neocortical focal epilepsy is usually nocturnal (Timofeev, 2011); therefore, below we describe the major differences in activities of cortical neurons exhibited during the three states of vigilance.

During the waking state the EEG is activated, the membrane potential of neurons is relatively stable (≈ -62 mV) (Steriade, Timofeev, & Grenier,

2001; Timofeev, Grenier & Steriade, 2000, 2001b) and the neurons occasionally fire action potentials. Extracellular unit recordings show that overall firing rates of unidentified cortical neurons during quiet wakefulness are higher than during sleep (Vyazovskiy et al., 2009). Intracellular recordings mainly from deep cortical neurons showed similar firing rates during quiet wakefulness, slow-wave, and REM sleep (Steriade et al., 2001). However, layer-specific analyses indicate that spontaneous firing during waking in upper layers is extremely low ($\ll$1 Hz) and that it is much higher in deeply lying neurons (Barth & Poulet, 2012; Chauvette, Volgushev, & Timofeev, 2010). Evidently, the firing is controlled by interactions of excitatory and inhibitory influences integrated over neuronal receptive fields. In slices and anesthetized preparations, the overall excitatory and inhibitory influences upon cortical neurons are balanced (Haider, Duque, Hasenstaub, & McCormick, 2006; Shu, Hasenstaub, & McCormick, 2003). During waking and sleep states, they are balanced too, but the balance is shifted and inhibition dominates at the level of neuronal soma (Haider, Hausser, & Carandini, 2013; Rudolph, Pospischil, Timofeev, & Destexhe, 2007); in these conditions action potentials are generated when inhibition decreases (Rudolph et al., 2007) and not when excitation increases as could be intuitively expected.

The state of sleep is composed of two main phases: REM and non-REM. Sleep is characterized by behavioral unresponsiveness. Similar to the waking state, during REM sleep the EEG is activated, and the membrane potential and firing of cortical neurons also are similar to the waking state (Steriade et al., 2001; Timofeev, Grenier & Steriade, 2001a). Cortical evoked potentials elicited by peripheral stimulation during REM sleep are similar to those evoked during wakefulness (Rector, Schei, & Rojas, 2009); however, evoked potentials elicited by electrical stimulation of prethalamic ascending fibers were much higher during REM as compared to the waking state (Chauvette, Seigneur, & Timofeev, 2012), suggesting a major blockage of peripheral information at the brainstem level and its significant amplification at the level of thalamocortical system. Indeed, the membrane potential of thalamocortical neurons during REM sleep is much more depolarized as than during slow-wave sleep (Hirsch, Fourment, & Marc, 1983), which might provide a mechanism for a very high responsiveness of the thalamocortical system during REM sleep.

Non-REM sleep is currently divided into three stages: stage 1, stage 2, and slow-wave sleep (Iber, Ancoli-Israel, Chesson, & Quan, 2007). Stage 1 is usually very short and it is characterized by a transition between alpha and theta activities. Theta rhythm is generated in hippocampal formation (Buzsaki, 2002) the structure that is out of scope of the current review. Stage 2 is dominated by K-complexes and spindles. Multiple previous investigations demonstrated the thalamic origin of spindles (Steriade, Deschenes, Domich, & Mulle, 1985; Steriade, Domich, Oakson, & Deschenes, 1987; von Krosigk, Bal, & McCormick, 1993). However, over the last several years a number of studies separated spindles into fast (12–15 Hz) and slow (9–12 Hz) (Ayoub et al., 2013; De Gennaro & Ferrara, 2003; Mölle, Bergmann, Marshall, & Born, 2011). Only fast spindles reveal properties similar to those of thalamic origin, while the mechanisms of slow spindles remain unknown (reviewed in Timofeev & Chauvette, 2013). Repeated stimulation of either cortex or thalamus at spindle frequencies can induce seizures with similar electrographic morphology (Steriade & Timofeev, 2001, 2003). This therefore suggests that these seizures originate as an interaction of cortex and thalamus without a specific cortical focus. As of today, we cannot exclude the possibility that slow spindles have a cortical origin. If that is true, then the mechanisms of generation of slow spindles and paroxysmal fast runs (Boucetta, Chauvette, Bazhenov, & Timofeev, 2008) might be similar. There are at least two factors that point to this potential similarity. (a) The frequency of fast runs (7–16 Hz, more commonly 9–12 Hz) overlaps with that of spindles. (b) The fast runs are usually recorded in frontal regions extending to parietal regions (Timofeev, Grenier, & Steriade, 1998). Slow spindles are recorded in the same spatial locations (Mölle et al., 2011).

Stage 3 or slow-wave sleep is the deepest part of sleep, dominating the first and the second cycle of sleep (Borbely, Baumann, Brandeis, Strauch, & Lehmann, 1981). The dominant activities of slow-wave sleep are slow oscillation and delta oscillation. These two oscillations show distinct dynamics during sleep. If delta oscillation declines from first to second sleep cycle, the slow oscillation does not (Achermann & Borbely, 1997). Both these types of activity are generated by synchronous de- and hyperpolarization of cortical neurons. The cellular basis of the slow oscillation was initially investigated in anesthetized animals (Contreras & Steriade, 1995; Metherate & Ashe, 1993; Steriade, Contreras, Dossi, & Nuñez, 1993; Steriade, Nuñez, & Amzica, 1993a, 1993b). Deep urethane or ketamine–xylazine anesthesia induces slow oscillation that is similar to sleep (Chauvette, Crochet, Volgushev, & Timofeev, 2011; Sharma, Wolansky, & Dickson, 2010). These studies demonstrated that during depth-positive (surface-negative) waves of local field potential (LFP), the cortical neurons are hyperpolarized and silent, while during depth-negative (surface-positive) waves, they are depolarized (active), display barrages of synaptic events, and fire action potentials. A set of intracellular investigations demonstrated that the main distinct feature of slow-wave sleep at the level of cortical network, as opposed to other states of vigilance, is the presence of silent states (Steriade et al., 2001; Timofeev, Grenier, Bazhenov, Sejnowski & Steriade, 2000; Timofeev, Grenier, & Steriade, 2000). The mechanisms of alternation between active and silent states of slow-wave sleep are not fully understood. A few demonstrated features of slow oscillation are the following:

a. Sleep slow oscillation is essentially a cortical phenomenon. It exists in isolated cortical preparations (Cossart, Aronov, & Yuste, 2003; Sanchez-Vives & McCormick, 2000; Sanchez-Vives, Reig, Winograd, & Descalzo, 2007; Timofeev, Grenier, Bazhenov, et al., 2000). However, the full extent of slow cortical oscillation can only be achieved with an active contribution from thalamus (Crunelli & Hughes, 2010; David et al., 2013; Lemieux, Chen, Lonjers, Bazhenov, & Timofeev, 2014).

b. Each slow wave starts in a particular location and from there it propagates to involve much if not the whole cortex (Massimini, Huber, Ferrarelli, Hill, & Tononi, 2004; Mohajerani, McVea, Fingas, & Murphy, 2010; Nir et al., 2011; Volgushev, Chauvette, Mukovski, & Timofeev, 2006). Intracellular recordings demonstrated that only the onset of active slow wave states propagate, while their termination (and the onset of silent states) occurs almost simultaneously (Sheroziya & Timofeev, 2014; Volgushev et al., 2006).

c. Active states originate in deep cortical layers, primarily layer 5 (Beltramo et al., 2013; Chauvette et al., 2010; Sanchez-Vives & McCormick, 2000). How exactly an active state starts remains unclear. Two main hypotheses exist: (i) Some intrinsic current in a subset of neurons leads to the generation of an active state in one cell, which then propagates across the whole cortex (Compte, Sanchez-Vives, McCormick, & Wang, 2003; Hill & Tononi, 2004; Sanchez-Vives & McCormick, 2000). During sleep and anesthesia, the duration of silent states is about 200–400 ms (Chauvette et al., 2011). Unfortunately, no current has yet been identified that is activated by a hyperpolarization to about −70 to −80 mV and can bring neurons to firing within 200–400 ms. Therefore, this hypothesis seems to be not supported by available knowledge. (ii) Some synaptic stochastic process (spike-independent miniature synaptic potentials, or similar) leads to occasional summation of depolarizing events in small groups of neurons that drive action potentials and trigger network active states (Chauvette et al., 2010; Cossart et al., 2003; Timofeev, Grenier, Bazhenov, et al., 2000). Considering mean miniature EPSP (mEPSP) amplitudes of 0.5 mV, their occurrence rates (12–13 s^{-1}) and membrane potentials of neurons during silent states (about 15 mV below firing threshold) an analytical model suggested that it would require about 100 million neurons to generate the slow oscillation with a frequency of 1 Hz (Timofeev, Grenier, Bazhenov, et al., 2000). Given a neuronal density of about 100,000 neurons per 1 mm^2, this would correspond to a surface of about 100 mm^2. This is a very realistic number for large animals (cats, monkey). However, a slow oscillation at similar frequencies is also present in smaller animals (mice), in which the total cortical area is less than 100 mm^2. Therefore, some other mechanisms are needed to trigger active states of slow oscillation.

d. Termination of active states or onset of silent states is also not well understood. Initial hypotheses suggested that short-term synaptic depression, activation of Ca^{2+}- and Na^{+}-activated K^{+} currents and similar cell-specific processes mediate termination of active states (Bazhenov, Timofeev, Steriade, & Sejnowski, 2002; Compte et al., 2003; Hill & Tononi, 2004; Milojkovic, Radojicic, & Antic, 2005). However, later measurements demonstrated that active states terminate nearly simultaneously over large cortical territories (Sheroziya & Timofeev, 2014; Volgushev et al., 2006). This points to the presence of a network mechanism. The most probable network mechanism is active inhibition. Indeed, (i) Cl^{-}-dependent inhibitory activities were detected in a small subset of cortical neurons at the beginning of silent states (Steriade et al., 1993a), (ii) a subset of cortical interneurons preferentially fires prior to silent state onset (Puig, Ushimaru, & Kawaguchi, 2008), (iii) modeling suggests that alteration of inhibitory activities disrupts synchronous termination of active states (Chen, Chauvette, Skorheim, Timofeev, & Bazhenov, 2012). The cortical interneurons most likely to be implicated in the synchronous termination of active states are short-axon cells. There should be a mechanism that induces their synchronous firing at the end of silent states. Our current study (Lemieux, Chauvette, & Timofeev, 2014) indicates that thalamus may be such a structure. Indeed, following surgical or functional disconnection of cortex from thalamus, the synchrony of termination of cortical silent states was lost.

e. Silent states are essentially periods of disfacilitation, i.e., periods in which spontaneous network activity is absent. Removal of synaptic activity results in neuronal hyperpolarization of several millivolts, a reduction or abolition of "synaptic noise," increased input resistance of neurons, and reduction of polysynaptic responsiveness (Contreras, Timofeev, & Steriade, 1996; Rosanova & Timofeev, 2005; Steriade et al., 2001; Timofeev, Contreras, & Steriade, 1996; Timofeev, Grenier, & Steriade, 2000; Timofeev, et al., 2001b). Intracellular Cs^{+} does not affect the duration of silent states but blocks the associated neuronal hyperpolarization, suggesting that K^{+} currents mediate this hyperpolarization (Timofeev, et al., 2001b). Interestingly, the extracellular Ca^{2+} concentration increases during silent states, mediating an increase in effectiveness of monosynaptic responses (Crochet, Chauvette, Boucetta, & Timofeev, 2005), but reducing intrinsic neuronal excitability (Boucetta, Crochet, Chauvette, Seigneur, & Timofeev, 2013). The increase in extracellular Ca^{2+} concentration increases neurotransmitter release and facilitates transition to the next active state.

3. NEOCORTICAL NEURONAL ACTIVITIES DURING SEIZURES

There are different forms of seizures. Those that originate in neocortex are composed of slower spike and wave components (1–3 Hz) and faster oscillations (fast runs) with frequencies 7–16 Hz (Boucetta et al., 2008; Steriade, Amzica, Neckelmann, & Timofeev, 1998; Timofeev et al., 1998; Timofeev & Steriade, 2004). LFP recordings point to the presence of high-frequency ripples (100–200 Hz) during active components of paroxysmal discharge, with the highest amplitude being recorded at the site of onset of each paroxysmal cycle (Grenier, Timofeev, Crochet, & Steriade, 2003; Grenier, Timofeev, & Steriade, 2003; Jefferys et al., 2012).

Cortically generated seizures evolve from slow oscillation. Figure 2.1A shows a typical electrographic seizure recorded in the suprasylvian gyrus (Fig. 2.1B) of an anesthetized cat using intracortical LFP and four intracellular electrodes. Prior to seizure, the neocortical activities were composed of slow oscillation (Fig. 2.1C). During LFP depth positivity, all four recorded cortical neurons were hyperpolarized, and during LFP depth negativity, they were depolarized, two of them fired action potentials. At the transition to seizure, the LFP recording showed a slight increase in frequency without any apparent increase in the amplitude (Fig. 2.1A and D). However, all the recorded neurons showed increases in amplitudes of depolarization and rates of firing. An increase in the amplitude of intracellular activities

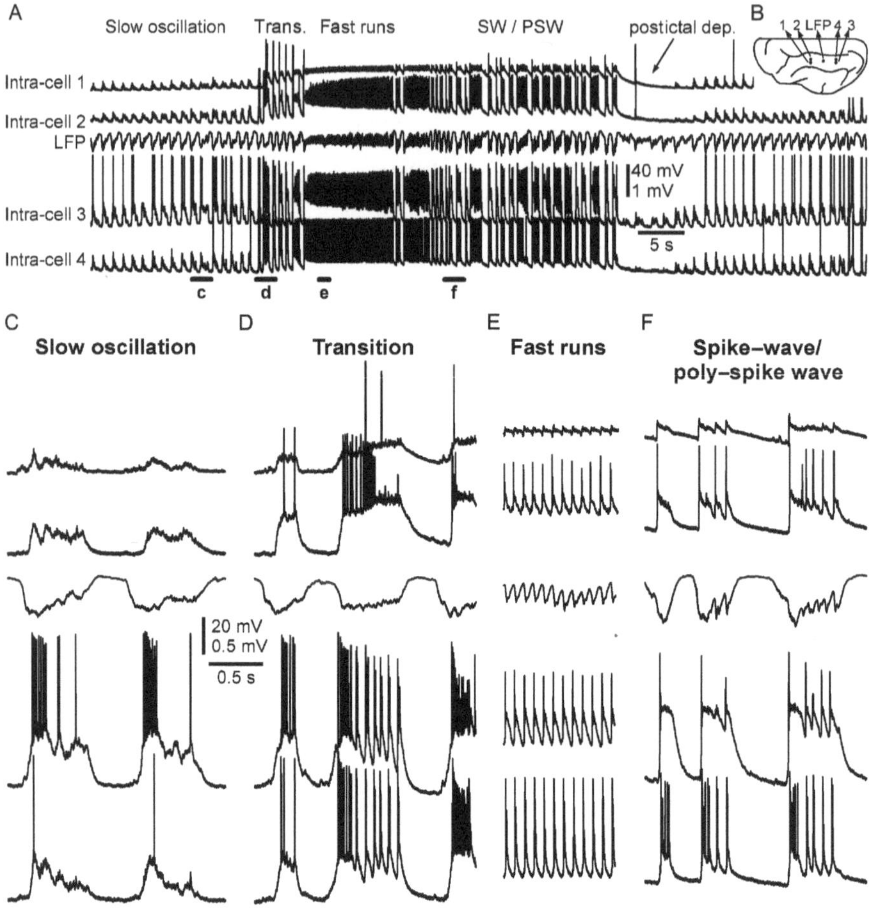

Figure 2.1 Electrographic activities during a spontaneous seizure in a cat anesthetized with ketamine–xylazine. (A) A segment of spontaneous local field potential (LFP) and quadruple intracellular recording before, during, and after seizure. (B) A drawing showing location of recording electrodes. (C) A segment of slow oscillation prior to the onset of seizure. (D) A segment of recording showing transition to seizure. (E) A segment of paroxysmal fast run. (F) LFP spike–wave/polyspike–wave component of seizure and corresponding intracellular paroxysmal depolarizing shifts.

without a parallel increase in the field potential amplitude suggests some extent of decrease in local synchrony, which contributes to the generation of LFP. The overt seizure started from a fast run (10–12 Hz) recorded with all available electrodes (Fig. 2.1A and E). Interestingly, in the neuron that was the most depolarized (recorded in the anterior part of suprasylvian gyrus), there was depolarizing block and that cell was unable to generate spikes, suggesting it could not excite its postsynaptic target neurons. The second half of this electrographic seizure was composed of mainly spike–wave/polyspike–wave complexes. As during the slow oscillation, during LFP depth negativity, all four recorded cortical neurons were depolarized, and during depth positivity, they were hyperpolarized. The depolarization of neurons during depth-negative LFP components was very strong and during each depolarizing event three out of four neurons showed either one spike or no spikes at all. Our previous studies show (Timofeev & Steriade, 2004) that the contribution of cortical neurons to seizure generation can roughly be divided into three groups: (a) leading neurons (high firing rates at the beginning of the seizure, Intra-cell 2 and 3 in Fig. 2.1), (b) maintaining neurons (high firing abilities during seizures, Intra-cell 4 in Fig. 2.1), and (c) reflecting neurons (occasional spikes at the middle/end of paroxysmal depolarizing shifts or no spikes at all, Intra-cell 1 in Fig. 2.1). The seizure terminated with postictal depression clearly seen in three out of four recorded neurons as a transient absence of synaptic activity (Fig. 2.1A).

4. ALTERATIONS OF SYNAPTIC AND INTRINSIC EXCITABILITY DURING SEIZURES

As mentioned above, during the EEG/LFP paroxysmal spike component, neurons are depolarized: this strong depolarization is called paroxysmal depolarizing shift. During the wave component, the neurons are hyperpolarized. How is synchronization of neuronal activities achieved during seizures? There are several mechanisms of neuronal synchronization: chemical synaptic interactions, electrical coupling via gap junctions, ephaptic interactions, and activity-dependent interactions between cells due to ionic changes.

Seizure activities are associated with an increase in extracellular K^+ concentration and a decrease in extracellular Ca^{2+} concentration (Somjen, 2004). During seizures, cellular swelling reduces extracellular space (Amzica & Neckelmann, 1999). These changes dramatically affect intrinsic and synaptic excitability of neurons. K^+ currents have a major impact in setting up membrane potentials of neurons. An increase in extracellular K^+ concentration shifts the reversal potential for these currents to more depolarized values. As a result, hyperpolarization-activated depolarizing current (I_h) becomes able to generate action potentials in a subset of cortical neurons at the end of wave component of paroxysmal spike (Timofeev, Bazhenov, Sejnowski, & Steriade, 2002). Firing of these neurons triggers the next paroxysmal cycle, which starts with some EPSPs and IPSPs. Due to high firing rates of inhibitory neurons and high extracellular levels of K^+, the chloride-dependent IPSPs in paroxysmal brain become depolarizing (Cohen, Navarro, Clemenceau, Baulac, & Miles, 2002; Fujiwara-Tsukamoto, Isomura, Nambu, & Takada, 2003; Payne, Rivera, Voipio, & Kaila, 2003; Timofeev, Grenier, & Steriade, 2002). This synaptic depolarization is amplified by persistent Na^+ current and high-threshold Ca^{2+} currents (Timofeev, Grenier, & Steriade, 2004). An increase in intracellular concentration of Na^+ and Ca^{2+} activates Na^+- and Ca^{2+}-activated K^+ currents that hyperpolarize neurons triggering the onset of the wave component of seizure (Timofeev et al., 2004). The absence of synaptic activities (disfacilitation) and activation of these two currents hyperpolarize neurons that activate h-current and trigger the next paroxysmal cycle (Timofeev, 2010; Timofeev & Steriade, 2004). High extracellular K^+ concentration and low extracellular Ca^{2+} concentration have another consequence. These conditions reduce or impair neurotransmitter release probability; therefore, the effectiveness of recurrent chemical synaptic interactions during seizures is impaired (Seigneur & Timofeev, 2010), and the efficacy of EPSPs and IPSPs triggering paroxysmal cycles is low. On the other hand, low Ca^{2+} concentrations open hemichannels forming electrical synapses (Thimm, Mechler, Lin, Rhee, & Lal, 2005). Increased effectiveness of electrical synapses increases local synchronization helping to generate large amplitude field potential paroxysmal activities (Timofeev et al., 2012). Electrical coupling is not effective to provide long-range synchronization; therefore, the synchrony of field potential discharges between different cortical regions during seizures is low (Boucetta et al., 2008; Meeren, Pijn, Van Luijtelaar, Coenen, & Lopes da Silva, 2002; Timofeev et al., 2002). Strong electrical fields generated by cortical activities also help to provide local synchronization via ephaptic interactions (Frohlich & McCormick, 2010; Jefferys, 1995).

5. UNBALANCE OF EXCITATORY AND INHIBITORY INFLUENCES LEADING TO SEIZURE GENERATION

It is widely accepted that the development of epileptiform activity results from a shift in the balance between excitation and inhibition toward excitation (Dichter & Ayala, 1987; Galarreta & Hestrin, 1998; Nelson & Turrigiano, 1998; Tasker & Dudek, 1991). The easiest way to elicit acute seizures experimentally is to block inhibition (Chagnac-Amitai &

Connors, 1989a, 1989b; Gutnick, Connors, & Prince, 1982; Matsumoto & Ajmone-Marsan, 1964a, 1964b; Prince, 1978; Steriade et al., 1998). Overt block of inhibition rarely if ever occurs outside experimental settings. Synaptic inhibition remains functional in many forms of paroxysmal activities (Davenport, Brown, & Babb, 1990; Esclapez, Hirsch, Khazipov, Ben-AriI, & Bernard, 1997; Higashima, 1988; Prince & Jacobs, 1998; Timofeev et al., 2002; Traub, Borck, Colling, & Jefferys, 1996). By contrast, multiple forms of acquired epilepsy depend on brain damage. These include brain trauma, stroke, tumor, and even diffuse damage, like Alzheimer disease (a high prevalence of epilepsy exists in "young" Alzheimer patients (Amatniek et al., 2006)). The common feature in all these patients is full or partial deafferentation of a particular cortical region and the main damage is done to excitatory cells. Indeed, excitatory cells constitute 80% of all cortical neurons, and all long-range connections are excitatory. Therefore, the reduced content of excitatory neurons is an epileptogenic factor.

Brain excitability is maintained at a normal level via homeostatic mechanisms. Silencing a cortical culture network for 2 days upregulates synaptic excitability, and an increase in network activity downregulates excitatory synaptic efficacy (Murthy, Schikorski, Stevens, & Zhu, 2001; Turrigiano, Leslie, Desai, Rutherford, & Nelson, 1998; Watt, van Rossum, MacLeod, Nelson, & Turrigiano, 2000). Conversely, prolonged enhanced activity induced by the blockade of synaptic inhibition or elevated $[K^+]_o$ reduces the size of mEPSCs (Leslie, Nelson, & Turrigiano, 2001; Lissin et al., 1998; Turrigiano et al., 1998). In addition, intrinsic excitability is regulated by activity. After a chronic blockade of activity, Na^+ currents increase and K^+ currents decrease in size, resulting in an enhanced responsiveness of pyramidal cells to current injection (Desai, Rutherford, & Turrigiano, 1999). Similar enhancement of intrinsic excitability was found during trauma-induced epileptogenesis (Avramescu & Timofeev, 2008). Thus, homeostatic plasticity controls the levels of neuronal activity through synaptic and intrinsic mechanisms (Murthy et al., 2001; Turrigiano et al., 1998). Our current study demonstrates that inactivation of lateral posterior thalamic nucleus disrupted slow oscillation in the target suprasylvian gyrus due to a functional removal of a small number of excitatory synapses. However, after 15–20 h, homeostatic upregulation of excitability was able to recover the slow oscillation (Lemieux, Chauvette, et al., 2014).

We recently demonstrated that (a) during trauma-induced epileptogenesis, cortical neurons undergo long-lasting silent periods during all states of vigilance (Nita, Cisse, Timofeev, & Steriade, 2007; Timofeev,

Bazhenov, Avramescu, & Nita, 2010), (b) in a kindling model of epilepsy, seizures are followed by continuing activity, called "outlasting," consisting of silent and active states (Nita, Cisse, Frohlich, & Timofeev, 2008; Nita, Cisse, & Timofeev, 2008). Therefore, silent periods are increased in both epileptogenesis models. These studies demonstrate that epileptogenesis is accompanied with increased network silence, which might induce upregulation of neuronal excitability.

Our collaborators and we developed network computational models in which partial cortical deafferentation led to upregulation of the neuronal excitability and the development of seizure-like activity (Frohlich, Bazhenov, & Sejnowski, 2008; Frohlich, Bazhenov, Timofeev, Steriade, & Sejnowski, 2006; Houweling, Bazhenov, Timofeev, Steriade, & Sejnowski, 2005; Volman, Bazhenov, & Sejnowski, 2011, 2012, 2013; Volman, Sejnowski, & Bazhenov, 2011). First, we found that only sufficiently strong deafferentation leads to the pathological network synchronization; after a weak deafferentation, homeostatic plasticity was able to recover the normal network activity (Houweling et al., 2005). Therefore, we predicted the existence of a critical degree of deafferentation for pathological network reorganization. Second, to be prone to paroxysmal bursting, the network should include a relatively large population of intact neurons and a population of deafferented cells with low spontaneous activity (Volman, Bazhenov, et al., 2011; Volman, Sejnowski, et al., 2011). This suggests that the epileptic activity should arise near the boundary of intact and deafferented areas and propagate from that point, as observed experimentally (Nita, Cissé, Timofeev, & Steriade, 2006; Nita et al., 2007; Topolnik, Steriade, & Timofeev, 2003a). Therefore, after deafferentation, even physiological level fluctuations of the input to the network may trigger transitions to recurrent epileptiform activity that would be impossible in the healthy network.

6. FOCUS OF EPILEPSY VERSUS SEIZURE AND INTERACTION OF EPILEPTIC FOCUS WITH INTACT CORTEX

The definitions of epileptic and seizure foci in the cortex are not simple but rather are dualistic. First, the epileptic focus can be described as a site in the cortex with some pathology that causes the generation of spontaneous seizures. Second, the seizure focus might be defined as a site at which seizures start and/or each paroxysmal discharge is initiated. As shown in Fig. 2.1, the neuron 1 obviously shows the most dramatic membrane depolarization in the seizure, but it almost fires no spikes (due to depolarization block). Such activity might be characteristic of the epileptic focus, but obviously not of the seizure focus, because a seizure generation requires active neuronal interactions, i.e., spikes. The neuron 2 was located close to neuron 1 (<1 mm). It showed intense firing at the beginning of seizure, but only one spike per paroxysmal cycle in the remaining part of the seizure. Therefore, such a neuron might belong to a group of neurons located in the epileptic focus, but likely not neurons in the seizure focus. The neuron 4 fired little prior to seizure, dramatically increased firing at the onset of seizure and fired multiple action potentials at each paroxysmal cycle; this neuron reveals a behavior expected for neurons recorded in the seizure focus. Interestingly, neuron 3, located close to neuron 4, fired many spikes prior to seizure as well as at the beginning of the seizure, but not during the main body of seizure. Intense neuronal firing excites target neurons and in addition, it increases local extracellular K^+ concentration that is one of the seizure-triggering factors (Bazhenov, Timofeev, Frohlich, & Sejnowski, 2008; Bazhenov, Timofeev, Steriade, & Sejnowski, 2004). Altogether, these data suggest that location that contained neurons 1 and 2 was likely the focus of epilepsy, but the location that contained neurons 3 and 4 corresponds to the requirements of the seizure focus.

Overall, the focus of epilepsy is a site in which a pathological process induced a major damage, and a large number of cellular elements cannot reliably contribute to network activities. The epileptic focus is surrounded by neurons that display abnormally high level of excitability and are ready to generate seizures. Therefore, the seizure focus is distributed around epileptic focus.

In previous and current studies, in our laboratory, we investigate the undercut model of epilepsy. In this model, a cut is done to transect white matter underneath an area of cortex (Nita et al., 2006; Prince & Tseng, 1993, Topolnik et al., 2003a; Topolnik, Steriade, & Timofeev, 2003b; Xiong, Ping, Gao, & Jin, 2011). The advantage of this model is that the epileptic focus is clearly defined, giving the ability to investigate where the seizures start in relation to this focus. This model closely matches clinical conditions of penetrating head wounds. Indeed, cortical undercut induces immediate depression of electrical activity in the affected area; acute seizures evolve within hours and can be observed for several hours in 70% of cats; they trigger epileptogenesis, which leads to development of epilepsy in 1.5–4 months from the undercut (Nita et al., 2007; Timofeev et al., 2010; Topolnik et al., 2003a). Postwar epidemiological studies from Vietnam and Croatia reported that about 50% of patients with penetrating cranial wounds develop recurring seizures 10–15 years after the trauma (Marcikic,

Melada, & Kovacevic, 1998; Salazar et al., 1985). Computational studies suggest that the undercut cortex model is also representative for other cases of epilepsy in which deafferentation is the main cause (Frohlich et al., 2008). In this model, the acute seizures started always at the border of undercut and relatively intact cortex (Topolnik et al., 2003a). The main factors causing these seizures are likely increased extracellular K^+, glutamate, etc. These changes are major at the site of the undercut and decrease with distance. However, at the site of the undercut, a cut of excitatory inputs abruptly reduced excitability. Therefore, the seizures start just outside that area.

In chronic conditions, the mechanisms of seizure generation are different. Epileptiform discharges can be elicited from slices obtained from the undercut cortex (Jacobs, Gutnick, & Prince, 1996; Prince & Tseng, 1993). However, spontaneous electrographic seizures first develop around the undercut cortex, with time they involve more cortical areas and when most of dorsal cortex shows electrographic seizure, the behavioral seizures started (Nita et al., 2007). Our current studies are done on mice in which spontaneous electrographic and behavioral seizures develop within the first week from the undercut. Like, in cats, the paroxysmal activities start around the undercut area (Fig. 2.2), never within the undercut, nor a long way from the undercut. Usually, some low-amplitude paroxysmal spikes may be recorded in minutes before the onset of seizure (Fig. 2.2, 1′, b, d). Progressively, they grow in amplitude and involve the undercut and other cortical areas, but still remaining isolated interictal spikes (Fig. 2.2, 2′). Prior to seizure (Fig. 2.2, 4′), the amplitude of interictal spikes increases in all channels, but in particular those remote from the undercut locations (Fig. 2.2, 3′, c, e). Despite the large amplitude of interictal spikes recorded in remote cortical areas, the earliest site showing the paroxysmal discharges was always located just in or close to the undercut area. The undercut cortex revealed the lowest amplitude of paroxysmal discharges and the onset of each paroxysmal discharge occurred after it started in the areas surrounding the undercut.

These results allow us to conclude that the obvious epileptic focus was the area of undercut. However, this area was never implicated in the onset of paroxysmal discharges; the focus of seizures was located just outside the undercut cortex.

7. NEURONAL PLASTICITY LEADING TO DEVELOPMENT OF EPILEPSY

What really happens in the undercut cortex and surrounding areas? As we described earlier, due to their stochastic nature, spontaneous active states

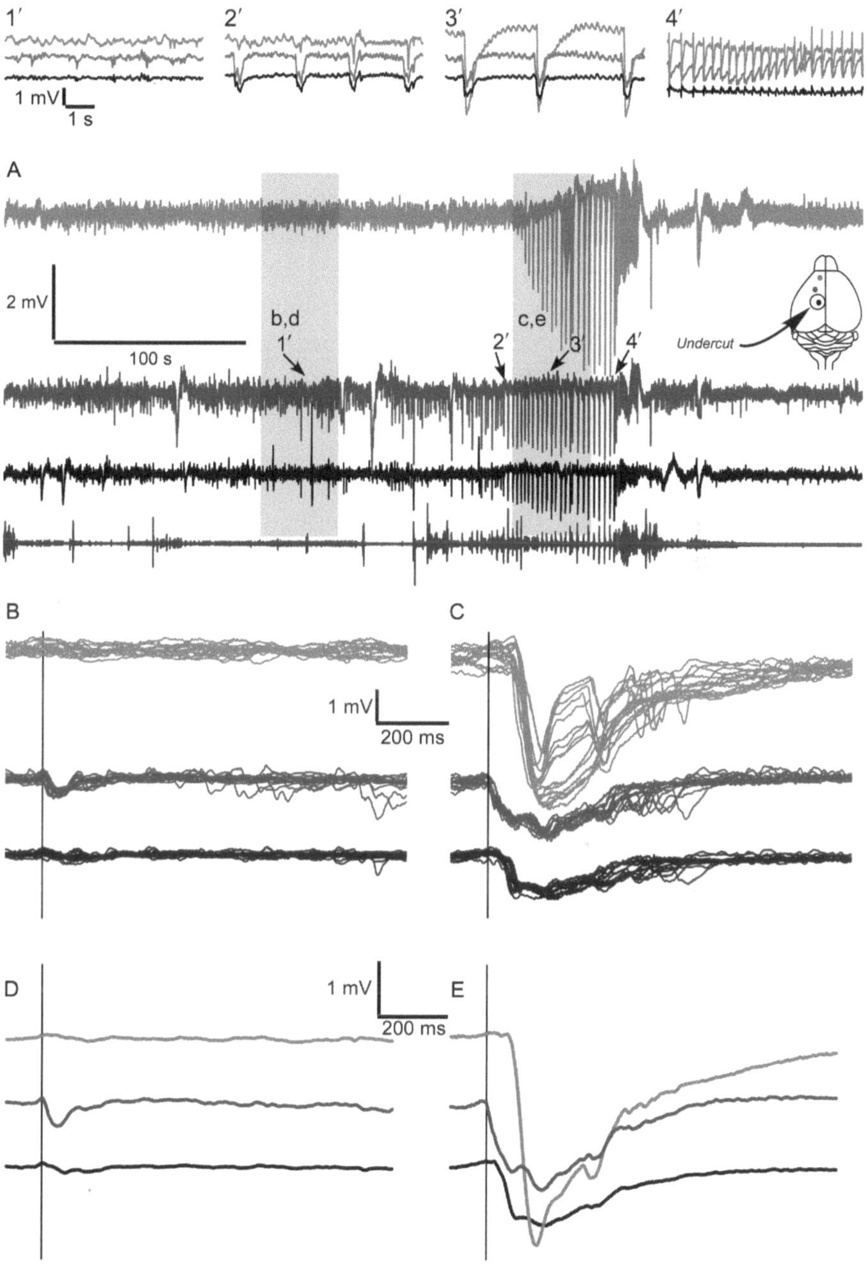

Figure 2.2 Full seizure recorded in a behaving mouse with chronic undercut. (A) Local field potentials in background condition and during paroxysmal activities. Position of electrodes is shown in drawing; undercut position is shown as circle. Different colors (grey shades in print version) indicating the location of electrodes used also to show electrographic

(Continued)

originate in different cortical locations and from there they propagate to involve all cortical regions. Decreased afferent drive to the undercut cortex results in an increase in the overall duration of silent states (Avramescu & Timofeev, 2008; Nita et al., 2007; Timofeev et al., 2010; Topolnik et al., 2003a). Therefore, the network remains silent for longer periods of time. Network silence upregulates neuronal excitability. In cell cultures (obtained from young animals), or in young animals *in vivo*, network silence triggers homeostatic plasticity processes (Turrigiano, 2011) which lead to upregulation of neuronal excitability till the target level of excitability is attained. This stands in agreement with the fact that neocortical seizures are generally nocturnal, and they evolve from sleep slow oscillation without discontinuity (Timofeev & Steriade, 2004) likely because disfacilitation during sleep silent states induce homeostatic upregulation of neuronal excitability. Sleep slow oscillation also induces long-term potentiation, which increases synaptic excitability and potentially can contribute to seizure generation (Chauvette et al., 2012). In adult animals, the neuronal excitability is also upregulated due to deafferentation (Avramescu & Timofeev, 2008; Lemieux, Chen, et al., 2014). Due to homeostatic plasticity processes, an increase in the overall silence leads to an increase in both intrinsic and synaptic excitability, and therefore, instantaneous firing rates in chronically deafferented cortex are much higher than in normal cortex (Avramescu & Timofeev, 2008). However, it appears that the upregulation of excitability in adult animals continues even when the "normal" levels of excitability are reached and the deafferented cortex becomes overexcitable. As a result, in the same experimental conditions of cortical undercut, young animals recover and they only display increased numbers of slow waves compared to control, but adult animals become epileptic (Timofeev, Sejnowski, Bazhenov, Chauvette, & Grand, 2013). This finding is congruent with available human data showing that

Figure 2.2—Cont'd activities at that location. Blue (bottom in print version) trace shows neck muscle activities. 1′, 2′, 3′, and 4′ are expended segments from indicated periods in (A). Low-amplitude interictal spikes (about 1 every 2 s) are seen in background state in front of the undercut cortex. Their amplitude progressively increases, involving other regions to become maximal just prior to seizure (4′). (B and C) Superposition of several interictal spikes obtained from indicated periods. Zero time is the onset of interictal spike in "green (second in print version)" electrode. (D and E) Averages of local field potential activities obtained from segments shown in (B) and (C). Note that in all occasions, the electrode located just outside undercut area (in green (the second trace from top in print version)) is the first one showing paroxysmal discharges, although the highest amplitude events were recorded in frontally located electrode.

posttraumatic seizures are a risk factor in adults, but it is certainly not true for infants (Angeleri et al., 1999) and that young children display mainly early seizures, but adolescents and adults, become epileptic after some latent period (Asikainen, Kaste, & Sarna, 1999).

Cortical undercut is accompanied with deafferentation and loss of neurons, in particular inhibitory neurons (Avramescu et al., 2009). Despite that the local connection probability within undercut cortex was increased (Avramescu & Timofeev, 2008; Jin, Prince, & Huguenard, 2006). We have now evidence that intracortical long-range excitatory connectivity is also increased between intact and deafferented cortex. (Kusmierczak, Lajeunesse, Grand, Timofeev, unpublished). Thus, a permanent loss of all connections coming via white matter in the undercut cortex is partially compensated by an increase in intracortical connectivity. The increase in intracortical connectivity is also present in surrounding undercut cortex areas. The number of neurons in areas surrounding undercut cortex is not different from control (Avramescu et al., 2009).

Normal active cortical states are generated via some stochastic process, which depends on the number of neurons and the number of connections (Chauvette et al., 2010; Timofeev, Grenier, Bazhenov, et al., 2000). At least at the beginning of seizure, paroxysmal active states are very similar to normal active states (Fig. 2.1C and D); therefore, they likely share common mechanisms of generation. As a result, areas with "normal" numbers of neurons and increased connectivity are the most probable sites to generate seizures.

Overall, cortical trauma induces major network reorganization. The site of trauma becomes the epileptic focus. In that site, the neurons are hyperexcitable and highly interconnected, but due to their low number, they cannot generate seizures. The areas immediately surrounding the undercut contain a normal number of neurons, but with increased connectivity. These areas become sites of seizure onset, that is, the main characteristic of seizure focus.

8. CONCLUSIONS

Acquired neocortical epilepsy is primarily associated with deafferentation induced either by direct cortical damage or indirect cortical damage resulting from stroke, tumor infections, and similar types of injury. The sites of damage become epileptic foci. This permanent deafferentaion upregulates the neuronal excitability. Silent periods of slow-wave sleep, periods of disfacilitation, further increase neuronal excitability promoting the development of seizures. The neuronal loss, altered connectivity in deafferented sites, and severe hyperexcitability in epileptic foci prevent these foci to trigger seizures. The neuronal density in the areas surrounding damaged cortical sites is normal, but their excitability and connectivity is increased. Therefore, the sites surrounding traumatized cortex trigger seizures. We propose that neocortical epileptic focus does not trigger seizures; the seizure foci are distributed around the epileptic focus.

ACKNOWLEDGMENTS

In recent years, our work was supported by CIHR, NIH-NINDS, NSERC, and FRSQ. We are thankful to Sergiu Ftomov for the technical assistance and Josée Seigneur for management assistance.

REFERENCES

Achermann, P., & Borbely, A. A. (1997). Low-frequency (<1 hz) oscillations in the human sleep electroencephalogram. *Neuroscience, 81*, 213–222.

Amatniek, J. C., Hauser, W. A., DelCastillo-Castaneda, C., Jacobs, D. M., Marder, K., Bell, K., et al. (2006). Incidence and predictors of seizures in patients with alzheimer's disease. *Epilepsia, 47*, 867–872.

Amzica, F., & Neckelmann, D. (1999). Membrane capacitance of cortical neurons and glia during sleep oscillations and spike-wave seizures. *Journal of Neurophysiology, 82*, 2731–2746.

Angeleri, F., Majkowski, J., Cacchio, G., Sobieszek, A., D'Acunto, S., Gesuita, R., et al. (1999). Posttraumatic epilepsy risk factors: One-year prospective study after head injury. *Epilepsia, 40*, 1222–1230.

Asikainen, I., Kaste, M., & Sarna, S. (1999). Early and late posttraumatic seizures in traumatic brain injury rehabilitation patients: Brain injury factors causing late seizures and influence of seizures on long-term outcome. *Epilepsia, 40*, 584–589.

Avramescu, S., Nita, D., & Timofeev, I. (2009). Neocortical post-traumatic epileptogenesis is associated with the loss of gabaergic neurons. *Journal of Neurotrauma, 26*, 799–812.

Avramescu, S., & Timofeev, I. (2008). Synaptic strength modulation following cortical trauma: A role in epileptogenesis. *Journal of Neuroscience, 28*, 6760–6772.

Ayoub, A., Aumann, D., Hörschelmann, A., Kouchekmanesch, A., Paul, P., Born, J., et al. (2013). Differential effects on fast and slow spindle activity, and the sleep slow oscillation in humans with carbamazepine and flunarizine to antagonize voltage-dependent na^{+} and ca^{2+} channel activity. *Sleep, 36*(6), 905–911.

Barth, A. L., & Poulet, J. F. A. (2012). Experimental evidence for sparse firing in the neocortex. *Trends in Neurosciences, 35*, 345–355.

Bazhenov, M., Timofeev, I., Frohlich, F., & Sejnowski, T. J. (2008). Cellular and network mechanisms of electrographic seizures. *Drug Discovery Today. Disease Models, 5*, 45–57.

Bazhenov, M., Timofeev, I., Steriade, M., & Sejnowski, T. J. (2002). Model of thalamocortical slow-wave sleep oscillations and transitions to activated states. *Journal of Neuroscience, 22*, 8691–8704.

Bazhenov, M., Timofeev, I., Steriade, M., & Sejnowski, T. J. (2004). Potassium model for slow (2–3 hz) in vivo neocortical paroxysmal oscillations. *Journal of Neurophysiology, 92*, 1116–1132.

Beltramo, R., D'Urso, G., Dal Maschio, M., Farisello, P., Bovetti, S., Clovis, Y., et al. (2013). Layer-specific excitatory circuits differentially control recurrent network dynamics in the neocortex. *Nature Neuroscience, 16*, 227–234.

Borbely, A. A., Baumann, F., Brandeis, D., Strauch, I., & Lehmann, D. (1981). Sleep deprivation: Effect on sleep stages and EEg power density in man. *Electroencephalography and Clinical Neurophysiology, 51*, 483–495.

Boucetta, S., Chauvette, S., Bazhenov, M., & Timofeev, I. (2008). Focal generation of paroxysmal fast runs during electrographic seizures. *Epilepsia, 49*, 1925–1940.

Boucetta, S., Crochet, S., Chauvette, S., Seigneur, J., & Timofeev, I. (2013). Extracellular ca^{2+} fluctuations in vivo affect afterhyperpolarization potential and modify firing patterns of neocortical neurons. *Experimental Neurology, 245*, 5–14.

Buzsaki, G. (2002). Theta oscillations in the hippocampus. *Neuron, 33*, 325–340.

Chagnac-Amitai, Y., & Connors, B. W. (1989a). Synchronized excitation and inhibition driven by intrinsically bursting neurons in neocortex. *Journal of Neurophysiology, 62*, 1149–1162.

Chagnac-Amitai, Y., & Connors, B. W. (1989b). Horizontal spread of synchronized activity in neocortex and its control by gaba-mediated inhibition. *Journal of Neurophysiology, 61*, 747–758.

Chauvette, S., Crochet, S., Volgushev, M., & Timofeev, I. (2011). Properties of slow oscillation during slow-wave sleep and anesthesia in cats. *Journal of Neuroscience, 31*, 14998–15008.

Chauvette, S., Seigneur, J., & Timofeev, I. (2012). Sleep oscillations in the thalamocortical system induce long-term neuronal plasticity. *Neuron, 75*, 1105–1113.

Chauvette, S., Volgushev, M., & Timofeev, I. (2010). Origin of active states in local neocortical networks during slow sleep oscillation. *Cerebral Cortex, 20*, 2660–2674.

Chen, J. Y., Chauvette, S., Skorheim, S., Timofeev, I., & Bazhenov, M. (2012). Interneuron-mediated inhibition synchronizes neuronal activity during slow oscillation. *Journal of Physiology, 590*, 3987–4010.

Cohen, I., Navarro, V., Clemenceau, S., Baulac, M., & Miles, R. (2002). On the origin of interictal activity in human temporal lobe epilepsy in vitro. *Science, 298*, 1418–1421.

Compte, A., Sanchez-Vives, M. V., McCormick, D. A., & Wang, X.-J. (2003). Cellular and network mechanisms of slow oscillatory activity (<1 hz) and wave propagations in a cortical network model. *Journal of Neurophysiology, 89*, 2707–2725.

Contreras, D., & Steriade, M. (1995). Cellular basis of eeg slow rhythms: A study of dynamic corticothalamic relationships. *Journal of Neuroscience, 15*, 604–622.

Contreras, D., Timofeev, I., & Steriade, M. (1996). Mechanisms of long-lasting hyperpolarizations underlying slow sleep oscillations in cat corticothalamic networks. *Journal of Physiology, 494*, 251–264.

Cossart, R., Aronov, D., & Yuste, R. (2003). Attractor dynamics of network up states in the neocortex. *Nature, 423*, 283–288.

Crochet, S., Chauvette, S., Boucetta, S., & Timofeev, I. (2005). Modulation of synaptic transmission in neocortex by network activities. *European Journal of Neuroscience, 21*, 1030–1044.

Crunelli, V., & Hughes, S. W. (2010). The slow (<1 hz) rhythm of non-rem sleep: A dialogue between three cardinal oscillators. *Nature Neuroscience, 13*, 9–17.

Davenport, C. J., Brown, W. J., & Babb, T. L. (1990). Gabaergic neurons are spared after intrahippocampal kainate in the rat. *Epilepsy Research, 5*, 28–42.

David, F., Schmiedt, J. T., Taylor, H. L., Orban, G., Di Giovanni, G., Uebele, V. N., et al. (2013). Essential thalamic contribution to slow waves of natural sleep. *Journal of Neuroscience, 33*, 19599–19610.

De Gennaro, L., & Ferrara, M. (2003). Sleep spindles: An overview. *Sleep Medicine Reviews, 7*, 423–440.

Desai, N. S., Rutherford, L. C., & Turrigiano, G. G. (1999). Plasticity in the intrinsic excitability of cortical pyramidal neurons. *Nature Neuroscience, 2*, 515–520.

Dichter, M. A., & Ayala, G. F. (1987). Cellular mechanisms of epilepsy: A status report. *Science, 237*, 157–164.

Epi K. Consortium Epilepsy Phenome/Genome P. (2013). De novo mutations in epileptic encephalopathies. *Nature, 501*, 217–221.

Esclapez, M., Hirsch, J. C., Khazipov, R., Ben-AriI, Y., & Bernard, C. (1997). Operative gabaergic inhibition in hippocampal ca1 pyramidal neurons in experimental epilepsy. *Proceedings of the National Academy of Sciences of the United States of America, 94*, 12151–12156.

Frohlich, F., Bazhenov, M., & Sejnowski, T. J. (2008). Pathological effect of homeostatic synaptic scaling on network dynamics in diseases of the cortex. *Journal of Neuroscience, 28*, 1709–1720.

Frohlich, F., Bazhenov, M., Timofeev, I., Steriade, M., & Sejnowski, T. J. (2006). Slow state transitions of sustained neural oscillations by activity-dependent modulation of intrinsic excitability. *Journal of Neuroscience, 26*, 6153–6162.

Frohlich, F., & McCormick, D. A. (2010). Endogenous electric fields may guide neocortical network activity. *Neuron, 67*, 129–143.

Fujiwara-Tsukamoto, Y., Isomura, Y., Nambu, A., & Takada, M. (2003). Excitatory gaba input directly drives seizure-like rhythmic synchronization in mature hippocampal ca1 pyramidal cells. *Neuroscience, 119*, 265–275.

Gabbott, P. L., & Somogyi, P. (1986). Quantitative distribution of gaba-immunoreactive neurons in the visual cortex (area 17) of the cat. *Experimental Brain Research, 61*, 323–331.

Galarreta, M., & Hestrin, S. (1998). Frequency-dependent synaptic depression and the balance of excitation and inhibition in the neocortex. *Nature Neuroscience, 1*, 587–594.

Grenier, F., Timofeev, I., Crochet, S., & Steriade, M. (2003). Spontaneous field potentials influence the activity of neocortical neurons during paroxysmal activities in vivo. *Neuroscience, 119*, 277–291.

Grenier, F., Timofeev, I., & Steriade, M. (2003). Neocortical very fast oscillations (ripples, 80–200 hz) during seizures: Intracellular correlates. *Journal of Neurophysiology, 89*, 841–852.

Gutnick, M. J., Connors, B. W., & Prince, D. A. (1982). Mechanisms of neocortical epileptogenesis in vitro. *Journal of Neurophysiology, 48*, 1321–1335.

Haider, B., Duque, A., Hasenstaub, A. R., & McCormick, D. A. (2006). Neocortical network activity in vivo is generated through a dynamic balance of excitation and inhibition. *Journal of Neuroscience, 26*, 4535–4545.

Haider, B., Hausser, M., & Carandini, M. (2013). Inhibition dominates sensory responses in the awake cortex. *Nature, 493*, 97–100.

Higashima, M. (1988). Inhibitory processes in development of seizure activity in hippocampal slices. *Experimental Brain Research, 72*, 37–44.

Hill, S. L., & Tononi, G. (2004). Modeling sleep and wakefulness in the thalamocortical system. *Journal of Neurophysiology, 93*, 1671–1698.

Hirsch, J. C., Fourment, A., & Marc, M. E. (1983). Sleep-related variations of membrane potential in the lateral geniculate body relay neurons of the cat. *Brain Research, 259*, 308–312.

Houweling, A. R., Bazhenov, M., Timofeev, I., Steriade, M., & Sejnowski, T. J. (2005). Homeostatic synaptic plasticity can explain post-traumatic epileptogenesis in chronically isolated neocortex. *Cerebral Cortex, 15*, 834–845.

Iber, C., Ancoli-Israel, S., Chesson, A., & Quan, S. (2007). *Manual for the scoring of sleep and associated events: Rules, terminology and technical specifications*. Westchester, NY: American Academy of Sleep Medicine.

Jacobs, K. M., Gutnick, M. J., & Prince, D. A. (1996). Hyperexcitability in a model of cortical maldevelopment. *Cerebral Cortex*, *6*, 514–523.

Jefferys, J. G. (1995). Nonsynaptic modulation of neuronal activity in the brain: Electric currents and extracellular ions. *Physiological Reviews*, *75*, 689–723.

Jefferys, J. G., Menendez de la Prida, L., Wendling, F., Bragin, A., Avoli, M., Timofeev, I., et al. (2012). Mechanisms of physiological and epileptic HFO generation. *Progress in Neurobiology*, *98*, 250–264.

Jin, X., Prince, D. A., & Huguenard, J. R. (2006). Enhanced excitatory synaptic connectivity in layer v pyramidal neurons of chronically injured epileptogenic neocortex in rats. *Journal of Neuroscience*, *26*, 4891–4900.

Lemieux, M., Chauvette, S., & Timofeev, I. (2014). Neocortical inhibitory activities and thalamocortical afferents contribute to the onset of silent states of the neocortical slow oscillation. *Journal of Neurophysiology*, (in review).

Lemieux, M., Chen, J.-Y., Lonjers, P., Bazhenov, M., & Timofeev, I. (2014). The impact of cortical deafferentation on the neocortical slow oscillation. *Journal of Neuroscience*, *34*, 5689–5703.

Leslie, K. R., Nelson, S. B., & Turrigiano, G. G. (2001). Postsynaptic depolarization scales quantal amplitude in cortical pyramidal neurons. *Journal of Neuroscience*, *21*, RC170.

Lissin, D. V., Gomperts, S. N., Carroll, R. C., Christine, C. W., Kalman, D., Kitamura, M., et al. (1998). Activity differentially regulates the surface expression of synaptic AMPA and NMDA glutamate receptors. *Proceedings of the National Academy of Sciences of the United States of America*, *95*, 7097–7102.

Marcikic, M., Melada, A., & Kovacevic, R. (1998). Management of war penetrating craniocerebral injuries during the war in croatia. *Injury*, *29*, 613–618.

Massimini, M., Huber, R., Ferrarelli, F., Hill, S., & Tononi, G. (2004). The sleep slow oscillation as a traveling wave. *Journal of Neuroscience*, *24*, 6862–6870.

Matsumoto, H., & Ajmone-Marsan, C. (1964a). Cortical cellular phenomena in experimental epilepsy: Interictal manifestations. *Experimental Neurology*, *9*, 286–304.

Matsumoto, H., & Ajmone-Marsan, C. (1964b). Cortical cellular phenomena in experimental epilepsy: Ictal manifestations. *Experimental Neurology*, *9*, 305–326.

Meeren, H. K., Pijn, J. P., Van Luijtelaar, E. L., Coenen, A. M., & Lopes da Silva, F. H. (2002). Cortical focus drives widespread corticothalamic networks during spontaneous absence seizures in rats. *Journal of Neuroscience*, *22*, 1480–1495.

Metherate, R., & Ashe, J. H. (1993). Ionic flux contributions to neocortical slow waves and nucleus basalis-mediated activation: Whole-cell recordings *in vivo*. *Journal of Neuroscience*, *13*, 5312–5323.

Milojkovic, B. A., Radojicic, M. S., & Antic, S. D. (2005). A strict correlation between dendritic and somatic plateau depolarizations in the rat prefrontal cortex pyramidal neurons. *Journal of Neuroscience*, *25*, 3940–3951.

Mohajerani, M. H., McVea, D. A., Fingas, M., & Murphy, T. H. (2010). Mirrored bilateral slow-wave cortical activity within local circuits revealed by fast bihemispheric voltage-sensitive dye imaging in anesthetized and awake mice. *Journal of Neuroscience*, *30*, 3745–3751.

Mölle, M., Bergmann, T. O., Marshall, L., & Born, J. (2011). Fast and slow spindles during the sleep slow oscillation: Disparate coalescence and engagement in memory processing. *Sleep*, *34*, 1411–1421.

Murthy, V. N., Schikorski, T., Stevens, C. F., & Zhu, Y. (2001). Inactivity produces increases in neurotransmitter release and synapse size. *Neuron*, *32*, 673–682.

Nelson, S. B., & Turrigiano, G. G. (1998). Synaptic depression: A key player in the cortical balancing act. *Nature Neuroscience*, *1*, 539–541.

Nir, Y., Staba, R. J., Andrillon, T., Vyazovskiy Vladyslav, V., Cirelli, C., Fried, I., et al. (2011). Regional slow waves and spindles in human sleep. *Neuron*, *70*, 153–169.

Nita, D., Cisse, Y., Frohlich, F., & Timofeev, I. (2008). Cortical and thalamic components of neocortical kindling-induced epileptogenesis in behaving cats. *Experimental Neurology, 211*, 518–528.
Nita, D., Cisse, Y., & Timofeev, I. (2008). State-dependent slow outlasting activities following neocortical kindling in cats. *Experimental Neurology, 211*, 456–468.
Nita, D., Cissé, Y., Timofeev, I., & Steriade, M. (2006). Increased propensity to seizures after chronic cortical deafferentation in vivo. *Journal of Neurophysiology, 95*, 902–913.
Nita, D. A., Cisse, Y., Timofeev, I., & Steriade, M. (2007). Waking-sleep modulation of paroxysmal activities induced by partial cortical deafferentation. *Cerebral Cortex, 17*, 272–283.
Payne, J. A., Rivera, C., Voipio, J., & Kaila, K. (2003). Cation-chloride co-transporters in neuronal communication, development and trauma. *Trends in Neurosciences, 26*, 199–206.
Prince, D. A. (1978). Neurophysiology of epilepsy. *Annual Review of Neuroscience, 1*, 395–415.
Prince, D. A., & Jacobs, K. (1998). Inhibitory function in two models of chronic epileptogenesis. *Epilepsy Research, 32*, 83–92.
Prince, D. A., & Tseng, G. F. (1993). Epileptogenesis in chronically injured cortex: In vitro studies. *Journal of Neurophysiology, 69*, 1276–1291.
Puig, M. V., Ushimaru, M., & Kawaguchi, Y. (2008). Two distinct activity patterns of fast-spiking interneurons during neocortical up states. *Proceedings of the National Academy of Sciences of the United States of America, 105*, 8428–8433.
Rector, D. M., Schei, J. L., & Rojas, M. J. (2009). Mechanisms underlying state dependent surface-evoked response patterns. *Neuroscience, 159*, 115–126.
Rosanova, M., & Timofeev, I. (2005). Neuronal mechanisms mediating the variability of somatosensory evoked potentials during sleep oscillations in cats. *Journal of Physiology, 562*(2), 569–582.
Rudolph, M., Pospischil, M., Timofeev, I., & Destexhe, A. (2007). Inhibition determines membrane potential dynamics and controls action potential generation in awake and sleeping cat cortex. *Journal of Neuroscience, 27*, 5280–5290.
Salazar, A., Jabbari, B., Vance, S., Grafman, J., Amin, D., & Dillon, J. (1985). Epilepsy after penetrating head injury. I. Clinical correlates: A report of the vietnam head injury study. *Neurology, 35*, 1406–1414.
Sanchez-Vives, M. V., & McCormick, D. A. (2000). Cellular and network mechanisms of rhythmic recurrent activity in neocortex. *Nature Neuroscience, 3*, 1027–1034.
Sanchez-Vives, M. V., Reig, R., Winograd, M., & Descalzo, V. F. (2007). An active cortical network in vitro. In I. Timofeev (Ed.), *Mechanisms of spontaneous active states in neocortex* (pp. 23–44). Kerala, India: Research Signpost.
Seigneur, J., & Timofeev, I. (2010). Synaptic impairment induced by paroxysmal ionic conditions in neocortex. *Epilepsia, 52*, 132–139.
Sharma, A. V., Wolansky, T., & Dickson, C. T. (2010). A comparison of sleeplike slow oscillations in the hippocampus under ketamine and urethane anesthesia. *Journal of Neurophysiology, 104*, 932–939.
Sheroziya, M. G., & Timofeev, I. (2014). Global intracellular slow wave dynamics of the thalamocortical system. *Journal of Neuroscience*, (in press).
Shu, Y., Hasenstaub, A., & McCormick, D. A. (2003). Turning on and off recurrent balanced cortical activity. *Nature, 423*, 288–293.
Somjen, G. C. (2004). *Ions in the brain: Normal function, seizures, and stroke*. Oxford, NY: Oxford University Press.
Steriade, M., Amzica, F., Neckelmann, D., & Timofeev, I. (1998). Spike-wave complexes and fast components of cortically generated seizures. II. Extra- and intracellular patterns. *Journal of Neurophysiology, 80*, 1456–1479.

Steriade, M., Contreras, D., Dossi, R. C., & Nuñez, A. (1993). The slow (<1 hz) oscillation in reticular thalamic and thalamo-cortical neurons: Scenario of sleep rhythm generation in interacting thalamic and neocortical networks. *Journal of Neuroscience*, *13*, 3284–3299.

Steriade, M., Deschenes, M., Domich, L., & Mulle, C. (1985). Abolition of spindle oscillations in thalamic neurons disconnected from nucleus reticularis thalami. *Journal of Neurophysiology*, *54*, 1473–1497.

Steriade, M., Domich, L., Oakson, G., & Deschenes, M. (1987). The deafferented reticular thalamic nucleus generates spindle rhythmicity. *Journal of Neurophysiology*, *57*, 260–273.

Steriade, M., Nuñez, A., & Amzica, F. (1993a). A novel slow (<1 hz) oscillation of neocortical neurons *in vivo*: Depolarizing and hyperpolarizing components. *Journal of Neuroscience*, *13*, 3252–3265.

Steriade, M., Nuñez, A., & Amzica, F. (1993b). Intracellular analysis of relations between the slow (<1 hz) neocortical oscillations and other sleep rhythms of electroencephalogram. *Journal of Neuroscience*, *13*, 3266–3283.

Steriade, M., & Timofeev, I. (2001). Corticothalamic operations through prevalent inhibition of thalamocortical neurons. *Thalamus & Related Systems*, *1*, 225–236.

Steriade, M., & Timofeev, I. (2003). Neuronal plasticity in thalamocortical networks during sleep and waking oscillations. *Neuron*, *37*, 563–576.

Steriade, M., Timofeev, I., & Grenier, F. (2001). Natural waking and sleep states: A view from inside neocortical neurons. *Journal of Neurophysiology*, *85*, 1969–1985.

Tasker, G. J., & Dudek, F. E. (1991). Electrophysiology of gaba-mediated synaptic transmission and possible roles in epilepsy. *Neurochemical Research*, *16*, 251–262.

Thimm, J., Mechler, A., Lin, H., Rhee, S., & Lal, R. (2005). Calcium-dependent open/closed conformations and interfacial energy maps of reconstituted hemichannels. *The Journal of Biological Chemistry*, *280*, 10646–10654.

Timofeev, I. (2010). Pathophysiology of neocortical seizures. In C. P. Panayiotopoulos (Ed.), *The atlas of epilepsies* (pp. 203–212). London: Springer-Verlag.

Timofeev, I. (2011). Injury induced epileptogenesis: Contribution of active inhibition, disfacilitation and deafferentation to seizure induction in thalamocortical system. In M. A. Woodin, & A. Maffei (Eds.), *Inhibitory synaptic plasticity* (pp. 107–122). New York, NY: Springer.

Timofeev, I., Bazhenov, M., Avramescu, S., & Nita, D. A. (2010). Posttraumatic epilepsy: The roles of synaptic plasticity. *The Neuroscientist*, *16*, 19–27.

Timofeev, I., Bazhenov, M., Seigneur, J., & Sejnowski, T. (2012). Neuronal synchronization and thalamocortical rhythms in sleep, wake, and epilepsy. In J. L. Noebels, M. Avoli, M. A. Rogawski, R. W. Olsen, & A. V. Delgado-Escueta (Eds.), *Jasper's basic mechanisms of the epilepsies* (pp. 157–175). New York, NY: Oxford University press.

Timofeev, I., Bazhenov, M., Sejnowski, T., & Steriade, M. (2002). Cortical hyperpolarization-activated depolarizing current takes part in the generation of focal paroxysmal activities. *Proceedings of the National Academy of Sciences of the United States of America*, *99*, 9533–9537.

Timofeev, I., & Chauvette, S. (2013). The spindles: Are they still thalamic? *Sleep*, *36*, 825–826.

Timofeev, I., Contreras, D., & Steriade, M. (1996). Synaptic responsiveness of cortical and thalamic neurones during various phases of slow sleep oscillation in cat. *Journal of Physiology*, *494*, 265–278.

Timofeev, I., Grenier, F., Bazhenov, M., Sejnowski, T. J., & Steriade, M. (2000). Origin of slow cortical oscillations in deafferented cortical slabs. *Cerebral Cortex*, *10*, 1185–1199.

Timofeev, I., Grenier, F., & Steriade, M. (1998). Spike-wave complexes and fast components of cortically generated seizures. IV. Paroxysmal fast runs in cortical and thalamic neurons. *Journal of Neurophysiology*, *80*, 1495–1513.

Timofeev, I., Grenier, F., & Steriade, M. (2000). Impact of intrinsic properties and synaptic factors on the activity of neocortical networks in vivo. *Journal of Physiology, Paris, 94*, 343–355.

Timofeev, I., Grenier, F., & Steriade, M. (2001a). Intrinsic and synaptic factors in activity of neocortical neurons during natural waking-sleep cycle: An intracellular study. In R. A. Velluti (Ed.), *Actas de fisiologia* (p. 110). Montevideo, Uruguay: Universidad de la Republica.

Timofeev, I., Grenier, F., & Steriade, M. (2001b). Disfacilitation and active inhibition in the neocortex during the natural sleep-wake cycle: An intracellular study. *Proceedings of the National Academy of Sciences of the United States of America, 98*, 1924–1929.

Timofeev, I., Grenier, F., & Steriade, M. (2002). The role of chloride-dependent inhibition and the activity of fast-spiking neurons during cortical spike-wave seizures. *Neuroscience, 114*, 1115–1132.

Timofeev, I., Grenier, F., & Steriade, M. (2004). Contribution of intrinsic neuronal factors in the generation of cortically driven electrographic seizures. *Journal of Neurophysiology, 92*, 1133–1143.

Timofeev, I., Sejnowski, T. J., Bazhenov, M., Chauvette, S., & Grand, L. B. (2013). Age dependency of trauma-induced neocortical epileptogenesis. *Frontiers in Cellular Neuroscience, 7*, 154. http://dx.doi.org/10.3389/fncel.2013.00154.

Timofeev, I., & Steriade, M. (2004). Neocortical seizures: Initiation, development and cessation. *Neuroscience, 123*, 299–336.

Topolnik, L., Steriade, M., & Timofeev, I. (2003a). Partial cortical deafferentation promotes development of paroxysmal activity. *Cerebral Cortex, 13*, 883–893.

Topolnik, L., Steriade, M., & Timofeev, I. (2003b). Hyperexcitability of intact neurons underlies acute development of trauma-related electrographic seizures in cats in vivo. *European Journal of Neuroscience, 18*, 486–496.

Traub, R. D., Borck, C., Colling, S. B., & Jefferys, J. G. (1996). On the structure of ictal events in vitro. *Epilepsia, 37*, 879–891.

Turrigiano, G. (2011). Too many cooks? Intrinsic and synaptic homeostatic mechanisms in cortical circuit refinement. *Annual Review of Neuroscience, 34*, 89–103.

Turrigiano, G. G., Leslie, K. R., Desai, N. S., Rutherford, L. C., & Nelson, S. B. (1998). Activity-dependent scaling of quantal amplitude in neocortical neurons. *Nature, 391*, 892–896.

Volgushev, M., Chauvette, S., Mukovski, M., & Timofeev, I. (2006). Precise long-range synchronization of activity and silence in neocortical neurons during slow-wave sleep. *Journal of Neuroscience, 26*, 5665–5672.

Volman, V., Bazhenov, M., & Sejnowski, T. J. (2011). Pattern of trauma determines the threshold for epileptic activity in a model of cortical deafferentation. *Proceedings of the National Academy of Sciences of the United States of America, 108*, 15402–15407.

Volman, V., Bazhenov, M., & Sejnowski, T. J. (2012). Computational models of neuron-astrocyte interaction in epilepsy. *Frontiers in Computational Neuroscience, 6*, 58.

Volman, V., Bazhenov, M., & Sejnowski, T. J. (2013). Divide and conquer: Functional segregation of synaptic inputs by astrocytic microdomains could alleviate paroxysmal activity following brain trauma. *PLoS Computational Biology, 9*, e1002856.

Volman, V., Sejnowski, T. J., & Bazhenov, M. (2011). Topological basis of epileptogenesis in a model of severe cortical trauma. *Journal of Neurophysiology, 106*, 1933–1942.

von Krosigk, M., Bal, T., & McCormick, D. A. (1993). Cellular mechanisms of a synchronized oscillation in the thalamus. *Science, 261*, 361–364.

Vyazovskiy, V. V., Olcese, U., Lazimy, Y. M., Faraguna, U., Esser, S. K., Williams, J. C., et al. (2009). Cortical firing and sleep homeostasis. *Neuron, 63*, 865–878.

Watt, A. J., van Rossum, M. C., MacLeod, K. M., Nelson, S. B., & Turrigiano, G. G. (2000). Activity coregulates quantal ampa and nmda currents at neocortical synapses. *Neuron*, *26*, 659–670.

Xiong, W., Ping, X., Gao, J., & Jin, X. (2011). Preparing undercut model of posttraumatic epileptogenesis in rodents. *Journal of Visualized Experiments*, e2840. http://dx.doi.org/10.3791/2840.

CHAPTER THREE

Malformations of Cortical Development and Neocortical Focus

Heiko J. Luhmann*[,1], Werner Kilb*, Hans Clusmann[†]
*Institute of Physiology, University Medical Center of the Johannes Gutenberg University, Mainz, Germany
[†]Department of Neurosurgery, RWTH Aachen University, Aachen, Germany
[1]Corresponding author: e-mail address: luhmann@uni-mainz.de

Contents

1. Introduction 36
2. Normal and Abnormal Development of the Cerebral Cortex 37
   2.1 Neurogenesis and apoptotic cell death 37
   2.2 Gliogenesis and myelination 40
   2.3 Neuronal migration 40
   2.4 Transient neurons and transient circuits 42
   2.5 The developmental excitatory–inhibitory shift of GABA 43
3. The (Un)identified Neocortical Focus 45
   3.1 Molecular, structural, and functional alterations in an epileptic focus 46
4. Removal of a Neocortical Focus 50
   4.1 Concept of presurgical evaluation in neocortical epilepsies 50
   4.2 Epilepsy surgery for cortical malformations 52
5. Conclusion 53
Acknowledgments 53
References 53

Abstract

Developmental neocortical malformations resulting from abnormal neurogenesis, disturbances in programmed cell death, or neuronal migration disorders may cause a long-term hyperexcitability. Early generated Cajal–Retzius and subplate neurons play important roles in transient cortical circuits, and structural/functional disorders in early cortical development may induce persistent network disturbances and epileptic disorders. In particular, depolarizing GABAergic responses are important for the regulation of neurodevelopmental events, like neurogenesis or migration, while pathophysiological alterations in chloride homeostasis may cause epileptic activity. Although modern imaging techniques may provide an estimate of the structural lesion, the site and extent of the cortical malformation may not correlate with the epileptogenic zone. The neocortical focus may be surrounded by widespread molecular, structural, and functional disturbances, which are difficult to recognize with imaging technologies. However,

International Review of Neurobiology, Volume 114
ISSN 0074-7742
http://dx.doi.org/10.1016/B978-0-12-418693-4.00003-0

modern imaging and electrophysiological techniques enable focused hypotheses of the neocortical epileptogenic zone, thus allowing more specific epilepsy surgery. Focal cortical malformation can be successfully removed with minimal rim, close to or even within eloquent cortex with a promising risk–benefit ratio.

1. INTRODUCTION

The neocortex (cerebral cortex) in all mammalian species, including humans, is a usually six-layered structure composed of pyramidal (projecting) neurons and local circuit (usually inhibitory GABAergic) interneurons, organized in vertical (radial) columns. Whereas the excitatory, glutamatergic neurons show a relatively uniform structure and function, inhibitory GABAergic interneurons differ in their anatomical, molecular, and electrophysiological function (for review, see DeFelipe et al., 2013). Besides these neuronal elements, the cortex, like other brain structures, consists of different types of glial cells, which fulfill a variety of important physiological functions.

The cerebral cortex gains its characteristic structure and function during pre- and early postnatal development by activity-independent and activity-dependent processes. Although the generation of neurons and glial cells, neuronal migration, and early cell death are controlled by a variety of molecular mechanisms, electrical activity plays an important regulatory function in these early developmental processes (for review, see Khazipov & Luhmann, 2006). Spontaneous and at later developmental stages sensory or movement-induced neuronal network activity patterns can be observed in neocortical regions of rodents and humans at surprisingly early stages of corticogenesis. These early activity patterns are generated in neuronal networks within the cortex (some of them, as the subplate, are only transiently expressed), in subcortical structures, or in the sensory periphery and have a strong impact on early developmental processes. Pathophysiological disturbances in these network activity patterns, as induced by perinatal hypoxia, inflammation, or traumatic brain injury, alter these developmental processes and may cause long-term alterations in neocortical structure and function (for review, see Kilb, Kirischuk, & Luhmann, 2011). Developmental disturbances in the generation of cortical layers, columns, and networks may result in the manifestation of epilepsy (Kilb et al., 2011).

During prenatal and early postnatal development, the cerebral cortex is highly vulnerable to internally or externally induced modifications of the homeostatic state, which may induce alterations in cell generation and programmed cell death or neuronal migration. Depending on the

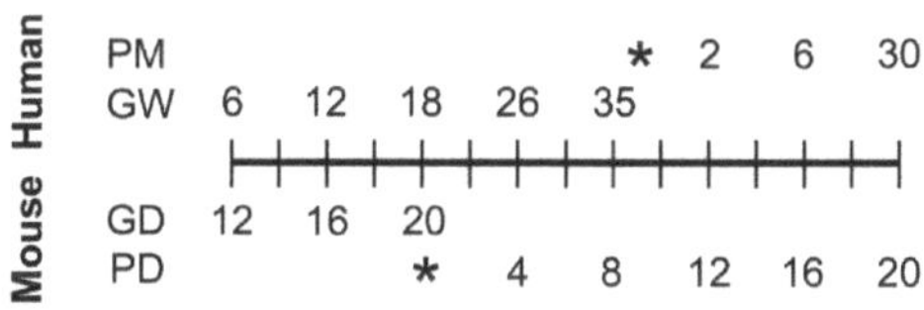

Figure 3.1 Comparative prenatal and early postnatal development of the human and mouse cerebral cortex. Comparative time points between both species are based on the calculations of the Web site http://www.translatingtime.net/ and on publications (Hagberg, Ichord, Palmer, Yager, & Vannucci, 2002; Romijn, Hofman, & Gramsbergen, 1991; Semple, Blomgren, Gimlin, Ferriero, & Noble-Haeusslein, 2013). PM, postnatal month; GW, gestational week; GD, gestational day; PD, postnatal day. Asterisks mark time point of birth in both species.

developmental state, different noxious stimuli may have the same pathological impact on early developmental processes and the same noxious stimulus may have a very different effect on corticogenesis during different developmental periods. A heterogeneous set of developmental disturbances may lead to neurological disorders including epilepsy.

Throughout this chapter, experimental results from rodents (mice and rats) and clinical data from humans are often compared. As far as we know, the early steps in neocortical development are rather similar in rodents and humans, although some differences have been reported between rodents and humans (for review, see Molnár & Clowry, 2012). Figure 3.1 gives a rough comparative estimate on the time course of neocortical development in rodents (specifically mice) and humans.

2. NORMAL AND ABNORMAL DEVELOPMENT OF THE CEREBRAL CORTEX

With the exception of the primary motor cortex (the agranular cortex lacking the granular layer 4), the cerebral cortex consists of six layers, which develop over the course of days in mice and months in humans (for review, see Rakic, 2009). The cortical layering is generated during early (mostly prenatal) development by newborn neurons that migrate into the cortical plate (CP). Disturbances in cell generation and neuronal migration have been implicated in diverse neurological and neuropsychiatric disorders including epilepsy, autism, and schizophrenia.

2.1. Neurogenesis and apoptotic cell death

In all mammalian species, neurons are not generated within the cerebral cortex itself, but rather in transient proliferative embryonic zones surrounding the cerebral lateral ventricle. Pyramidal cells are generated in the ventricular zone (VZ) directly underneath the telencephalic wall and migrate in an inside first outside last pattern radially along radial glial cells into the CP. In contrast, GABAergic interneurons are generated in the lateral and medial ganglionic eminences (future striatum and basal ganglia) and migrate laterally into the cerebral cortex.

The VZ contains neuroepithelial progenitor cells, which divide symmetrically to produce two identical daughter cells, thereby increasing the pool of neuronal progenitors. In mammals, a subset of neuroepithelial cells will differentiate into radial glial cells, which divide asymmetrically to produce another radial glial cell and either a mature neocortical neuron that migrates to its appropriate layer or an intermediate progenitor cell that translocates to the adjacent subventricular zone (SVZ). Here, intermediate progenitor cells undergo one to three symmetric cell divisions to produce either two identical neurons destined for the same cortical layer or two daughter cells that continue the cycle, thereby amplifying the neuronal population.

A number of transcription factors are expressed in specific cortical regions and during specific early developmental periods, forming different spatiotemporal molecular patterns, which influence neurogenesis and promote neuronal diversity (for review, see Hevner, Hodge, Daza, & Englund, 2006). These morphogenetic gradients control the topographic ingrowth of thalamocortical connections and thereby specify neocortical areas (for review, see Vanderhaeghen & Polleux, 2004). Furthermore, neurogenesis also seems to be under the control of the dense and elaborate vasculature expressed in the VZ and SVZ, where dividing neuronal cells are often located close to the vessels.

Although neurogenesis occurs mostly during early stages of corticogenesis, this process is already influenced by electrical activity (for review, see Kilb et al., 2011). Cell proliferation in the visual cortex is controlled by spontaneous activity patterns in the retina, so-called retinal waves (Bonetti & Surace, 2010). Activation of GABA or glutamate receptors on neocortical progenitor cells reduced their proliferation in the SVZ, while in the VZ neurogenesis was enhanced (Haydar, Wang, Schwartz, & Rakic, 2000; LoTurco, Owens, Heath, Davis, & Kriegstein, 1995). A modulatory influence on cell proliferation has been also described for other neurotransmitters such as dopamine, noradrenaline, and serotonin (for review, see Kilb et al., 2011).

Certain pathological conditions during early development may interfere with neurogenesis and thereby may cause long-term neurological disorders, such as epilepsy. One prevalent factor is perinatal hypoxia. In 1–8 births per 1000, oxygen levels drop below optimal levels and about half of the children suffering from perinatal hypoxic ischemia develop neurological deficits. Autopsy materials from preterm infants of gestational week (GW) 16–35 and experimental studies in rabbits have demonstrated that neurogenesis is suppressed by premature birth (Malik et al., 2013). Rodents exposed to neonatal hypoxia show an increased susceptibility to develop epileptic seizures (Jensen, Applegate, Holtzman, Belin, & Burchfiel, 1991), and in humans, hypoxic ischemic encephalopathy accounts for 60% of all neonatal seizures.

The mammalian target of rapamycin (mTOR) is a serine/threonine protein kinase with a central role in the regulation of cell proliferation. Hyperactivity of mTOR, as observed in tuberous sclerosis complex (TSC), may lead to abnormal brain development associated with epileptic seizures and other neurological symptoms. TSC patients show neocortical malformations that are generated during prenatal development. A number of experimental studies in mice have demonstrated that activation of mTOR leads to an overproduction of newborn neurons and subcortical heterotopia, a pathological malformation (Feliciano & Bordey, 2013), which is often associated with pharmacoresistant epilepsy.

Early cortical development is also characterized by a delicate balance between the generation of newborn neurons and simultaneously the death of immature neurons, a process called programmed cell death or apoptosis. Caspases are key elements in the mammalian cell death machinery, and experimental knockout of only one of these caspases causes a markedly enlarged and malformed cerebrum in consequence of a reduced apoptosis during development (Kuida et al., 1998). As other developmental processes, programmed cell death is also controlled by early electrical activity patterns (for review, see Kilb et al., 2011). In neocortical slice cultures, blockade of spontaneous activity by TTX application or blockade of ionotropic GABA or glutamate receptors causes a significant increase in apoptosis (Heck et al., 2008), demonstrating a direct impact of neuronal activity on the control of programmed cell death during early stages of cortical development. These data also indicate that a certain spontaneous electrical activity pattern, as also observed *in vivo* in the neonatal cerebral cortex of rodents (Yang et al., 2013; Yang, Hanganu-Opatz, Sun, & Luhmann, 2009) and in preterm and full-term human babies (Milh et al., 2007; Tolonen, Palva, Andersson, & Vanhatalo, 2007; Vanhatalo et al., 2005, 2002), is essential to control the balance of neurogenesis and cell death. Physiologically relevant burst patterns play an essential role in this process (Golbs, Nimmervoll, Sun,

Sava, & Luhmann, 2011) and the essential role of the PI3K pathway in this process has been recently demonstrated (Wagner-Golbs & Luhmann, 2012).

These experimental data also suggest that any disturbances in early activity patterns, as in perinatal hypoxia–ischemia or inflammation, may also cause alterations in programmed cell death. This has been recently demonstrated using *in vitro* and *in vivo* rodent models of neocortical development, in which inflammation caused rapid alterations in the pattern of spontaneous burst activities, which subsequently led to an increase in apoptosis (Nimmervoll et al., 2013).

2.2. Gliogenesis and myelination

Besides neurons, glial cells play a most prominent role in early brain development and in adulthood. During early postnatal weeks, the SVZ is also a source of newly generated astrocytes and oligodendrocytes. In humans, glial cells are generated during fetal and postnatal periods with a peak that coincides with the growth of blood vessels. In humans, oligodendrocyte development begins *in utero* and myelination in certain brain regions continues well into childhood and adolescence. Injury to the periventricular white matter due to hypoxia–ischemia, infections, or inflammation *in utero* or in premature infants represents the most common cause of cerebral palsy (Volpe, 2001). In the premature infant, white matter is more vulnerable than gray matter, and the period of greatest vulnerability is between GW23 and GW32. The majority of the major fiber tracts are myelinated during early childhood, but axons continue to myelinate during postnatal years. These developmental changes in myelination are identifiable by magnetic resonance (MR) images, which highlight fat (myelin) and gray matter. With maturation, the increase in myelination is accompanied by a decrease in brain water content.

White matter in humans prior to GW32, equivalent to rodents before P6, is particularly sensitive to damage from hypoxia–ischemia and inflammation, most likely because of the enhanced vulnerability of immature oligodendrocytes to oxidative stress and NMDA receptor overstimulation (Back et al., 2001).

2.3. Neuronal migration

Migrating pyramidal neurons use the radial process of radial glial cells, which extend to the upper CP, as a guiding structure to reach in an inside-out manner their final layer destination (for review, see Rakic, Ayoub, Breunig, & Dominguez, 2009). Although in full-term human babies neocortical neuronal migration is mostly terminated around birth, recent findings indicate that migration in human prefrontal cortex continues until postnatal month 18 (Sanai et al., 2011). Whereas pyramidal neurons show an almost exclusively vertical migration from their birthplace in the VZ and SVZ to their final cortical layer position, GABAergic interneurons generated in the lateral and medial ganglionic eminences have to migrate first tangentially over long distances until they also follow a radial migration path at their final cortical position (for review, see Kriegstein, 2005). The transition from tangential to radial migration seems to be also controlled by early spontaneous network activity. GABAergic interneurons terminate their tangential migration when they become engaged in synchronous network activity (De Lima, Gieseler, & Voigt, 2009).

Experimental evidence from rodent *in vitro* and *in vivo* models strongly indicates that neuronal migration is controlled by the action of neurotransmitters (for review, see Kilb et al., 2011). In the cerebral cortex and also in the hippocampus, neuronal migration of pyramidal neurons and GABAergic interneurons is influenced by the neurotransmitters GABA and glutamate (for review, see Manent, Beguin, Ganay, & Represa, 2011). Both transmitters act in a paracrine manner on migrating neurons, which are not yet integrated in a synaptic network. Transient blockade of NMDA receptors in rat parietal cortex *in vivo* causes disturbances in cortical layering and heterotopic cell clusters in layer I, which are accompanied by a synaptic hyperexcitability (Reiprich, Kilb, & Luhmann, 2005). Similar neuronal migration disorders have been also demonstrated in neocortical organotypic slice cultures and *in vivo* in the rat parietal cortex with compounds acting on GABA-A receptors (Heck et al., 2007). Furthermore, recent *in vitro* experiments have shown that GABA-C receptors also control neuronal migration in neocortical organotypic slice cultures (Denter et al., 2010). Interestingly, activation of high-affinity GABA-C receptors in the intermediate zone acts as a "go-signal" for migration, whereas activation of low-affinity GABA-A receptors in the CP acts as a "stop signal" for migrating neurons (Denter et al., 2010). Finally, glycine receptors also modulate neuronal migration (Nimmervoll, Denter, Sava, Kilb, & Luhmann, 2011). It is not surprising that compounds acting on NMDA or ionotropic GABA receptors (e.g., alcohol, anesthetics, and antiepileptics) may have a profound influence on neuronal migration and may cause structural abnormalities in the cerebral cortex (for review, see Manent et al., 2011).

Neuronal migrations disorders in the cerebral cortex represent a large group of neuropathological malformations (mostly focal cortical dysplasia, heterotopia, and polymicrogyria), which are caused during pre- and perinatal periods by internal pathophysiological states (e.g., *in utero* hypoxia, inflammation, and infection), external noxious stimuli (e.g., maternal drug abuse during pregnancy and X-ray irradiation), or genetic mutations (for review, see Guerrini & Parrini, 2010; Palmini, Andermann, & Andermann, 1994; Ross & Walsh, 2001). A large number of experimental and clinical data strongly indicate that neuronal migration disorders may cause epileptic seizures (see Sections 3 and 4).

2.4. Transient neurons and transient circuits

Besides the important and physiologically relevant process of programmed cell death during early corticogenesis (see Section 2.1), two distinct subpopulations of transient and early generated neurons play a central role in neocortical development: Cajal–Retzius cells and subplate neurons (for review, see Luhmann, 2013; Luhmann, Hanganu, & Kilb, 2003, Luhmann, Kirischuk, Sinning, & Kilb, 2014). The glutamatergic Cajal–Retzius neurons are located in the most superficial layer of the cerebral cortex, the marginal zone, which later becomes the cell sparse layer I, and produce the extracellular matrix protein reelin, which is essential for radial neuronal migration. When radial migration is terminated, almost all Cajal–Retzius cells disappear by programmed cell death. Cajal–Retzius neurons are an integral part of the early neocortical network and receive excitatory GABAergic inputs (Achilles et al., 2007; Kolbaev, Achilles, Luhmann, & Kilb, 2011; Kolbaev, Luhmann, & Kilb, 2011). A pathophysiological consequence of a Cajal–Retzius cell dysfunction is the reeler mutant mouse, which shows prominent neuronal migration disorders in the cerebellum, hippocampus, and cerebral cortex. It has been suggested that reelin may be also implicated in synaptic transmission and in neurodevelopmental disorders, such as epilepsy and neuropsychiatric disorders (for review, see Herz & Chen, 2006). The high number of Cajal–Retzius cells in layer I of the temporal cortex from epileptic patients with focal cortical dysplasia may indicate that these cells did not undergo apoptosis and contributed as surviving network elements to the manifestation of epileptic seizures (Garbelli et al., 2001). A close correlation between reelin deficiency and focal migration defects has been demonstrated in the hippocampus from patients suffering from temporal lobe epilepsy and in animal models (for review, see Frotscher, 2010).

Glutamatergic and GABAergic subplate cells represent another population of early neurons, which are transiently expressed in all mammalian species during early cortical development between the intermediate zone and the CP (for review, see Kanold & Luhmann, 2010). Subplate cells represent a very heterogeneous population of neurons with relatively mature structural and functional properties, which receive prominent synaptic inputs from specific thalamic nuclei (Hanganu, Kilb, & Luhmann, 2002; Hirsch & Luhmann, 2008) and play an important role in the generation and amplification of early synchronized network activity (Dupont, Hanganu, Kilb, Hirsch, & Luhmann, 2006; Hanganu, Okabe, Lessmann, & Luhmann, 2009; Yang et al., 2013, 2009; for review, see Luhmann, Kilb, & Hanganu-Opatz, 2009). Furthermore, subplate cells are crucial for the path finding of corticopetal and corticofugal axonal projections, in the formation of the neocortical columnar architecture, in developmental plasticity, and in the maturation of neocortical GABAergic inhibition (for review, see Kanold & Luhmann, 2010). It has been suggested that cortical dysplasia associated with pharmacoresistant epilepsy could be the consequence of postnatal retention of subplate neurons, which did not undergo programmed cell death (Cepeda et al., 2007).

2.5. The developmental excitatory–inhibitory shift of GABA

The observation that the activation of GABA-A receptors induces a membrane depolarizing and in many cases excitatory postsynaptic responses in the developing nervous system (Achilles et al., 2007; Ben-Ari, Cherubini, Corradetti, & Gaiarsa, 1989; Luhmann & Prince, 1991; Owens, Boyce, Davis, & Kriegstein, 1996) is of particular relevance to understand the role of GABA in controlling neurodevelopmental events like proliferation, migration, cell death, and early network activity (for review, see Kilb et al., 2011; Wang & Kriegstein, 2009). Depolarizing responses are mainly caused by an altered Cl^- homeostasis during early development (Fig. 3.2A) (Ben-Ari et al., 2012), although HCO_3^- fluxes also considerably contribute to depolarizing driving forces in GABAergic responses (Rivera, Voipio, & Kaila, 2005). In immature neurons, the intracellular Cl^- concentration ($[Cl^-]_i$) is elevated, as compared to most mature neurons in the central nervous system. This elevated $[Cl^-]_i$ is caused by the dominance of transport proteins mediating Cl^- uptake, in particular by isoform 1 of the Na^+–K^+–Cl^- cotransporter (NKCC1), while the neuron-specific K^+–Cl^- cotransporter (KCC2) is expressed at lower levels (Rivera et al., 1999; Yamada et al., 2004). In rodents, the adult-like low $[Cl^-]_i$ is generally reached at the end of the second postnatal week (Owens & Kriegstein,

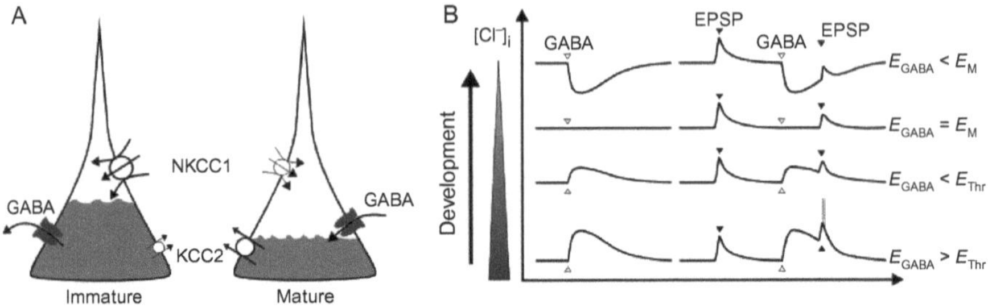

Figure 3.2 Chloride homeostasis in immature and mature neurons and its influence on GABAergic function. (A) In immature neurons, the expression of the Cl^- extruder KCC2 is low; thus Cl^- can be efficiently accumulated by the Cl^- loader NKCC1. In mature neurons, the expression and membrane insertion of KCC2 are high, enabling the maintenance of a low intracellular Cl^- concentration ($[Cl^-]_i$). (B) Influence of $[Cl^-]_i$ on GABAergic membrane responses. In low $[Cl^-]_i$, as in mature neurons, the GABA reversal potential (E_{GABA}) is below the resting membrane potential (E_M), leading to hyperpolarizing GABAergic membrane responses (upper trace). Under this condition, GABA reduced the peak amplitude of excitatory postsynaptic potentials (EPSP). If $E_{GABA} = E_M$ (second trace from top), application of GABA does not change E_M, but still reduces the EPSP amplitude by shunting inhibition. If E_{GABA} is above E_M, as in immature neurons, but below the action potential threshold (E_{Thr}), GABA mediates depolarizing responses, but the shunting effect on EPSP overpowers the depolarizing effect, leading to an inhibitory action (third trace from top). Only if E_{GABA} is above E_{Thr}, as in very immature neurons, the GABAergic membrane response will facilitate action potential generation by the EPSP (lower trace).

2002), however, with substantial cell-type, layer, and even compartment-specific diversity (Ikeda et al., 2003; Khirug et al., 2008; Shimizu-Okabe et al., 2002). Interestingly, a shift to elevated $[Cl^-]_i$ and to depolarizing GABAergic responses has also been observed after traumatic brain injuries and in neurons from epileptic patients (Huberfeld et al., 2007; Nabekura et al., 2002; Palma et al., 2006), indicating that maladaptive changes in $[Cl^-]_i$ homeostasis can contribute to neurological symptoms.

The status of $[Cl^-]_i$ homeostasis in human interm babies is currently a matter of debate (for review, see Löscher, Puskarjov, & Kaila, 2013). Based on the expression levels of NKCC1 and KCC2, it has been proposed that in human babies, a considerable maturation of KCC2 expression and $[Cl^-]_i$ may occur in the first postnatal weeks, suggesting depolarizing responses in early postnatal periods. On the other hand, a recent study demonstrated in the prefrontal cortex and hippocampus of early postnatal animals only slightly reduced KCC2 expression levels and KCC2/NKCC1 expression ratio that is comparable to the adult situation (Hyde et al., 2011).

For the assessment of the E–I shift in GABAergic action, it must also be considered that GABA-A receptor-induced depolarizations are not necessarily excitatory, but can also mediate inhibitory actions by shunting membrane currents (Edwards, 1990), activating voltage-dependent K^+ currents (Monsivais & Rubel, 2001), or inactivating voltage-dependent Na^+ currents (Rojas, Akrouh, Eisenman, & Mennerick, 2011). In general, GABAergic depolarizations are inhibitory if the GABA reversal potential is negative to the action potential threshold (Fig. 3.2B) (Kolbaev, Achilles, et al., 2011; Owens & Kriegstein, 2002). However, even smaller $[Cl^-]_i$ can reliably mediate depolarizing actions by recruitment of subthreshold voltage-gated currents, like persistent Na^+ currents (Valeeva, Valiullina, & Khazipov, 2013).

The depolarizing and potentially excitatory action of GABA has been considered as one reason for the poor pharmacological responsiveness of antiepileptic drugs in children (Ben-Ari & Holmes, 2005). Accordingly, it was proposed that inhibition of NKCC1-mediated Cl^- accumulation can alleviate seizures and/or enhance the anticonvulsive efficiency of GABA-acting antiepileptic drugs (Dzhala, Brumback, & Staley, 2008; for critical review, see Löscher et al., 2013).

3. THE (UN)IDENTIFIED NEOCORTICAL FOCUS

The molecular, anatomical, and electrophysiological changes in a cortical epileptic focus generating a hyperexcitable network have been studied in a number of animal models and in human epileptic tissue. However, it is still unclear whether every cortical malformation causes epilepsy and whether the location (e.g., cortical area and layer) and size of the malformation correlates with the severity of the seizures (for review, see Schwartzkroin & Wenzel, 2012)? The epileptic focus or epileptic zone, defined as the region where seizures originate, cannot always unequivocally be identified and does not necessarily correlate with the site of the structural lesion with more or less obvious morphological changes as recognized by imaging techniques (Fig. 3.3). For presurgical evaluation, a precise identification of the epileptic focus is highly admirable to reach best outcomes for the patients with minimal neurological side effects (see Section 4). This section describes the molecular, anatomical, and electrophysiological changes in a cortical epileptic focus as identified in various animal models and in experimental studies of human neocortical tissue obtained during surgery for the treatment of severe seizures (for review, see also Najm, Tilelli, & Oghlakian, 2007).

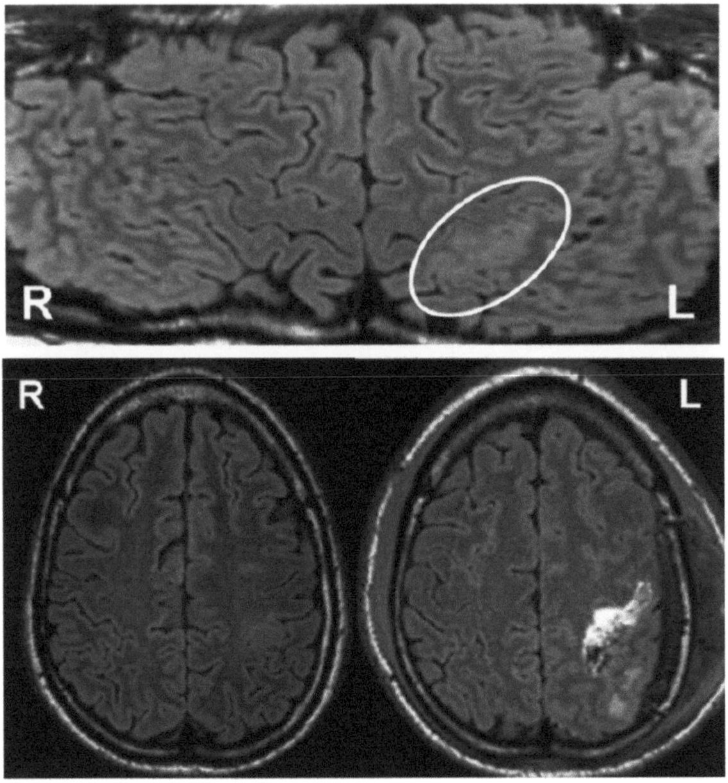

Figure 3.3 Identification of the neocortical structural lesion in a 22-year-old male patient suffering from longstanding epilepsy with simple and complex partial seizures and secondary generalization. There is no neurological deficit and no mental disorder. The epilepsy is refractory to different kinds of medication so that the patient underwent presurgical evaluation. Upper part shows the MR FLAIR image of the patient: the planar reformatting (pancake-view) clearly shows a structural alteration in the left postcentral gyrus, suggestive of focal cortical dysplasia. Lower part compares in regular axial orientation pre- and early postoperative findings, documenting the completeness of lesionectomy. Histology: Focal cortical dysplasia Type IIb (Palmini & Lüders, 2002).

3.1. Molecular, structural, and functional alterations in an epileptic focus

3.1.1 Animal models

The primary causes of a cortical malformation resulting in an epileptic focus are disturbances in cell generation, programmed cell death, and neuronal migration. These developmental disturbances are accompanied by the presence of atypical cell types, aberrant local and long-range connectivity, imbalance in excitatory and inhibitory synaptic transmission, unusual gap junctional coupling, and changes in passive and active neuronal membrane properties.

Experimental models of neocortical malformations resembling to some extent the human neuropathology in cortical epilepsies are the *in utero* irradiation model (Xiang, Chen, Yu, King, & Roper, 2006), the model of *in utero* injection of methylazoxymethanol acetate (mostly hippocampal) also used as a "double-hit" model in combination with pilocarpine (Battaglia, Colciaghi, Finardi, & Nobili, 2013; Colacitti et al., 1999; Colciaghi et al., 2011; Tschuluun, Wenzel, Katleba, & Schwartzkroin, 2005), the 1-3-bis-chloroethyl-nitrosurea model (Moroni et al., 2011, 2009), the model of *in utero* RNA interference of double cortex (Dcx) (Lapray et al., 2010; Ramos, Bai, & LoTurco, 2006), the mutant rat model of telencephalic internal structural heterotopia (Trotter, Kapur, Anzivino, & Lee, 2006), the mutant mouse model of subcortical band heterotopia (Croquelois et al., 2009), spontaneous neocortical malformations in inbred lines as the NXSM-D/Ei mice (Gabel & LoTurco, 2002), and the freeze lesion model (for review, see Luhmann, 2006). Spontaneous seizures have been demonstrated in only some of these models. Often the threshold for seizure induction is reduced in these animals. In other models, modifications in the experimental protocol for the induction of neocortical malformation may lead to clinically more relevant models. For example, in the cortical freeze lesion model (Jacobs, Gutnick, & Prince, 1996; Luhmann & Raabe, 1996), lesion induction at postnatal ages causes an intracortical hyperexcitability *in vitro*, but no spontaneous epileptic seizures *in vivo* (for review, see Luhmann, 2006). However, when multiple lesions are induced at late prenatal stages, the animals may develop spontaneous seizures (Kamada et al., 2013), indicating that the time point of lesion induction has a strong influence on the outcome.

A number of studies have demonstrated an imbalance between excitatory and inhibitory synaptic function. In the *in utero* irradiation and freeze lesion model, an impairment in GABAergic function has been reported. This loss in inhibition may result from a decrease in the density (Rosen, Jacobs, & Prince, 1998) or in the excitatory synaptic drive of GABAergic interneurons (Luhmann, Karpuk, Qü, & Zilles, 1998; Zhou & Roper, 2011), a loss or reorganization of inhibitory axonal projections (Jacobs et al., 1996), a reduction in GABA receptors (Zilles, Qü, Schleicher, & Luhmann, 1998), changes in the subunit composition of GABA receptors (Redecker, Luhmann, Hagemann, Fritschy, & Witte, 2000), and disturbances in intracellular chloride regulation (Shimizu-Okabe et al., 2007).

Besides a decrease in GABAergic inhibition, an increase in glutamatergic excitation has been also reported in various experimental models. Pyramidal neurons may receive a stronger excitatory synaptic drive (Brill & Huguenard, 2010; Jacobs & Prince, 2005), presynaptic glutamate release may be enhanced (Chen, Xiang, & Roper, 2007), thalamocortical and intracortical axonal projections may be disorganized (Jacobs, Hwang, & Prince, 1999; Jacobs, Mogensen, Warren, & Prince, 1999), and glutamatergic receptors may be increased (Kamada et al., 2013; Zilles et al., 1998) or their subunit composition may be disturbed (Hagemann, Kluska, Redecker, Luhmann, & Witte, 2003). This imbalance between excitatory and inhibitory synaptic function is accompanied by an impairment in synaptic plasticity (Peters et al., 2004).

A consistent finding in different experimental models was the surprisingly widespread structural and functional changes around the cortical lesion. Not the structural lesion itself but rather the perilesional site was the epileptogenic zone (EZ), which spontaneously or upon stimulation generated epileptiform activity propagating over large neocortical regions (Jacobs et al., 1996; Luhmann & Raabe, 1996; Luhmann, Raabe, Qü, & Zilles, 1998). In the rat freeze lesion model, GABA and glutamate receptors are changed in their density and subunit composition over neocortical regions extending several millimeters from the site of the structural lesion (Hagemann et al., 2003; Redecker et al., 2000; Zilles et al., 1998). These findings may have important clinical implications because in epileptic patients the presurgical evaluation and definition of the epileptic focus are often difficult (see Section 4). *In vitro* studies in the NXSM-D/Ei mutant mouse model of neocortical ectopias have shown that even a single ectopia is associated with a higher excitability, but that epileptiform activity is not generated within the malformation itself (Gabel & LoTurco, 2002). Furthermore, removal of the cortical malformation (structural lesion) did not restore normal excitability (Gabel & LoTurco, 2002).

However, a direct causal relationship between the cortical malformation and the severity of epileptic activity has been demonstrated in the rat model of subcortical band heterotopia generated by *in utero* RNA interference of the Dcx gene. When aberrantly positioned neurons were stimulated to migrate by reexpressing the Dcx after birth, migration restarted and neocortical malformations were reduced. This repair resulted in a reduction of the convulsant-induced seizure threshold to normal levels (Manent, Wang, Chang, Paramasivam, & LoTurco, 2009). These elegant experiments also suggest that neuronal migration disorders may be treatable by restarting developmental programs again.

3.1.2 Experimental studies in human tissue

As in animal models of neocortical malformations associated with epilepsy, neuropathological studies of tissue from epileptic patients also indicate that focal cortical dysplasia may be the result of neuronal migration disorders, an exuberant production of neuroblasts, and/or a disruption of mechanisms for naturally occurring cell death (Spreafico et al., 1998). Focal defects in cortical organization often involve deeper cortical neurons and the white matter (Rossini et al., 2011), supporting the hypothesis that surviving subplate neurons may be involved in the pathophysiology of some forms of human focal epilepsies. Quantitative receptor autoradiography in the neocortex of patients with focal epilepsy revealed a prominent and consistent increase in the density of AMPA receptor-binding sites, but no significant changes in NMDA or GABA-A receptors (Palomero-Gallagher et al., 2012; Zilles, Qü, Köhling, & Speckmann, 1999).

In cortical tissue obtained from children with intractable focal epilepsy, Jansen, Peugh, Roden, and Ojemann (2010) demonstrated a disruption of the normal maturation of cortical GABA-A receptor subunit expression. Focal decreases in GABA-A receptor binding even remote from the neocortical epileptic focus have been reported with PET imaging in young patients with medically refractory epilepsy (Juhasz et al., 2009). *In vitro* electrophysiological recordings in neocortical slices obtained from epileptic patients who underwent surgery also indicate prominent changes in GABAergic function in the epileptic focus and in the surrounding cortical regions. GABA-A receptor-mediated inhibition was substantially altered in regions of dysplasia and GABA transporter expression was markedly reduced in patients with focal cortical dysplasia (Calcagnotto, Paredes, Tihan, Barbaro, & Baraban, 2005). Neocortical slices from tissue resected for the treatment of pharmacoresistant epilepsy in children revealed spontaneous GABA-A-mediated depolarizations, which even elicited action potentials (Cepeda et al., 2007). Recent anatomical and electrophysiological studies in slices obtained from pharmacoresistant pediatric epilepsy patients reported abnormal, often immature-like cell morphologies and spontaneous rhythmic activity, which was action potential dependent, mediated by GABA-A receptors and unaffected by glutamate receptor antagonists (Cepeda et al., 2014). The expression of excitatory GABA-A mediated activity may result from long-term alterations in the expression of the neuron-specific K^+–Cl^- cotransporter KCC2, as demonstrated in human cortical dysplasia (Munakata et al., 2007). In contrast to phasic (synaptic) GABAergic inhibition, tonic (extrasynaptic) inhibition seems to be preserved in animal models and in human epileptic tissue (for review, see Pavlov & Walker, 2013).

An intrinsic ability to generate ictal-like epileptiform events has been demonstrated in human neocortical slices with focal cortical dysplasia treated with 4-aminopyridine (Avoli, Bernasconi, Mattia, Olivier, & Hwa, 1999). Two distinct patterns of spontaneous epileptiform events, which were sensitive to glutamate or GABA-A receptor antagonists, could be observed under these experimental conditions (Avoli et al., 1999). Acutely dissociated neurons from pediatric cortical dysplasia tissue showed a decrease in NR2B subunit expression and a reduced magnesium sensitivity of the NMDA receptor, which may contribute to the hyperexcitability in or near the epileptic focus (André et al., 2004).

4. REMOVAL OF A NEOCORTICAL FOCUS

4.1. Concept of presurgical evaluation in neocortical epilepsies

In selected cases epilepsy surgery, i.e., removing a neocortical malformation and surrounding epileptogenic cortex may lead to partial or complete seizure control. Success largely depends not only on the type and extent of malformation but also on preoperative evaluation and transfer of relevant data to decision making and resulting surgical strategy.

Former work resulted in a theory of different zones involved in the epilepsies. In spite of some definition differences, these concepts turned out to be important because they are the fundamental of applying multiple diagnostic means to approach the epileptic focus or "epileptogenic zone" (Palmini, 2006; Rosenow & Lüders, 2001). Six cortical zones that can be defined in the presurgical evaluation of candidates for epilepsy surgery are (1) the symptomatogenic zone (initial ictal clinical symptoms); (2) the *irritative* zone (interictal EEG spikes); (3) the *ictal onset* zone (initial EEG ictal seizure activity); (4) the *epileptogenic lesion* (MRI); (5) the *functional deficit zone* (neurological or neuropsychological deficits); and (6) the *eloquent cortex* (normal adjacent functional cortex). Different diagnostic techniques are used in the delineation of these cortical zones, such as video-EEG monitoring and structural as well as functional imaging. Presurgical diagnostics should result in an estimate of the "epileptogenic zone," defined as the neocortical tissue, which is inevitably necessary for the generation of clinical epileptic seizures. There are no diagnostic tools to measure directly which and which amount of tissue is necessary to generate epileptic seizures. The more circumscribed a cortical malformation, the potentially more restricted the EZ. This holds especially true for focal cortical dysplasia Type IIb

(Palmini & Lüders, 2002), where the EZ is mainly located in the immediate vicinity, so that strictly limited resections are successful in most cases (Wagner et al., 2011). As there are so far no tools to directly demonstrate the EZ, new developments aim at biomarkers which are closely related to the EZ (Gnatkovsky et al., 2014). Thus, at present, the EZ can only be proven by a circumscribed neocortical resection or disconnection leading to seizure freedom (Schramm & Clusmann, 2008).

The concept of the EZ is still predominantly a domain of simultaneous video-EEG monitoring and seizure semiology analysis (Cascino, 2002; Janszky et al., 2001). Invasive recording via chronically implanted electrodes is often indicated if,

1. nonconclusive or even discordant results from noninvasive procedures, especially from interictal and ictal EEG, are recorded, or
2. nonlesional high-resolution MRI not clearly distinguishable from normal tissue, questionable, diffuse, or multiple lesions are found, or
3. localization of the assumed epileptogenic lesion is close to or overlapping with eloquent areas, thus requiring electrical stimulation for cortical mapping (as an alternative to awake craniotomy) (Schramm & Clusmann, 2008).

Multiple different electrode types and procedures may be used for chronic invasive monitoring: depth electrodes with few or as stereo-EEG-recording with multiple depth electrodes (Talairach & Szikla, 1980), subdural electrodes (Wyler, Ojemann, Lettich, & Ward, 1984), epidural electrodes, peg electrodes inserted in the bone, sphenoidal, and foramen ovale electrodes. These electrodes allow interictal and also ictal recordings and are therefore able, if positioned adequately, to provide closer hints to the epileptogenic focus.

Originally, invasive monitoring was intraoperative electrocorticography (ECoG) or cortical mapping in the awake patient (Jasper, Pertuisset, & Flanigin, 1951). Intraoperative ECoG is often performed in neocortical epilepsy surgery after noninvasive evaluation to determine the border of extended lesionectomy in patients with neocortical lesions, as, for example, illustrated in Fig. 3.3. The disadvantages of this method are the poorly defined influence of anesthetics, the short recording time, and the lack of seizure recording. Basically, intraoperative ECoG is an interictal recording. Therefore, intraoperative ECoG is restricted to the definition of the irritative zone and has thus limitations for sufficiently delineating the EZ or eloquent cortices (Zentner et al., 1997). Intraoperative spikes or bursts may indicate the irritative zone. Cessation of such activity after resection can be taken as an indirect hint, that parts of the epileptogenic focus have been resected or disconnected successfully. However, correlation with postoperative seizure control has been weak (Kuruvilla & Flink, 2003). High-frequency oscillations (HFO: ripples, 80–250 Hz; fast ripples, >250 Hz) have been suggested as a marker closer correlated to the EZ: with removal of HFO-generating areas, good surgical outcome was reached in most cases. Thus, HFOs may be used as a marker of epileptogenicity, potentially more accurate than spike-generating areas. In patients in whom the majority of HFO-generating tissue remained, a poor surgical outcome occurred (Jacobs et al., 2010). A recent study described seven characteristic HFO seizure-onset patterns: low-voltage fast activity (43%); low-frequency high-amplitude periodic spikes (21%); sharp activity at ≤13 Hz (15%); spike-and-wave activity (9%); burst of high-amplitude polyspikes (6%); burst suppression (4%); and delta brush (4%). The latter occurred only in cases of cortical dysplasia, but not always with this pathology. Compared to other patterns, low-voltage fast activity was associated with a larger seizure-onset zone (Perucca, Dubeau, & Gotman, 2014).

4.2. Epilepsy surgery for cortical malformations

Several surgery types are used at present: lobectomy, lesionectomy (pure or with variable rim of cortex), and corticectomy. Complete removal of the EZ can be compromised by the overlap with eloquent cortex, e.g., the primary motor cortex, cortex areas representing speech function, and visual cortex. Established measures to reliably assess the eloquence of certain cortical areas are cortical mapping via chronically implanted electrodes and intraoperative mapping during "awake craniotomy" (Berger, Kincaid, Ojemann, & Lettich, 1989; Ojemann, Ojemann, Lettich, & Berger, 1989). The aim of these measures is to remove as much tissue as thought to be necessary to provide complete seizure relief, without causing inacceptable permanent neurological damage. The depth of resection should include the whole cortical surface, including the deep cortical folds. A resection of 2.5–3 cm in depth is usually sufficient. It is more difficult to define the horizontal extent of resection, and this definition may also depend on the pathology. A lesionectomy with rim is preferable whenever possible, although in many cases of Type IIb focal cortical dysplasia (Palmini & Lüders, 2002) nearly pure lesionectomies and leaving the deep "root" are thought to be sufficient, as illustrated in Fig. 3.3 (Urbach et al., 2002; Wagner et al., 2011).

Complete yet safe resection close to motor areas in medically intractable epilepsy requires functional information. New deficits may occur despite preservation of motor cortex, e.g., through vascular compromise. Continuous motor-evoked potential (MEP) monitoring in focal epilepsy surgery may provide additional safety and should be used in cases close to the motor strip as an alternative or additionally to extraoperative motor cortex mapping via chronically implanted subdural electrodes. It has been shown that MEP changes predict occurrence and permanence of new pareses. Successful MEP monitoring correlates with unimpaired motor outcome and full seizure control (Neuloh, Bien, Clusmann, von Lehe, & Schramm, 2010). There are some hints toward an increasing role for intraoperative MRI in immediate resection control of epileptogenic lesions. However, applicability and superiority have not been proven yet (Buchfelder, Fahlbusch, Ganslandt, Stefan, & Nimsky, 2002).

5. CONCLUSION

A large number of developmental malformations may not fulfill the criteria of an epileptic "focus" in a strict sense, although clinical imaging techniques reveal a focal lesion. Molecular, anatomical, and electrophysiological changes surrounding this lesion may be rather widespread, and it is currently unclear whether these global modifications are the cause or the consequence of epileptic seizures.

Although modern diagnostic techniques enable more stringent hypotheses of the neocortical EZ, thus enabling more precisely directed resection, the exact location and size of the EZ are still difficult to determine. Future methods and studies combining additional parameters, e.g., intraoperative molecular imaging, may lead to a further improvement and better identification of the epileptic focus.

ACKNOWLEDGMENTS

We are most thankful to our coworkers, who over many years contributed substantially to the experimental and clinical research cited in this chapter. The work of the authors has been supported by grants of the DFG.

REFERENCES

Achilles, K., Okabe, A., Ikeda, M., Shimizu-Okabe, C., Yamada, J., Fukuda, A., et al. (2007). Kinetic properties of Cl uptake mediated by Na+-dependent K+-2Cl cotransport in immature rat neocortical neurons. *Journal of Neuroscience*, *27*, 8616–8627.

André, V. M., Flores-Hernandez, J., Cepeda, C., Starling, A. J., Nguyen, S., Lobo, M. K., et al. (2004). NMDA receptor alterations in neurons from pediatric cortical dysplasia tissue. *Cerebral Cortex*, *14*, 634–646.

Avoli, M., Bernasconi, A., Mattia, D., Olivier, A., & Hwa, G. G. C. (1999). Epileptiform discharges in the human dysplastic neocortex: In vitro physiology and pharmacology. *Annals of Neurology*, *46*, 816–826.

Back, S. A., Luo, N. L., Borenstein, N. S., Levine, J. M., Volpe, J. J., & Kinney, H. C. (2001). Late oligodendrocyte progenitors coincide with the developmental window of vulnerability for human perinatal white matter injury. *Journal of Neuroscience*, *21*, 1302–1312.

Battaglia, G., Colciaghi, F., Finardi, A., & Nobili, P. (2013). Intrinsic epileptogenicity of dysplastic cortex: Converging data from experimental models and human patients. *Epilepsia*, *54*, 33–36.

Ben-Ari, Y., Cherubini, E., Corradetti, R., & Gaiarsa, J.-L. (1989). Giant synaptic potentials in immature rat CA3 hippocampal neurones. *Journal of Physiology (London)*, *416*, 303–325.

Ben-Ari, Y., & Holmes, G. L. (2005). The multiple facets of gamma-aminobutyric acid dysfunction in epilepsy. *Current Opinion in Neurology*, *18*, 141–145.

Ben-Ari, Y., Woodin, M. A., Sernagor, E., Cancedda, L., Vinay, L., Rivera, C., et al. (2012). Refuting the challenges of the developmental shift of polarity of GABA actions: GABA more exciting than ever! *Frontiers in Cellular Neuroscience*, *6*, 35. http://dx.doi.org/10.3389/fncel.2012.00035.

Berger, M. S., Kincaid, J., Ojemann, G. A., & Lettich, E. (1989). Brain mapping techniques to maximize resection, safety, and seizure control in children with brain tumors. *Neurosurgery*, *25*, 786–792.

Bonetti, C., & Surace, E. M. (2010). Mouse embryonic retina delivers information controlling cortical neurogenesis. *PLoS One*, *5*(12), e15211.

Brill, J., & Huguenard, J. R. (2010). Enhanced infragranular and supragranular synaptic input onto layer 5 pyramidal neurons in a rat model of cortical dysplasia. *Cerebral Cortex*, *20*, 2926–2938.

Buchfelder, M., Fahlbusch, R., Ganslandt, O., Stefan, H., & Nimsky, C. (2002). Use of intraoperative magnetic resonance imaging in tailored temporal lobe surgeries for epilepsy. *Epilepsia*, *43*, 864–873.

Calcagnotto, M. E., Paredes, M. F., Tihan, T., Barbaro, N. M., & Baraban, S. C. (2005). Dysfunction of synaptic inhibition in epilepsy associated with focal cortical dysplasia. *Journal of Neuroscience*, *25*, 9649–9657.

Cascino, G. D. (2002). Video-EEG monitoring in adults. *Epilepsia*, *43*(Suppl. 3), 80–93.

Cepeda, C., Andre, V. M., Wu, N., Yamazaki, I., Uzgil, B., Vinters, H. V., et al. (2007). Immature neurons and GABA networks may contribute to epileptogenesis in pediatric cortical dysplasia. *Epilepsia*, *48*(Suppl. 5), 79–85.

Cepeda, C., Chen, J. Y., Wu, J. Y., Fisher, R. S., Vinters, H. V., Mathern, G. W., et al. (2014). Pacemaker GABA synaptic activity may contribute to network synchronization in pediatric cortical dysplasia. *Neurobiology of Disease*, *62*, 208–217.

Chen, H. X., Xiang, H., & Roper, S. N. (2007). Impaired developmental switch of short-term plasticity in pyramidal cells of dysplastic cortex. *Epilepsia*, *48*, 141–148.

Colacitti, C., Sancini, G., DeBiasi, S., Franceschetti, S., Caputi, A., Frassoni, C., et al. (1999). Prenatal methylazoxymethanol treatment in rats produces brain abnormalities with morphological similarities to human developmental brain dysgeneses. *Journal of Neuropathology & Experimental Neurology*, *58*, 92–106.

Colciaghi, F., Finardi, A., Frasca, A., Balosso, S., Nobili, P., Carriero, G., et al. (2011). Status epilepticus-induced pathologic plasticity in a rat model of focal cortical dysplasia. *Brain*, *134*, 2828–2843.

Croquelois, A., Giuliani, F., Savary, C., Kielar, M., Amiot, C., Schenk, F., et al. (2009). Characterization of the HeCo mutant mouse: A new model of subcortical band heterotopia associated with seizures and behavioral deficits. *Cerebral Cortex*, *19*, 563–575.

DeFelipe, J., Lopez-Cruz, P. L., Benavides-Piccione, R., Bielza, C., Larranaga, P., Anderson, S., et al. (2013). New insights into the classification and nomenclature of cortical GABAergic interneurons. *Nature Reviews Neuroscience*, *14*, 202–216.

De Lima, A. D., Gieseler, A., & Voigt, T. (2009). Relationship between GABAergic interneurons migration and early neocortical network activity. *Developmental Neurobiology*, *69*, 105–123.

Denter, D. G., Heck, N., Riedemann, T., White, R., Kilb, W., & Luhmann, H. J. (2010). GABA(C) receptors are functionally expressed in the intermediate zone and regulate radial migration in the embryonic mouse neocortex. *Neuroscience*, *167*, 124–134.

Dupont, E., Hanganu, I. L., Kilb, W., Hirsch, S., & Luhmann, H. J. (2006). Rapid developmental switch in the mechanisms driving early cortical columnar networks. *Nature*, *439*, 79–83.

Dzhala, V. I., Brumback, A. C., & Staley, K. J. (2008). Bumetanide enhances phenobarbital efficacy in a neonatal seizure model. *Annals of Neurology*, *63*, 222–235.

Edwards, D. H. (1990). Mechanisms of depolarizing inhibition at the crayfish giant motor synapse. I. Electrophysiology. *Journal of Neurophysiology*, *64*, 532–540.

Feliciano, D. M., & Bordey, A. (2013). Newborn cortical neurons: Only for neonates? *Trends in Neurosciences*, *36*, 51–61.

Frotscher, M. (2010). Role for Reelin in stabilizing cortical architecture. *Trends in Neurosciences*, *33*, 407–414.

Gabel, L. A., & LoTurco, J. J. (2002). Layer I ectopias and increased excitability in murine neocortex. *Journal of Neurophysiology*, *87*, 2471–2479.

Garbelli, R., Frassoni, C., Ferrario, A., Tassi, L., Bramerio, M., & Spreafico, R. (2001). Cajal–Retzius cell density as marker of type of focal cortical dysplasia. *Neuroreport*, *12*, 2767–2771.

Gnatkovsky, V., de Curtis, M., Pastori, C., Cardinale, F., Lo Russo, G., Mai, R., et al. (2014). Biomarkers of epileptogenic zone defined by quantified stereo-EEG analysis. *Epilepsia*, *55*(2), 296–305. http://dx.doi.org/10.1111/epi.12507, Article first published online: 13 January 2014.

Golbs, A., Nimmervoll, B., Sun, J. J., Sava, I. E., & Luhmann, H. J. (2011). Control of programmed cell death by distinct electrical activity patterns. *Cerebral Cortex*, *21*, 1192–1202.

Guerrini, R., & Parrini, E. (2010). Neuronal migration disorders. *Neurobiology of Disease*, *38*, 154–166.

Hagberg, H., Ichord, R., Palmer, C., Yager, J. Y., & Vannucci, S. J. (2002). Animal models of developmental brain injury: Relevance to human disease—A summary of the panel discussion from the third Hershey conference on developmental cerebral blood flow and metabolism. *Developmental Neuroscience*, *24*, 364–366.

Hagemann, G., Kluska, M. M., Redecker, C., Luhmann, H. J., & Witte, O. W. (2003). Distribution of glutamate receptor subunits in experimentally induced cortical malformations. *Neuroscience*, *117*, 991–1002.

Hanganu, I. L., Kilb, W., & Luhmann, H. J. (2002). Functional synaptic projections onto subplate neurons in neonatal rat somatosensory cortex. *Journal of Neuroscience*, *22*, 7165–7176.

Hanganu, I. L., Okabe, A., Lessmann, V., & Luhmann, H. J. (2009). Cellular mechanisms of subplate-driven and cholinergic input-dependent network activity in the neonatal rat somatosensory cortex. *Cerebral Cortex*, *19*, 89–105.

Haydar, T. F., Wang, F., Schwartz, M. L., & Rakic, P. (2000). Differential modulation of proliferation in the neocortical ventricular and subventricular zones. *Journal of Neuroscience*, *20*, 5764–5774.

Heck, N., Golbs, A., Riedemann, T., Sun, J. J., Lessmann, V., & Luhmann, H. J. (2008). Activity-dependent regulation of neuronal apoptosis in neonatal mouse cerebral cortex. *Cerebral Cortex, 18*, 1335–1349.

Heck, N., Kilb, W., Reiprich, P., Kubota, H., Furukawa, T., Fukuda, A., et al. (2007). GABA-A receptors regulate neocortical neuronal migration in vitro and in vivo. *Cerebral Cortex, 17*, 138–148.

Herz, J., & Chen, Y. (2006). Reelin, lipoprotein receptors and synaptic plasticity. *Nature Reviews Neuroscience, 7*, 850–859.

Hevner, R. F., Hodge, R. D., Daza, R. A., & Englund, C. (2006). Transcription factors in glutamatergic neurogenesis: Conserved programs in neocortex, cerebellum, and adult hippocampus. *Neuroscience Research, 55*, 223–233.

Hirsch, S., & Luhmann, H. J. (2008). Pathway-specificity in N-methyl-D-aspartate receptor-mediated synaptic inputs onto subplate neurons. *Neuroscience, 153*, 1092–1102.

Huberfeld, G., Wittner, L., Clemenceau, S., Baulac, M., Kaila, K., Miles, R., et al. (2007). Perturbed chloride homeostasis and GABAergic signaling in human temporal lobe epilepsy. *Journal of Neuroscience, 27*, 9866–9873.

Hyde, T. M., Lipska, B. K., Ali, T., Mathew, S. V., Law, A. J., Metitiri, O. E., et al. (2011). Expression of GABA signaling molecules KCC2, NKCC1, and GAD1 in cortical development and schizophrenia. *Journal of Neuroscience, 31*, 11088–11095.

Ikeda, M., Toyoda, H., Yamada, J., Okabe, A., Sato, K., Hotta, Y., et al. (2003). Differential development of cation-chloride cotransporters and Cl − homeostasis contributes to differential GABAergic actions between developing rat visual cortex and dorsal lateral geniculate nucleus. *Brain Research, 984*, 149–159.

Jacobs, K. M., Gutnick, M. J., & Prince, D. A. (1996). Hyperexcitability in a model of cortical maldevelopment. *Cerebral Cortex, 6*, 514–523.

Jacobs, K. M., Hwang, B. J., & Prince, D. A. (1999). Focal epileptogenesis in a rat model of polymicrogyria. *Journal of Neurophysiology, 81*, 159–173.

Jacobs, K. M., Mogensen, M., Warren, E., & Prince, D. A. (1999). Experimental microgyri disrupt the barrel field pattern in rat somatosensory cortex. *Cerebral Cortex, 9*, 733–744.

Jacobs, K. M., & Prince, D. A. (2005). Excitatory and inhibitory postsynaptic currents in a rat model of epileptogenic microgyria. *Journal of Neurophysiology, 93*, 687–696.

Jacobs, J., Zijlmans, M., Zelmann, R., Chatillon, C. E., Hall, J., Olivier, A., et al. (2010). High-frequency electroencephalographic oscillations correlate with outcome of epilepsy surgery. *Annals of Neurology, 67*, 209–220.

Jansen, L. A., Peugh, L. D., Roden, W. H., & Ojemann, J. G. (2010). Impaired maturation of cortical GABA(A) receptor expression in pediatric epilepsy. *Epilepsia, 51*, 1456–1467.

Janszky, J., Fogarasi, A., Jokeit, H., Schulz, R., Hoppe, M., & Ebner, A. (2001). Spatiotemporal relationship between seizure activity and interictal spikes in temporal lobe epilepsy. *Epilepsy Research, 47*, 179–188.

Jasper, H. H., Pertuisset, B., & Flanigin, H. F. (1951). EEG and cortical electrograms in patients with temporal lobe seizures. *Archives of Neurology and Psychiatry, 65*, 272–290.

Jensen, F. E., Applegate, C. D., Holtzman, D., Belin, T. R., & Burchfiel, J. L. (1991). Epileptogenic effect of hypoxia in the immature rodent brain. *Annals of Neurology, 29*, 629–637.

Juhasz, C., Asano, E., Shah, A., Chugani, D. C., Batista, C. E., Muzik, O., et al. (2009). Focal decreases of cortical GABAA receptor binding remote from the primary seizure focus: What do they indicate? *Epilepsia, 50*, 240–250.

Kamada, T., Sun, W., Takase, K., Shigeto, H., Suzuki, S. O., Ohyagi, Y., et al. (2013). Spontaneous seizures in a rat model of multiple prenatal freeze lesioning. *Epilepsy Research, 105*, 280–291.

Kanold, P. O., & Luhmann, H. J. (2010). The subplate and early cortical circuits. *Annual Review of Neuroscience, 33*, 23–48.

Khazipov, R., & Luhmann, H. J. (2006). Early patterns of electrical activity in the developing cerebral cortex of human and rodents. *Trends in Neurosciences, 29*, 414–418.
Khirug, S., Yamada, J., Afzalov, R., Voipio, J., Khiroug, L., & Kaila, K. (2008). GABAergic depolarization of the axon initial segment in cortical principal neurons is caused by the Na-K-2Cl cotransporter NKCC1. *Journal of Neuroscience, 28*, 4635–4639.
Kilb, W., Kirischuk, S., & Luhmann, H. J. (2011). Electrical activity patterns and the functional maturation of the neocortex. *European Journal of Neuroscience, 34*, 1677–1686.
Kolbaev, S. N., Achilles, K., Luhmann, H. J., & Kilb, W. (2011). Effect of depolarizing GABA(A)-mediated membrane responses on the excitability of Cajal–Retzius cells in the immature rat neocortex. *Journal of Neurophysiology, 106*, 2034–2044.
Kolbaev, S. N., Luhmann, H. J., & Kilb, W. (2011). Activity-dependent scaling of GABAergic excitation by dynamic Cl(−) changes in Cajal–Retzius cells. *Pflügers Archiv, 461*, 557–565.
Kriegstein, A. R. (2005). Constructing circuits: Neurogenesis and migration in the developing neocortex. *Epilepsia, 46*(Suppl. 7), 15–21.
Kuida, K., Haydar, T. F., Kuan, C. Y., Gu, Y., Taya, C., Karasuyama, H., et al. (1998). Reduced apoptosis and cytochrome *c*-mediated caspase activation in mice lacking caspase 9. *Cell, 94*, 325–337.
Kuruvilla, A., & Flink, R. (2003). Intraoperative electrocorticography in epilepsy surgery: Useful or not? *Seizure, 12*, 577–584.
Lapray, D., Popova, I. Y., Kindler, J., Jorquera, I., Becq, H., Manent, J. B., et al. (2010). Spontaneous epileptic manifestations in a DCX knockdown model of human double cortex. *Cerebral Cortex, 20*, 2694–2701.
Löscher, W., Puskarjov, M., & Kaila, K. (2013). Cation-chloride cotransporters NKCC1 and KCC2 as potential targets for novel antiepileptic and antiepileptogenic treatments. *Neuropharmacology, 69*, 62–74.
LoTurco, J. J., Owens, D. F., Heath, M. J. S., Davis, M. B. E., & Kriegstein, A. R. (1995). GABA and glutamate depolarize cortical progenitor cells and inhibit DNA synthesis. *Neuron, 15*, 1287–1298.
Luhmann, H. J. (2006). The cortical freeze lesion model. In A. Pitkänen, P. A. Schwartzkroin, & S. L. Moshe (Eds.), *Models of seizures and epilepsy* (pp. 295–303). Amsterdam: Elsevier.
Luhmann, H. J. (2013). Cajal–Retzius and subplate cells—Transient cortical neurons and circuits. In J. L. R. Rubenstein, & P. Rakic (Eds.), *Comprehensive developmental neuroscience: Cellular migration and formation of neuronal connections* (pp. 843–856). Amsterdam: Academic Press.
Luhmann, H. J., Hanganu, I. L., & Kilb, W. (2003). Cellular physiology of the neonatal rat cerebral cortex. *Brain Research Bulletin, 60*, 345–353.
Luhmann, H. J., Karpuk, N., Qü, M., & Zilles, K. (1998). Characterization of neuronal migration disorders in neocortical structures: II. Intracellular in vitro recordings. *Journal of Neurophysiology, 80*, 92–102.
Luhmann, H. J., Kilb, W., & Hanganu-Opatz, I. L. (2009). Subplate cells: Amplifiers of neuronal activity in the developing cerebral cortex. *Frontiers in Neuroanatomy, 3*, 19ff.
Luhmann, H. J., Kirischuk, S., Sinning, A., & Kilb, W. (2014). Early GABAergic circuitry in the cerebral cortex. *Current Opinion in Neurobiology, 26*, 72–78.
Luhmann, H. J., & Prince, D. A. (1991). Postnatal maturation of the GABAergic system in rat neocortex. *Journal of Neurophysiology, 65*, 247–263.
Luhmann, H. J., & Raabe, K. (1996). Characterization of neuronal migration disorders in neocortical structures. I. Expression of epileptiform activity in an animal model. *Epilepsy Research, 26*, 67–74.

Luhmann, H. J., Raabe, K., Qü, M., & Zilles, K. (1998). Characterization of neuronal migration disorders in neocortical structures: Extracellular *in vitro* recordings. *European Journal of Neuroscience, 10*, 3085–3094.

Malik, S., Vinukonda, G., Vose, L. R., Diamond, D., Bhimavarapu, B. B. R., Hu, F. R., et al. (2013). Neurogenesis continues in the third trimester of pregnancy and is suppressed by premature birth. *Journal of Neuroscience, 33*, 411–423.

Manent, J. B., Beguin, S., Ganay, T., & Represa, A. (2011). Cell-autonomous and cell-to-cell signalling events in normal and altered neuronal migration. *European Journal of Neuroscience, 34*, 1595–1608.

Manent, J. B., Wang, Y., Chang, Y., Paramasivam, M., & LoTurco, J. J. (2009). Dcx reexpression reduces subcortical band heterotopia and seizure threshold in an animal model of neuronal migration disorder. *Nature Medicine, 15*, 84–90.

Milh, M., Kaminska, A., Huon, C., Lapillonne, A., Ben-Ari, Y., & Khazipov, R. (2007). Rapid cortical oscillations and early motor activity in premature human neonate. *Cerebral Cortex, 17*, 1582–1594.

Molnár, Z., & Clowry, G. (2012). Cerebral cortical development in rodents and primates. *Evolution of the Primate Brain: From Neuron to Behavior, 195*, 45–70.

Monsivais, P., & Rubel, E. W. (2001). Accommodation enhances depolarizing inhibition in central neurons. *Journal of Neuroscience, 21*, 7823–7830.

Moroni, R. F., Cipelletti, B., Inverardi, F., Regondi, M. C., Spreafico, R., & Frassoni, C. (2011). Development of cortical malformations in BCNU-treated rat, model of cortical dysplasia. *Neuroscience, 175*, 380–393.

Moroni, R. F., Inverardi, F., Regondi, M. C., Watakabe, A., Yamamori, T., Spreafico, R., et al. (2009). Expression of layer-specific markers in the adult neocortex of BCNU-treated rat, a model of cortical dysplasia. *Neuroscience, 159*, 682–691.

Munakata, M., Watanabe, M., Otsuki, T., Nakama, H., Arima, K., Itoh, M., et al. (2007). Altered distribution of KCC2 in cortical dysplasia in patients with intractable epilepsy. *Epilepsia, 48*, 837–844.

Nabekura, J., Ueno, T., Okabe, A., Furuta, A., Iwaki, T., Shimizu-Okabe, C., et al. (2002). Reduction of KCC2 expression and GABAA receptor-mediated excitation after in vivo axonal injury. *Journal of Neuroscience, 22*, 4412–4417.

Najm, I. M., Tilelli, C. Q., & Oghlakian, R. (2007). Pathophysiological mechanisms of focal cortical dysplasia: A critical review of human tissue studies and animal models. *Epilepsia, 48*(Suppl. 2), 21–32.

Neuloh, G., Bien, C. G., Clusmann, H., von Lehe, M., & Schramm, J. (2010). Continuous motor monitoring enhances functional preservation and seizure-free outcome in surgery for intractable focal epilepsy. *Acta Neurochirurgica (Wien), 152*, 1307–1314.

Nimmervoll, B., Denter, D. G., Sava, I., Kilb, W., & Luhmann, H. J. (2011). Glycine receptors influence radial migration in the embryonic mouse neocortex. *Neuroreport, 22*, 509–513.

Nimmervoll, B., White, R., Yang, J. W., An, S., Henn, C., Sun, J. J., et al. (2013). LPS-induced microglial secretion of TNF-alpha increases activity-dependent neuronal apoptosis in neonatal cerebral cortex. *Cerebral Cortex, 23*, 1742–1755.

Ojemann, G. A., Ojemann, J., Lettich, E., & Berger, M. (1989). Cortical language localization in left, dominant hemisphere. An electrical stimulation mapping investigation in 117 patients. *Journal of Neurosurgery, 71*, 316–326.

Owens, D. F., Boyce, L. H., Davis, M. B., & Kriegstein, A. R. (1996). Excitatory GABA responses in embryonic and neonatal cortical slices demonstrated by gramicidin perforated-patch recordings and calcium imaging. *Journal of Neuroscience, 16*, 6414–6423.

Owens, D. F., & Kriegstein, A. R. (2002). Is there more to GABA than synaptic inhibition? *Nature Reviews Neuroscience, 3*, 715–727.

Palma, E., Amici, M., Sobrero, F., Spinelli, G., Di Angelantonio, S., Ragozzino, D., et al. (2006). Anomalous levels of Cl^- transporters in the hippocampal subiculum from temporal lobe epilepsy patients make GABA excitatory. *Proceedings of the National Academy of Sciences of the United States of America*, *103*, 8465–8468.
Palmini, A. (2006). The concept of the epileptogenic zone: A modern look at Penfield and Jasper's views on the role of interictal spikes. *Epileptic Disorders*, *8*(Suppl. 2), 10–15.
Palmini, A., Andermann, E., & Andermann, F. (1994). Prenatal events and genetic factors in epileptic patients with neuronal migration disorders. *Epilepsia*, *35*, 965–973.
Palmini, A., & Lüders, H. O. (2002). Classification issues in malformations caused by abnormalities of cortical development. *Neurosurgery Clinics of North America*, *13*, 1–16.
Palomero-Gallagher, N., Schleicher, A., Bidmon, H. J., Pannek, H. W., Hans, V., Gorji, A., et al. (2012). Multireceptor analysis in human neocortex reveals complex alterations of receptor ligand binding in focal epilepsies. *Epilepsia*, *53*, 1987–1997.
Pavlov, I., & Walker, M. C. (2013). Tonic GABAA receptor-mediated signalling in temporal lobe epilepsy. *Neuropharmacology*, *69*, 55–61.
Perucca, P., Dubeau, F., & Gotman, J. (2014). Intracranial electroencephalographic seizure-onset patterns: Effect of underlying pathology. *Brain*, *137*, 183–196.
Peters, O., Redecker, C., Hagemann, G., Bruehl, C., Luhmann, H. J., & Witte, O. W. (2004). Impaired synaptic plasticity in the surround of perinatally acquired dysplasia in rat cerebral cortex. *Cerebral Cortex*, *14*, 1081–1087.
Rakic, P. (2009). Evolution of the neocortex: A perspective from developmental biology. *Nature Reviews Neuroscience*, *10*, 724–735.
Rakic, P., Ayoub, A. E., Breunig, J. J., & Dominguez, M. H. (2009). Decision by division: Making cortical maps. *Trends in Neurosciences*, *32*, 291–301.
Ramos, R. L., Bai, J., & LoTurco, J. J. (2006). Heterotopia formation in rat but not mouse neocortex after RNA interference knockdown of DCX. *Cerebral Cortex*, *16*, 1323–1331.
Redecker, C., Luhmann, H. J., Hagemann, G., Fritschy, J. M., & Witte, O. W. (2000). Differential downregulation of $GABA_A$ receptor subunits in widespread brain regions in the freeze-lesion model of focal cortical malformations. *Journal of Neuroscience*, *20*, 5045–5053.
Reiprich, P., Kilb, W., & Luhmann, H. J. (2005). Neonatal NMDA receptor blockade disturbs neuronal migration in rat somatosensory cortex in vivo. *Cerebral Cortex*, *15*, 349–358.
Rivera, C., Voipio, J., & Kaila, K. (2005). Two developmental switches in GABAergic signalling: The K^+-Cl^- cotransporter KCC2 and carbonic anhydrase CAVII. *Journal of Physiology*, *562*, 27–36.
Rivera, C., Voipio, J., Payne, J. A., Ruusuvuori, E., Lahtinen, H., Lamsa, K., et al. (1999). The K^+/Cl^- co-transporter KCC2 renders GABA hyperpolarizing during neuronal maturation. *Nature*, *397*, 251–255.
Rojas, P., Akrouh, A., Eisenman, L. N., & Mennerick, S. (2011). Differential effects of axon initial segment and somatodendritic GABA(A) receptors on excitability measures in rat dentate granule neurons. *Journal of Neurophysiology*, *105*, 366–379.
Romijn, H. J., Hofman, M. A., & Gramsbergen, A. (1991). At what age is the developing cerebral cortex of the rat comparable to that of the full-term newborn human baby? *Early Human Development*, *26*, 61–67.
Rosen, G. D., Jacobs, K. M., & Prince, D. A. (1998). Effects of neonatal freeze lesions on expression of parvalbumin in rat neocortex. *Cerebral Cortex*, *8*, 753–761.
Rosenow, F., & Lüders, H. (2001). Presurgical evaluation of epilepsy. *Brain*, *124*, 1683–1700.
Ross, M. E., & Walsh, C. A. (2001). Human brain malformations and their lessons for neuronal migration. *Annual Review of Neuroscience*, *24*, 1041–1070.

Rossini, L., Moroni, R. F., Tassi, L., Watakabe, A., Yamamori, T., Spreafico, R., et al. (2011). Altered layer-specific gene expression in cortical samples from patients with temporal lobe epilepsy. *Epilepsia, 52*, 1928–1937.

Sanai, N., Nguyen, T., Ihrie, R. A., Mirzadeh, Z., Tsai, H. H., Wong, M., et al. (2011). Corridors of migrating neurons in the human brain and their decline during infancy. *Nature, 478*, 382–387.

Schramm, J., & Clusmann, H. (2008). The surgery of epilepsy. *Neurosurgery, 62*(Suppl. 2), 463–481.

Schwartzkroin, P. A., & Wenzel, H. J. (2012). Are developmental dysplastic lesions epileptogenic? *Epilepsia, 53*, 35–44.

Semple, B. D., Blomgren, K., Gimlin, K., Ferriero, D. M., & Noble-Haeusslein, L. J. (2013). Brain development in rodents and humans: Identifying benchmarks of maturation and vulnerability to injury across species. *Progress in Neurobiology, 106*, 1–16.

Shimizu-Okabe, C., Okabe, A., Kilb, W., Sato, K., Luhmann, H. J., & Fukuda, A. (2007). Changes in the expression of cation-Cl(−) cotransporters, NKCC1 and KCC2, during cortical malformation induced by neonatal freeze-lesion. *Neuroscience Research, 59*, 288–295.

Shimizu-Okabe, C., Yokokura, M., Okabe, A., Ikeda, M., Sato, K., Kilb, W., et al. (2002). Layer-specific expression of Cl^- transporters and differential $[Cl^-]_i$ in newborn rat cortex. *Neuroreport, 13*, 2433–2437.

Spreafico, R., Pasquier, B., Minotti, L., Garbelli, R., Kahane, P., Grand, S., et al. (1998). Immunocytochemical investigation on dysplastic human tissue from epileptic patients. *Epilepsy Research, 32*, 34–48.

Talairach, J., & Szikla, G. (1980). Application of stereotactic concepts to the surgery of epilepsy. *Acta Neurochirurgica Supplement (Wien), 30*, 35–54.

Tolonen, M., Palva, J. M., Andersson, S., & Vanhatalo, S. (2007). Development of the spontaneous activity transients and ongoing cortical activity in human preterm babies. *Neuroscience, 145*, 997–1006.

Trotter, S. A., Kapur, J., Anzivino, M. J., & Lee, K. S. (2006). GABAergic synaptic inhibition is reduced before seizure onset in a genetic model of cortical malformation. *Journal of Neuroscience, 26*, 10756–10767.

Tschuluun, N., Wenzel, J. H., Katleba, K., & Schwartzkroin, P. A. (2005). Initiation and spread of epileptiform discharges in the methylazoxymethanol acetate rat model of cortical dysplasia: Functional and structural connectivity between CA1 heterotopia and hippocampus/neocortex. *Neuroscience, 133*, 327–342.

Urbach, H., Scheffler, B., Heinrichsmeier, T., von Oertzen, J., Kral, T., Wellmer, J., et al. (2002). Focal cortical dysplasia of Taylor's balloon cell type: A clinicopathological entity with characteristic neuroimaging and histopathological features, and favorable postsurgical outcome. *Epilepsia, 43*, 33–40.

Valeeva, G., Valiullina, F., & Khazipov, R. (2013). Excitatory actions of GABA in the intact neonatal rodent hippocampus in vitro. *Frontiers in Cellular Neuroscience*, 7, 20.

Vanderhaeghen, P., & Polleux, F. (2004). Developmental mechanisms patterning thalamocortical projections: Intrinsic, extrinsic and in between. *Trends in Neurosciences, 27*, 384–391.

Vanhatalo, S., Palva, J. M., Andersson, S., Rivera, C., Voipio, J., & Kaila, K. (2005). Slow endogenous activity transients and developmental expression of K^+-Cl^- cotransporter 2 in the immature human cortex. *European Journal of Neuroscience, 22*, 2799–2804.

Vanhatalo, S., Tallgren, P., Andersson, S., Sainio, K., Voipio, J., & Kaila, K. (2002). DC-EEG discloses prominent, very slow activity patterns during sleep in preterm infants. *Clinical Neurophysiology, 113*, 1822–1825.

Volpe, J. J. (2001). Perinatal brain injury: From pathogenesis to neuroprotection. *Mental Retardation and Developmental Disabilities Research Reviews*, 7, 56–64.

Wagner, J., Urbach, H., Niehusmann, P., von Lehe, M., Elger, C. E., & Wellmer, J. (2011). Focal cortical dysplasia type IIb: Completeness of cortical, not subcortical, resection is necessary for seizure freedom. *Epilepsia*, *52*, 1418–1424.

Wagner-Golbs, A., & Luhmann, H. J. (2012). Activity-dependent survival of developing neocortical neurons depends on PI3K signalling. *Journal of Neurochemistry*, *120*, 495–501.

Wang, D. D., & Kriegstein, A. R. (2009). Defining the role of GABA in cortical development. *Journal of Physiology (London)*, *587*, 1873–1879.

Wyler, A. R., Ojemann, G. A., Lettich, E., & Ward, A. A., Jr. (1984). Subdural strip electrodes for localizing epileptogenic foci. *Journal of Neurosurgery*, *60*, 1195–1200.

Xiang, H., Chen, H. X., Yu, X. X., King, M. A., & Roper, S. N. (2006). Reduced excitatory drive in interneurons in an animal model of cortical dysplasia. *Journal of Neurophysiology*, *96*, 569–578.

Yamada, J., Okabe, A., Toyoda, H., Kilb, W., Luhmann, H. J., & Fukuda, A. (2004). Cl^- uptake promoting depolarizing GABA actions in immature rat neocortical neurones is mediated by NKCC1. *Journal of Physiology (London)*, *557*, 829–841.

Yang, J. W., An, S., Sun, J. J., Reyes-Puerta, V., Kindler, J., Berger, T., et al. (2013). Thalamic network oscillations synchronize ontogenetic columns in the newborn rat barrel cortex. *Cerebral Cortex*, *23*, 1299–1316.

Yang, J. W., Hanganu-Opatz, I. L., Sun, J. J., & Luhmann, H. J. (2009). Three patterns of oscillatory activity differentially synchronize developing neocortical networks *in vivo*. *Journal of Neuroscience*, *29*, 9011–9025.

Zentner, J., Hufnagel, A., Wolf, H. K., Ostertun, B., Behrens, E., Campos, M. G., et al. (1997). Surgical treatment of neoplasms associated with medically intractable epilepsy. *Neurosurgery*, *41*, 378–386.

Zhou, F. W., & Roper, S. N. (2011). Altered firing rates and patterns in interneurons in experimental cortical dysplasia. *Cerebral Cortex*, *21*, 1645–1658.

Zilles, K., Qü, M. S., Köhling, R., & Speckmann, E. J. (1999). Ionotropic glutamate and GABA receptors in human epileptic neocortical tissue: Quantitative *in vitro* receptor autoradiography. *Neuroscience*, *94*, 1051–1061.

Zilles, K., Qü, M., Schleicher, A., & Luhmann, H. J. (1998). Characterization of neuronal migration disorders in neocortical structures: Quantitative receptor autoradiography of ionotropic glutamate, $GABA_A$, and $GABA_B$ receptors. *European Journal of Neuroscience*, *10*, 3095–3106.

CHAPTER FOUR

Limbic Networks and Epileptiform Synchronization: The View from the Experimental Side

Charles Behr*, Margherita D'Antuono*, Shabnam Hamidi*, Rochelle Herrington*, Maxime Lévesque*, Pariya Salami*, Zahra Shiri*, Rüdiger Köhling†, Massimo Avoli*,‡,1

*Department of Neurology, Neurosurgery and Physiology, Montréal Neurological Institute, Montréal, Québec, Canada
†Institute of Physiology, University of Rostock, Rostock, Germany
‡Department of Experimental Medicine, Facoltà di Medicina e Odontoiatria, Sapienza Università di Roma, Roma, Italy
[1]Corresponding author: e-mail address: massimo.avoli@mcgill.ca

Contents

1. Background 64
2. Mechanisms Underlying Epileptiform Synchronization 66
   2.1 Interictal spikes 67
   2.2 High-frequency oscillations 69
   2.3 Ictal discharges 70
3. Interactions Between Limbic Areas 73
4. Seizure-Onset Types 75
   4.1 Human background 75
   4.2 Experimental evidence *in vitro* 76
   4.3 Chronic models of MTLE 77
5. Modulation of Epileptiform Synchronization by Neurosteroids 78
6. Conclusions 81
Acknowledgments 81
References 81

Abstract

In this review, we summarize findings obtained in acute and chronic epilepsy models and in particular experiments that have revealed how neuronal networks in the limbic system—which is closely involved in the pathophysiogenesis of mesial temporal lobe epilepsy (MTLE)—produce hypersynchronous discharges. MTLE is often associated with a typical pattern of brain damage known as mesial temporal sclerosis, and it is one of the most refractory forms of partial epilepsy in adults. Specifically, we will address the cellular and pharmacological features of abnormal electrographic events that, as in MTLE patients, can occur in *in vivo* and *in vitro* animal models; these include interictal and

International Review of Neurobiology, Volume 114
ISSN 0074-7742
http://dx.doi.org/10.1016/B978-0-12-418693-4.00004-2

ictal discharges along with high-frequency oscillations. In addition, we will consider how different limbic structures made hyperexcitable by acute pharmacological manipulations interact during epileptiform discharge generation. We will also review the electrographic characteristics of two types of seizure onsets that are most commonly seen in human and experimental MTLE as well as in *in vitro* models of epileptiform synchronization. Finally, we will address the role played by neurosteroids in reducing epileptiform synchronization and in modulating epileptogenesis.

ABBREVIATIONS

4AP 4-aminopyridine
HFOs high-frequency oscillations
HYP hypersynchronous onset
IPSPs inhibitory postsynaptic potentials
LVF low-voltage fast onset

1. BACKGROUND

The limbic system—as first described by Broca, followed by Papez and McLeans—includes the hippocampus proper and several parahippocampal areas such as the subiculum, the entorhinal and perirhinal cortices, and the amygdala complex. It is involved not only in physiological processes such as learning and memory (Gloor, 1997) but also in pathological conditions such as MTLE. With up to 80% of patients not achieving adequate seizure control, MTLE is one of the most refractory adult forms of epilepsy (Spencer, 2002; Wiebe, Blume, Girvin, Eliasziw, & Effectiveness and Efficiency of Surgery for Temporal Lobe Epilepsy Study Group, 2001). It consists of seizures that originate from the hippocampus, entorhinal cortex, or amygdala, often many years after an initial brain insult such as status epilepticus (SE), encephalitis, or febrile convulsions, but also subacutely as a result of autoimmune encephalitis. MTLE is often associated with a pattern of brain damage known as mesial temporal (also termed *Ammon's horn*) sclerosis, which is characterized by a selective neuronal loss in the CA3/CA1 region of the hippocampus and the hilus, along with granule cell dispersion and aberrant mossy fiber sprouting in the molecular layer of the dentate gyrus (Berkovic et al., 1991; Buckmaster, 2012; Du, Eid, Lothman, Kohler, & Schwarcz, 1995; Gloor, 1997). Anterior temporal resection will reduce seizure occurrence, but still approximately

30% of patients are not seizure-free after surgery, most certainly due to insufficient resection of the epileptic tissue.

MTLE can be reproduced in laboratory animals by topical or systemic injection of chemoconvulsants, such as kainic acid or pilocarpine, as well as by repetitive electrical stimulation (Ben-Ari, 1985; Curia, Longo, Biagini, Jones, & Avoli, 2008; Gorter, van Vliet, Aronica, & Lopes da Silva, 2001; Lévesque & Avoli, 2013). In adult animals, these experimental procedures induce an initial SE and lead to a chronic condition of recurrent seizures within 1–4 weeks; remarkably, a similar, seizure-free, latent period can be identified in MTLE patients who develop partial seizures in adolescence or early adulthood after having suffered an initial insult (e.g., complex febrile convulsions) in childhood (French et al., 1993; Salanova, Markand, & Worth, 1994). As illustrated in Fig. 4.1, similar electrographic patterns of epileptiform activity can also be induced *in vitro* in brain slices or in whole brain preparations following "acute" pharmacological manipulations that aim at increasing neuronal excitability (Avoli & de Curtis, 2011; Avoli, Panuccio, et al., 2013); it is important to emphasize that these "acute" experiments, which were originally made in *in vivo* models during 1960s, have provided relevant information regarding the fundamental mechanisms underlying epileptiform synchronization. However, their involvement in the pathophysiogenesis of seizures occurring in MTLE must be taken with caution as they were obtained from normal, control brain tissue.

Understanding the pathophysiogenesis of limbic seizures and thus of MTLE remains an open challenge in epilepsy research. In this review, we will mainly focus on the brain processes that lead to the initiation of seizure activity (a process that is also termed ictogenesis). Specifically, we will summarize the cellular and pharmacological features of synchronous, abnormal events that, as in MTLE patients, can occur in *in vivo* and *in vitro* animal models; these include interictal and ictal discharges along with high-frequency oscillations (HFOs). In addition, we will consider how different limbic structures made hyperexcitable by acute pharmacological manipulations do interact during epileptiform discharge generation. We will also review the electrographic characteristics of two types of seizure onsets that are most commonly seen in human and experimental MTLE as well as in *in vitro* models of epileptiform synchronization. Finally, we will address the role played by neurosteroids in reducing epileptiform synchronization and in modulating epileptogenesis.

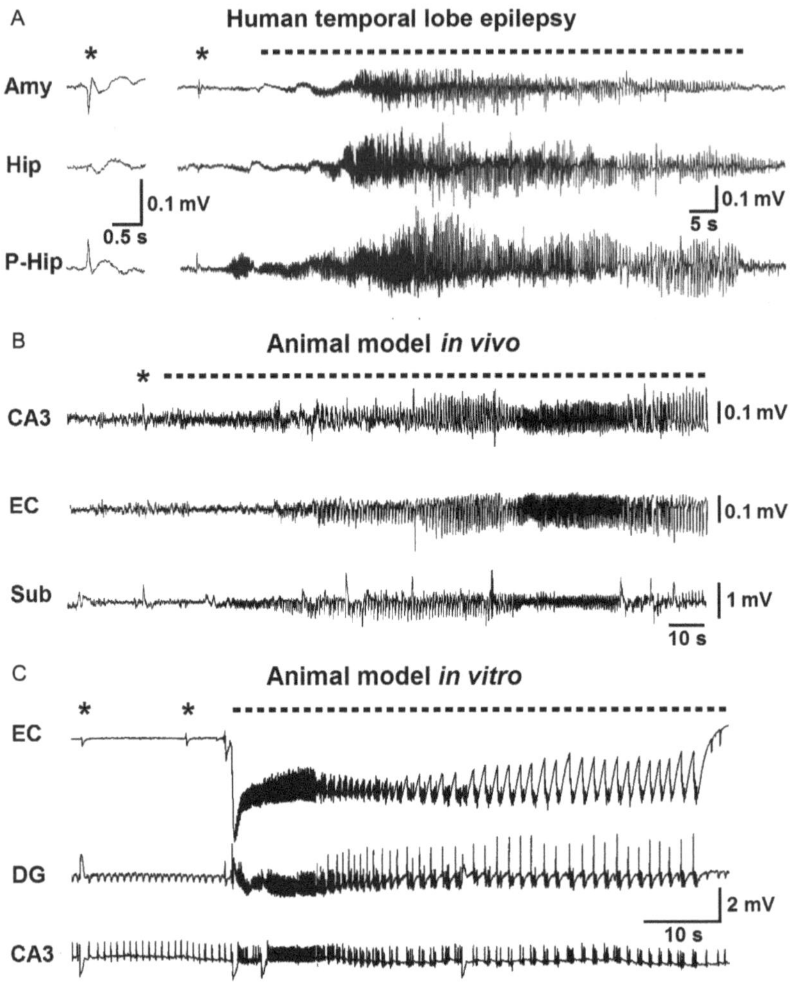

Figure 4.1 Interictal (asterisks) and ictal (dotted lines) epileptiform discharges recorded with depth EEG electrodes from a patient with MTLE (A), from a pilocarpine-treated epileptic rat (B), and with an extracellular microelectrode from the EC area in a rat brain slice treated with 4AP (C). *EEG recordings shown in (A) were kindly provided by Drs. P. Perucca, F. Dubeau, and J. Gotman at the Montreal Neurological Institute & Hospital.*

2. MECHANISMS UNDERLYING EPILEPTIFORM SYNCHRONIZATION

For almost a century, interictal spikes (Fig. 4.1, asterisks) have been a key feature in the noninvasive diagnosis of partial epilepsy. They also contribute to the presurgical evaluation, helping to define the epileptogenic zone by delineating the so-called "irritative zone" (Rosenow & Lüders, 2001). MTLE, however, differs from other partial epilepsies due to the deep topography of limbic structures. As a result, despite being often associated with interictal spikes, sharp waves, SW complexes, as well as slow activity, electrophysiological localization of the "irritative zone" in MTLE can be equivocal (Niedermeyer, 2004). Additionally, bilateral interictal findings are not rare but do not always preclude a lateralized underlying pathology.

Ictal activity on scalp EEG is usually characterized by rhythmic discharges, at a frequency of around 5 Hz, limited to temporal regions or involving more widespread networks, and starting before the occurrence of clinical symptoms (Fish, 1996; Fig. 4.1, dotted lines). In some cases, stereo-EEG is required to properly lateralize the seizure-onset zone (Diehl & Lüders, 2000) and when recorded using this method, mesial temporal seizures can be divided into two onset types: low-voltage fast onset (LVF) and hypersynchronous onset (HYP) types (Spencer, Kim, & Spencer, 1992, Velasco, Wilson, Babb, & Engel, 2000; see also Section 4). More recently, HFOs have been recorded in MTLE patients using microelectrodes (Staba, Wilson, Bragin, Fried, & Engel, 2002) and macroelectrodes (Crépon et al., 2010; Jacobs et al., 2008), and they were proposed as better markers of the epileptogenic zone than interictal spikes (Jacobs et al., 2008).

2.1. Interictal spikes

Interictal spikes consist of a large-amplitude, rapid component usually followed by a slow wave. Intracellular recordings obtained from cortical neurons in acute *in vivo* and *in vitro* brain slice models have demonstrated that interictal spikes induced by $GABA_A$ receptor antagonists correspond to paroxysmal depolarizing shifts of the membrane potential with sustained action potential firing that can be followed by hyperpolarization (Ayala, Dichter, Gumnit, Matsumoto, & Spencer, 1973; Dingledine & Gjerstad, 1980; Schwartzkroin & Prince, 1980; Fig. 4.2A). These studies have also shown that interictal synchronization results from enhancement of glutamatergic excitation (presumably caused by the pharmacological weakening of inhibition), high-threshold Ca^{2+} currents leading to action potential bursting, and nonsynaptic interactions such as intercellular gap junctions and ephaptic interactions (De Curtis, Jefferys, & Avoli, 2012; Jefferys, Jiruska, de Curtis, & Avoli, 2012).

Successive studies have, however, demonstrated that interictal activity induced by applying K^+ channel blockers or medium containing low Mg^{2+}, low Cl^-, or high K^+ are also contributed by $GABA_A$ receptor

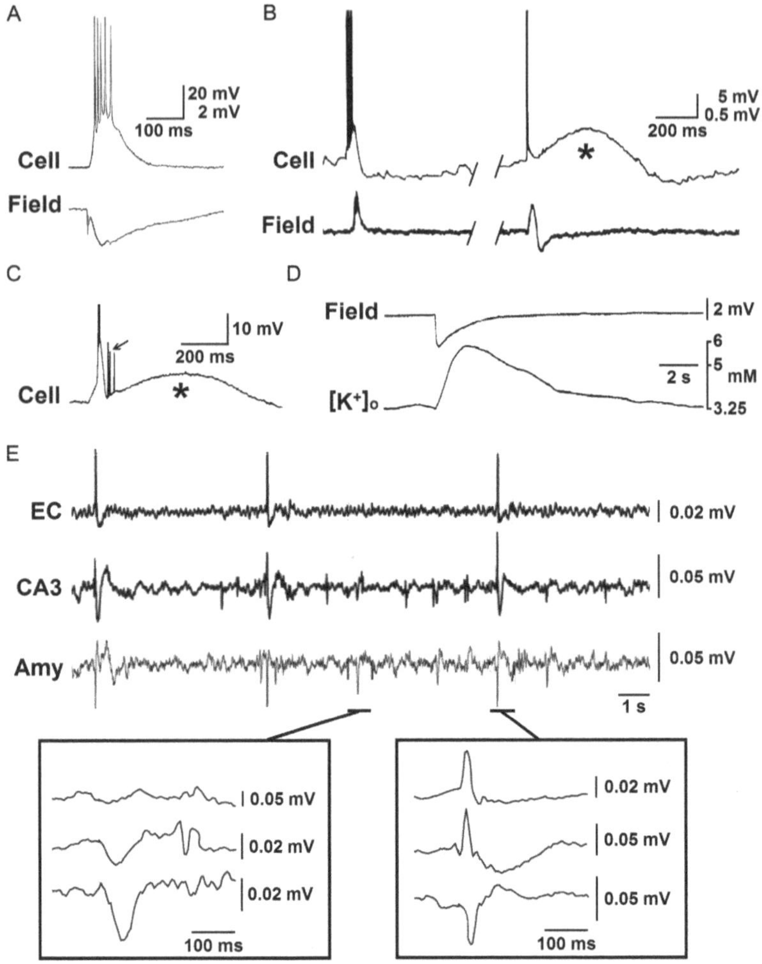

Figure 4.2 (A)_ Intracellular (K-acetate-filled microelectrode, Cell) and field potential (Field) recordings obtained in a neocortical slice during application of medium containing the $GABA_A$ receptor antagonist bicuculline methiodide. (B) Intracellular (K-acetate-filled microelectrode, Cell) and field potential (Field) recordings obtained from the CA3 subfield during 4AP application. Note that the intracellular counterpart of the "fast" interictal discharges recorded from a CA3 pyramidal cell consists of depolarizations that trigger bursts of fast action potentials, while the counterpart of the "slow" interictal spike is characterized by a slow depolarization with a single fast action potential. (C) Ectopic spikes (arrow) occur during the slow depolarization associated with the slow interictal discharge recorded during 4AP application; intracellular recording was obtained with a K-acetate-filled microelectrode. (D) Simultaneous field potential and extracellular $[K^+]$ recordings of the glutamatergic-independent interictal spike generated by CA3 neuronal networks during concomitant application of 4AP and glutamatergic antagonists; note that the extracellular $[K^+]$ increases up to approximately 5.0 mM from a resting concentration of 3.2 mM shortly after the negative peak of the field potential. (E) Interictal activity recorded with depth electrodes from the EC, CA3, and amygdala in a pilocarpine-treated animal during the chronic epileptic period. The expanded traces show in detail the two types of interictal spikes.

mechanisms (Avoli et al., 2002; Avoli & de Curtis, 2011). In particular, application of the K^+ channel blocker 4-aminopyridine (4AP) induces, in addition to frequently occurring interictal events that are driven by the CA3 network and are abolished by AMPA receptor antagonists (Fig. 4.2B, left sample), "slow" interictal discharges that (i) continue to occur and to propagate even in the presence of ionotropic glutamatergic receptor blockers and (ii) are abolished by drugs that interfere with $GABA_A$ receptor signaling (Fig. 4.2B, right sample). These "slow" interictal spikes are associated with a long-lasting depolarization that can be preceded by a fast hyperpolarizing component (depending upon the resting membrane potential of the recorded cell; Fig. 4.2B, right sample) and often trigger ectopic action potentials that are presumably due to local, transient elevations in extracellular $[K^+]$ (Fig. 4.2C; Avoli, Methot, & Kawasaki, 1998). Slow interictal spikes are indeed accompanied by sizeable increases in extracellular $[K^+]$ even during blockade of ionotropic glutamatergic transmission (Fig. 4.2D). Several studies have identified 4AP-induced glutamatergic-independent spikes in several structures (e.g., neocortex, amygdala, entorhinal, perirhinal, insular, and piriform cortices) maintained *in vitro* both in the slice and in the isolated guinea pig brain preparation (Avoli, de Curtis, & Köhling, 2013). In addition, similar network-driven events have been reported to occur following application of the muscarinic agonist carbachol (Dickson & Alonso, 1997).

The diverse contribution of distinct mechanisms to interictal spikes as demonstrated in *in vitro* experimental models of epileptiform synchronization is in line with the diversity in shape that is seen during *in vivo* EEG recordings of MTLE patients or of chronically epileptic animals (Fig. 4.2E; Bortel, Lévesque, Biagini, Gotman, & Avoli, 2010; Chauvière et al., 2012). In addition, by recording single unit activity from epileptic patients, Keller et al. (2010) have reported high levels of heterogeneity in the firing patterns of single units during interictal spikes. Single units in the seizure-onset zone could either show an increase or a decrease in firing rates at different time points before, during, and after the interictal spikes. It should be emphasized that interictal spikes may play divergent functional roles with respect to ictogenesis (see Section 3) as well as epileptogenesis as recently suggested by Chauvière et al. (2012).

2.2. High-frequency oscillations

HFOs (80–500 Hz) represent a new type of EEG activity that is captivating neuroscientists, as well as clinical and fundamental epileptologists. Oscillations at frequencies higher than what is normally considered standard in

EEG recordings (i.e., <60 Hz) were initially reported in physiological studies that have demonstrated the presence of gamma (up to 80 Hz) and ripples (80–200 Hz) during visual perception in the neocortex, during consummatory behaviors, slow-wave sleep, and memory consolidation in the hippocampus (Buzsáki & da Silva, 2012; Buzsaki, Horvath, Urioste, Hetke, & Wise, 1992; Jefferys, Menendez de la Prida, et al., 2012). Approximately 15 years ago, HFOs (ripples: 80–200 Hz, fast ripples: 250–500 Hz) were recorded from patients with MTLE and in animal models mimicking this condition; these findings have suggested that these oscillatory events represent biomarkers of epileptogenic neuronal networks (Staba, 2012).

HFOs are not visible in standard intracranial EEG recordings and can only be extracted by amplifying the appropriately filtered signal. As it will be discussed in Section 4, ripples and fast ripples have been recorded with depth electrodes in chronic epileptic rodents and in epileptic patients both in coincidence with interictal events and in their absence (Staba, 2012). These pathological HFOs presumably reflect the activity of dysfunctional neural networks sustaining epileptogenesis, and they are now considered better markers than interictal spikes to identify seizure-onset zones (Staba, 2012). Experimental studies have highlighted several mechanisms that may contribute to physiological and pathological HFOs. These include kinetics of intrinsic neuronal currents such as after-hyperpolarizing currents (I_{AHP}) or fast potassium currents (I_A), as well as excitatory currents and recurrent inhibitory connectivity in combination with fast timescales of inhibitory postsynaptic potentials (IPSPs), out-of-phase firing in neuronal clusters, and interneuronal coupling through gap junctions (Jefferys, Jiruska, et al., 2012; Jefferys, Menendez de la Prida, et al., 2012, Kohling & Staley, 2011). The exact role played by these mechanisms in pathologic HFOs remains undefined, but it has been proposed that ripples may represent population IPSPs generated by principal neurons entrained by synchronously active interneuron networks (Buzsáki & Chrobak, 1995; Ylinen et al., 1995); in contrast, fast ripples would mirror synchronous in-phase or out-of-phase firing of abnormally active principal cells, and would be independent of inhibitory neurotransmission (Bragin, Benassi, Kheiri, & Engel, 2011; Dzhala & Staley, 2004; Engel, Bragin, Staba, & Mody, 2009; Foffani, Uzcategui, Gal, & Menendez de la Prida, 2007; Ibarz, Foffani, Cid, Inostroza, & Menendez de la Prida, 2010).

2.3. Ictal discharges

The process leading to seizure generation is termed ictogenesis; the underlying mechanisms make neuronal networks generate prolonged periods of hypersynchronous activity that disrupt normal brain function. Seizure termination and postictal state complete the sequence of ictogenesis. As shown in Figs. 4.1 and 4.3, the electrographic counterpart of a seizure (or ictal discharge) consists of a prolonged series of fast EEG transients that can vary in amplitude, and is easily distinguished from the background activity as well as from the interictal spikes. In the case of a prototypical tonic–clonic seizure, the clonic phase will follow the tonic one, mirrored on the EEG by an increasing amplitude and a decreasing frequency. Interspersed periods of "suppression" of increasing duration will occur and diffuse until finally reaching a typical postictal depression state (Engel, Dichter, & Schwartzkroin, 1997). Intracellular recordings have shown that ictal discharges are associated with sustained depolarizations capped by sustained action potential firing (Fig. 4.3A). It has also been established that neuronal network synchronization during ictal activity depends on glutamatergic ionotropic neurotransmission (particularly that mediated by the NMDA receptor; Köhr & Heinemann, 1989; Löscher and Hönack, 1991; Traynelis & Dingledine, 1988) and on persistent Na^+ currents that are fundamental to maintain the sustained action potential firing (Mantegazza, Curia, Biagini, Ragsdale, & Avoli, 2010). In line with this view, many of the most widely used antiepileptic drugs, including phenytoin, carbamazepine, lamotrigine, and, perhaps, topiramate, inhibit voltage-gated Na^+ channels (Mantegazza et al., 2010).

However, $GABA_A$ receptor signaling is also required for ictogenesis and under specific *in vitro* pharmacological manipulations it can be the sole synaptic mechanism responsible for the occurrence of seizure-like events (Uusisaari, Smirnov, Voipio, & Kaila, 2002). The role of $GABA_A$ receptor signaling in initiating and sustaining ictal discharges has been clearly defined in several studies performed in the *in vitro* 4AP model (Avoli & de Curtis, 2011). These experiments have shown that ictal activity disappears in several limbic and extralimbic structures during bath application of $GABA_A$ receptor antagonists or of a μ-opioid receptor agonist. In addition, it has been found that inhibitory currents at the onset of and/or during *in vitro* ictal activity may overload the ability of the neuronal Cl^- transporter KCC2 to maintain a low intracellular Cl^- concentration thus disclosing depolarizing, and potentially excitatory, $GABA_A$ receptor-mediated conductances. These events lead to transient increases in extracellular $[K^+]$ that are caused by the activity of transporters responsible for reuptake of GABA and Cl^- extrusion (Viitanen, Ruusuvuori, Kaila, & Voipio, 2010) and that are larger in size during preictal spikes as compared to those seen during interictal discharges (Fig. 4.3B, arrow). However, the role played by

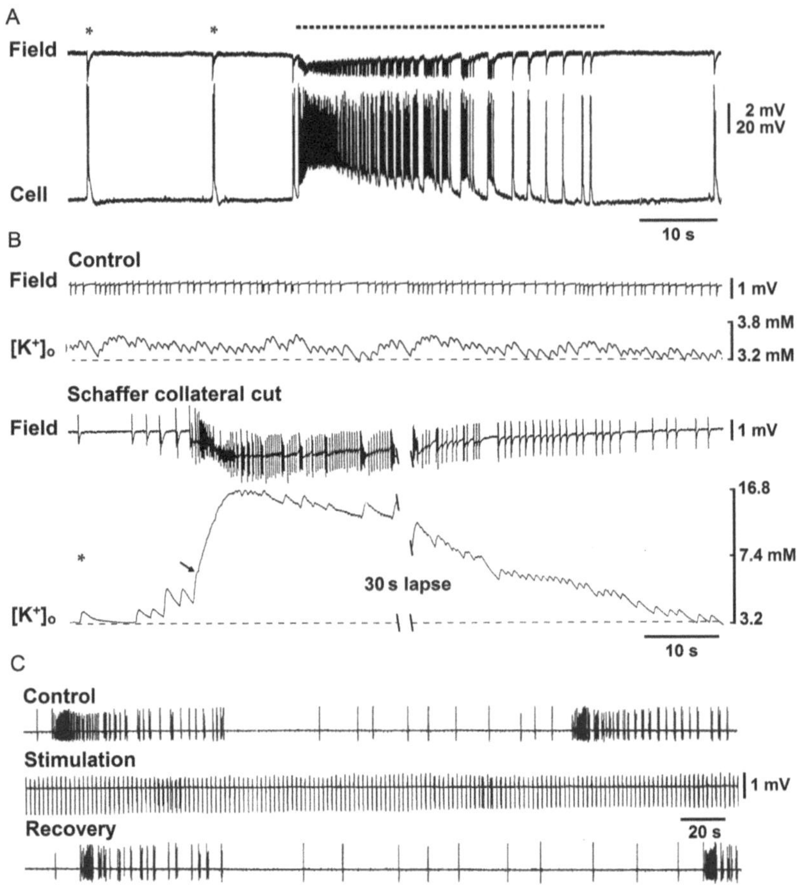

Figure 4.3 (A) Field and intracellular (Cell) recordings obtained from a mouse EC neuron during application of Mg^{2+}-free medium. Both interictal (asterisks) and ictal (dotted line) discharges occurred during this experimental procedure. (B) Changes induced by Schaffer collateral cut on the epileptiform activity recorded from the EC in a combined mouse brain slice. Under control conditions, fast CA3-driven interictal activity is generated by EC neuronal networks; however, cutting the Schaffer collateral prevents CA3-driven interictal activity to propagate to the EC and uncovers slow interictal events along with ictal discharge in this area. Note that the slow interictal spike (asterisk) is associated with an increase in extracellular [K^+] that is smaller than what occurring in coincidence with the initial spike leading to ictal activity (arrow). Note also the much larger elevation in extracellular [K^+] occurring during the overt ictal discharge. (C) Ictal discharges recorded *in vitro* from the EC during application of 4AP do not occur during local electrical stimulation implemented at 0.5 Hz. Note also that ictal activity reappears following termination of the stimulating protocol.

synaptic mechanisms during seizure discharges should be considered with caution as epileptiform activity is accompanied by decreases in extracellular [Ca^{2+}] that are not compatible with efficient transmitter release (Pumain, Menini, Heinemann, Louvel, & Silva-Barrat, 1985; Timofeev & Steriade, 2004). Nevertheless, several studies have shown that these seizure-like discharges are abolished by specific neurotransmitter receptor antagonists.

3. INTERACTIONS BETWEEN LIMBIC AREAS

Limbic structures display a high level of feedforward as well as feedback connectivity (McIntyre & Schwartzkroin, 2008). Temporal lobectomy, first achieved by Horsley in 1886, was progressively refined based on anatomopathological examination (Falconer, 1974; Margerison & Corsellis, 1966) and intracranial stereo-EEG evaluation (Lieb, Engel, Brown, Gevins, & Crandall, 1981; Talairach et al., 1974). Nowadays, anterior mesial temporal resection remains the most common procedure for MTLE surgery (Vives, Lee, Doyle, & Spencer, 2008). Pathogenicity of medial temporal structures was progressively recognized and resection evolved in that sense; however, precise interictal and ictal interactions of limbic structures in human remains poorly understood, partly due to the constraints in terms of spatial sampling during depth EEG studies (Spencer & Spencer, 1994). Developed at the Montreal Neurological Institute, intraoperative electrocorticography helped in identifying the pathological nature of the amygdalo-hippocampus complex, eventually leading to the selective amygdalo-hippocampectomy procedure as an attempt to reduce functional postoperative deficit (Olivier, 2010; Wieser & Yaşargil, 1982). However, tractography studies using diffusion-tensor imaging suggest that the main neocorticolimbic fiber tracts may still be affected in selective approach, underlying the role played by neocorticolimbic connectivity in MTLE epileptogenicity (Colnat-Coulbois et al., 2010). Another crucial aspect of limbic connectivity with regard to epileptogenicity and seizure spread concerns the interhemispheric connections through anterior and posterior commissures. The classical view of a vestigial anterior commissure (Gloor, Salanova, Olivier, & Quesney, 1993) has been somewhat challenged by recent DTI studies, showing an anterior interamygdalian commissural connectivity (Colnat-Coulbois et al., 2010). Despite this evidence, the question of whether different areas of the limbic system interact during epileptiform synchronization and, above all, how they interact remains unclear.

Over a decade ago, we reported that during application of 4AP or medium containing low Mg^{2+}, cutting the Schaffer collaterals in mouse brain slices comprising interconnected hippocampus and entorhinal cortex abolishes the propagation of CA3-driven interictal discharges to the entorhinal cortex and discloses slow interictal spikes and seizure-like events in the latter structure (Fig. 4.3B; Barbarosie & Avoli, 1997; Barbarosie, Louvel, Kurcewicz, & Avoli, 2000). Therefore these findings—which have later been replicated in rat brain slices by extending the analysis to other limbic areas such as the perirhinal cortex, the amygdala, or the insular cortex—indicate that CA3 output activity controls rather than reinforces the expression of ictal epileptiform activity in the entorhinal cortex and other parahippocampal areas (Avoli & de Curtis, 2011).

Although a definitive answer to what determines the control exerted by hippocampal output activity on parahippocampal ictogenesis remains to be found, experimental evidence suggests that a main player relates to the ability of CA3-driven interictal spikes to hinder the transient elevations in extracellular [K^+] that occur during the slow interictal events and may be instrumental for ictal discharge onset (Fig. 4.3B). Therefore, CA3-driven interictal events should insure that GABA release during the slow interictal discharge is downregulated, thus controlling the associated elevations in extracellular [K^+].

This hypothesis is further supported by experiments in which parahippocampal structures such as the EC, the amygdala, or the insular cortex are stimulated at frequencies similar to those of CA3-driven interictal spikes (i.e., at 0.5–1.0 Hz; Fig. 4.3C; Avoli & de Curtis, 2011; Avoli, de Curtis, et al., 2013). Clinical studies have indeed shown that low-frequency stimulation—delivered through transcranial magnetic or deep-brain electrical procedures—can reduce seizures in epileptic patients not responding to conventional antiepileptic therapy (Tergau, Naumann, Paulus, & Steinhoff, 1999; Theodore & Fisher, 2004; Vonck, Boon, Achten, De Reuck, & Caemaert, 2002).

The ability of hippocampal outputs to control ictogenesis may also be relevant in MTLE, a disorder that is associated with CA3 and CA1 subfield damage (Gloor, 1997); according to this view, a decrease in hippocampal output activity secondary to cell damage may facilitate the generation of ictal discharges in parahippocampal structures (Barbarosie & Avoli, 1997, D'Antuono et al., 2002, Panuccio et al., 2010). The sclerotic hippocampus presents with almost complete loss of CA1 neurons (the Sommer sector), especially at the subicular boundary, as well as disappearance of hilar cells in the dentate gyrus (endfolium sclerosis). Neuronal cell loss is also found in the medial EC (especially in layer III) and in the amygdala. Recent evidence obtained from studies performed in animal models mimicking MTLE suggests that the pattern of neuronal damage found in Ammon's horn sclerosis cuts off the normal physiological route of hippocampal network activity (the trisynaptic hippocampal pathway), thus reinforcing the activity of direct connections between EC–CA1 (when spared) and EC–subiculum (D'Antuono et al., 2002, Panuccio et al., 2010). These findings are challenging the traditional view that has emphasized the role played by mossy fiber sprouting in the dentate gyrus in epileptogenesis.

4. SEIZURE-ONSET TYPES

4.1. Human background

Depth electrode recordings in MTLE patients have opened a new window to the analysis of seizure-onset patterns. With surface electrodes, it is difficult to perform a detailed analysis of seizure onsets because signals originating from deep regions are attenuated before reaching the scalp and noise introduced by muscle artifacts further compounds the problem. This induces low reliability between independent observers when trying to identify regions of seizure onset and when classifying seizures based on their onset EEG pattern (Spencer et al., 1985). However, the use of depth electrodes implanted in temporal and extratemporal regions of MTLE patients has led to the identification of two main seizure-onset types (Bragin, Wilson, Fields, Fried and Engel, 2005; Ogren et al., 2009; Spencer et al., 1992; Velasco et al., 2000). HYP seizures, which are characterized by rhythmic high-amplitude spikes followed by fast activity in the 10–20 Hz frequency range, mainly originate from hippocampal regions and rarely spread to areas adjacent or contralateral to the seizure-onset zone (Velasco et al., 2000); in addition, they are often associated with mesial temporal lobe sclerosis (Spanedda, Cendes, & Gotman, 1997; Spencer et al., 1992; Townsend, 1991).

Further anatomopathological and magnetic resonance imaging studies have shown that patients presenting with HYP seizures have extensive cell loss in the dentate gyrus and in all hippocampal areas except CA2 and presubiculum (Spencer et al., 1992; Velasco et al., 2000). The second type of seizure onset, termed LVF, is characterized by the occurrence of a positive- or negative-going spike that is followed by the appearance of low-amplitude. high-frequency activity. Compared to HYP seizures, the site of origin of LVF seizures is more diffuse and often extrahippocampal;

moreover, these seizures spread more rapidly and more widely to regions ipsilateral and contralateral to the side of onset (Engel et al., 1990; Spencer et al., 1992; Townsend, 1991; Velasco et al., 2000). Patients with LVF seizures also show more extensive and diffuse neuronal loss with bilateral hippocampal atrophy associated with cell loss in the CA2/CA4 regions (Ogren et al., 2009; Velasco et al., 2000). Thus, taken together, the differences between the two seizure-onset types indicate that they may arise through different mechanisms as well as through the predominant involvement of specific types of neurons (i.e., glutamatergic and inhibitory cells).

4.2. Experimental evidence *in vitro*

LVF and HYP seizures can be reproduced in *in vitro* brain slice preparations. As shown in Fig. 4.4A, during application of 4AP, EC networks commonly generate LVF seizures (panel a); however, in few cases, HYP can occur (panel b; Avoli, Panuccio, et al., 2013). Interestingly, in combined brain slices comprising the EC and the perirhinal cortex, both seizure-onset types can co-occur: LVF seizures are usually recorded in the EC while HYP seizures predominate in the perirhinal area (Köhling et al., unpublished results; Fig. 4.4B). Indeed, HYP seizures could reflect the synchronous firing or principal (glutamatergic) cells, whereas LVF seizures could be related to an inhibitory process with enhancement of interneuronal activity that leads to synchronous $GABA_A$ receptor-mediated signaling at the seizure onset. Results supporting this hypothesis also come from *in vitro* studies showing that in the human epileptic tissue recorded *in vitro*, preictal discharges of HYP-like ictal events induced by low Mg^{2+} are not suppressed by $GABA_A$ receptor antagonists (Huberfeld et al., 2011). Moreover, ripples, with the virtual absence of fast ripples, were shown to be related to LVF ictal events induced by 4AP *in vitro* in the piriform cortex (Panuccio et al., 2012) and in the EC (Avoli, Panuccio, et al., 2013).

We have also reported that in the 4AP model of *in vitro* limbic epileptiform synchronization both EC and perirhinal networks can be seizure-onset zone but with a clear predominance of the latter structure (De Guzman, D'Antuono, & Avoli, 2004). Finally, it has been shown that ictal discharges in combined rat hippocampus–EC slices bathed with Mg^{2+}-free medium initiate and propagate in variable ways with about 35% of these events starting in hippocampal areas, 25% in the EC, and the remaining 40% of unclear origin (Schulte-Weß S. & Köhling R., unpublished observations; Fig. 4.4C).

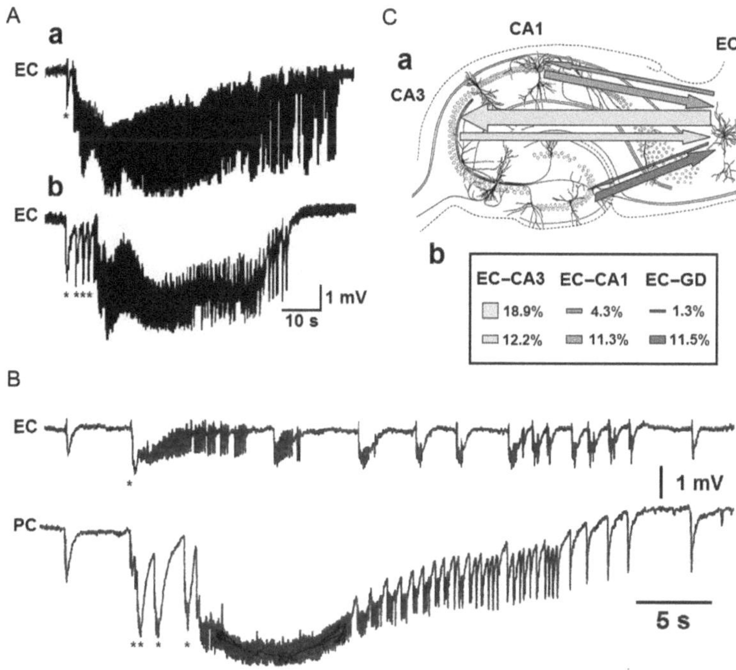

Figure 4.4 (A) LVF and HYP ictal discharges recorded *in vitro* from the EC in two different experiments during application of 4AP. Asterisks highlight the interictal-like events that precede the beginning of the ictal activity. (B) Field potential recordings obtained in a rat brain slice during application of 4AP show that EC and perirhinal cortex (PC) can generate at the same time LVF and HYP seizures, respectively. (C) Ictal activity in an EC–hippocampus brain slice during application of Mg^{2+}-free medium can initiate in different areas with variable propagation. Schematic diagram of propagation directions are shown in (a) where arrows represent direction (arrowheads) and percentages (arrow width) of propagation directions. Data used for this analysis rest on simultaneous field recordings obtained from areas CA1, CA3, DG of the hippocampus and EC in 51 combined rat brain slices (b) Percentages of propagation directions corresponding to arrow widths shown in (a). Note that 35% of the ictal discharges arose in hippocampus and propagated to EC, 24.5% of them started in EC and propagated to hippocampal areas, with latencies of 60–175 ms, while 40.5% of them had latencies of less than 5 ms, which hindered the identification of any definite initiation site. *From: Schulte-Wess, S. Iktogenese und Propagation epileptischer Aktivität im entorhinalen Cortex und Hippocampus. Dissertation, Muenster, 2007***.

4.3. Chronic models of MTLE

Results obtained to date in animals models of MTLE support clinical findings on seizure-onset types. Bragin, Azizyan, Almajano, Wilson and Engel (2005) reported that an SE induced by the local administration of kainic acid in the CA3 region of the hippocampus is followed by the occurrence of both

HYP and LVF seizures, within a week after the injection. In the same study, recordings with depth electrodes in hippocampal and para-hippocampal regions demonstrated that HYP seizures mostly originated from hippocampal structures ipsilateral to the injection of kainic acid. These seizures also occurred in most cases when the animal was sleeping or immobile and they were rarely associated with behavioral symptoms, suggesting that they presumably remained focal. On the other hand, the seizure-onset zones of LVF seizures were located in both hippocampus and EC and they frequently propagated to other brain structures. Interestingly, during HYP seizures, a transition to an LVF pattern was associated with propagation of seizures to other brain regions outside of the hippocampus. Similar results have been obtained with the pilocarpine model of MTLE, since HYP seizures mostly originate from the CA3 region of the hippocampus whereas the seizure-onset zones of LVF seizures are located in both the hippocampus and the entorhinal cortex or are located outside of the hippocampus (Lévesque, Salami, Gotman, Avoli, 2012).

As shown in Fig. 4.5A, the analysis of HFOs occurring during the preictal and ictal periods has revealed distinct patterns of HFO occurrence: HYP seizures were mostly associated with fast ripple occurrence, whereas fast ripples occurred rarely during LVF seizures (Bragin, Azizyan, et al., 2005; Lévesque et al., 2012). On the other hand, LVF seizures were mainly associated with an increase of activity in the beta–gamma range (15–20 Hz; Bragin, Azizyan, et al., 2005) or in ripple occurrence (Lévesque et al., 2012). These increases in occurrence of fast ripples during HYP seizures and ripples during LVF seizures are mostly occurring in seizure-onset zones, suggesting that they could reflect distinct mechanisms of ictogenesis (Lévesque et al., 2012).

5. MODULATION OF EPILEPTIFORM SYNCHRONIZATION BY NEUROSTEROIDS

Neurosteroids are a class of compounds that can modulate neuronal excitability at the level of ion channels and membrane receptors (Akk et al., 2009; Mellon & Griffin, 2002). They are endogenously synthesized, mainly in glial cells and principal neurons, from circulating steroidal precursors but exert no known genomic effects (Agís-Balboa et al., 2006; Reddy, 2010). There are three main classes: the pregnane, androstane, and sulfated neurosteroids (Reddy, 2010). The former two exert mainly inhibitory effects and the later, excitatory (Majewska, 1992; Reddy, 2010). More specifically, pregnane neurosteroids are positive allosteric modulators of

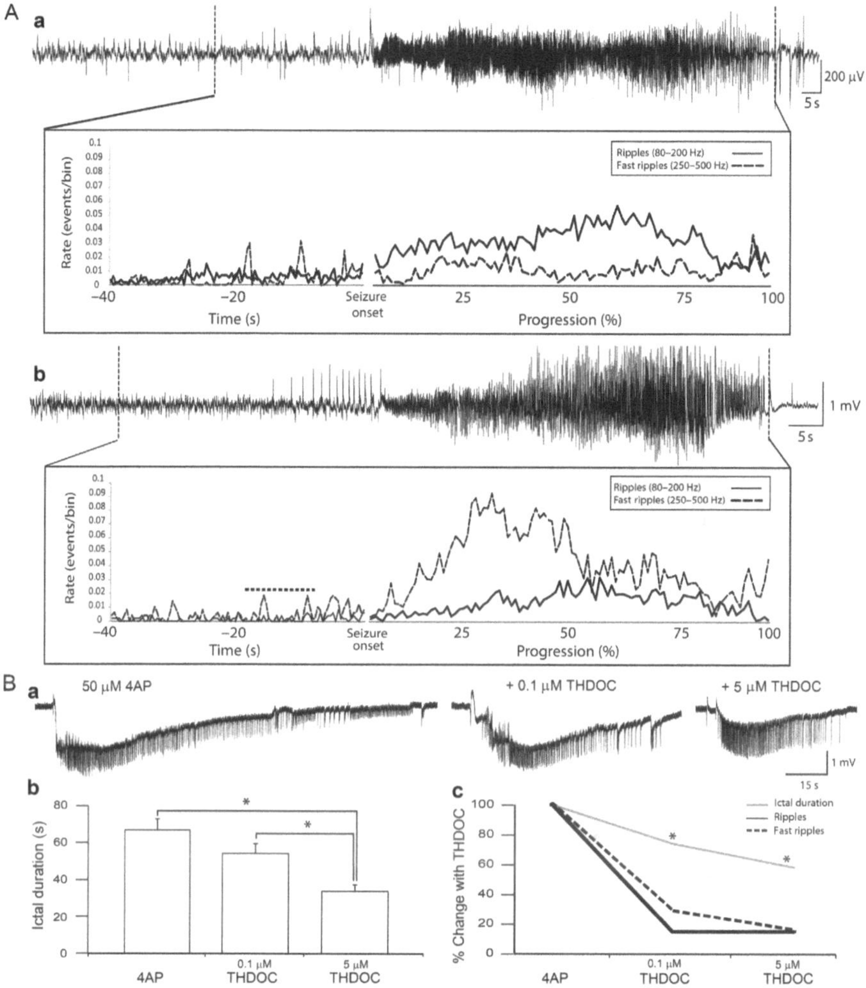

Figure 4.5 (A) EEG recordings of spontaneous seizures obtained from the CA3 region of the hippocampus in a pilocarpine-treated rat. Panel (a) shows an example of an LVF seizure in the CA3 region of the hippocampus. Note the occurrence of a single positive-going spike at the onset of the seizure, followed by activity in the 5–20 Hz range. The analysis of HFOs shows that high rates of ripples compared to fast ripples are associated with this type of seizure onset ($n = 18$ LVF seizures). Panel (b) shows an example of a HYP seizure recorded in the same animal in the CA3 region of the hippocampus ($n = 21$ HYP seizures). Note the occurrence of multiple ictal spikes followed by activity in the 10–20 Hz range. Compared to LVF seizures, HYP seizures are mostly associated with the occurrence of fast ripples. (B) Ictal events recorded with an extracellular microelectrode from the piriform cortex during 4AP application show a concentration-dependent decrease in duration with increasing concentrations of THDOC (a and b). Note in (c) that THDOC reduces HFO occurrence to a greater extent than it reduces duration of ictal events.

$GABA_A$ receptor function and have been shown to provide some anti-ictogenic effect (Reddy & Rogawski, 2012).

Allotetrahydrodeoxycorticosterone (THDOC) and allopregnanolone are examples of pregnane neurosteroids that can be found endogenously at nanomolar concentrations (Reddy, 2011). By increasing the mean open time of $GABA_A$ receptor channels, THDOC effectively potentiates both phasic and tonic currents (Reddy, 2011; Wohlfarth, Bianchi, & Macdonald, 2002); however, it is worth noting that the δ subunit of the $GABA_A$ receptor confers a greater transduction of the neurosteroid signal upon binding of THDOC (Wohlfarth et al., 2002). $GABA_A$ receptors containing the δ subunit are mainly found extrasynaptically and are more resistant to desensitization but have a low affinity to GABA, even at saturating concentrations (Glykys & Mody, 2007; Wohlfarth et al., 2002). In animal models of MTLE, subunits of $GABA_A$ receptors undergo alterations that render them less sensitive to neurosteroid modulation (Mtchedlishvili, Bertram, & Kapur, 2001). In the epileptic rat, a downregulation of the δ subunit as well as an upregulation of the α4 subunit have been reported; these changes have been proposed to contribute to increased excitability (Peng, Huang, Stell, Mody, & Houser, 2004; Smith et al., 1998). THDOC can enhance the tonic current generated by these $GABA_A$ receptors in a way that increasing GABA alone cannot (Wohlfarth et al., 2002).

By modulating the efficacy of $GABA_A$ receptor function, THDOC acts as a broad spectrum anticonvulsant, enhancing the inhibitory tone of the brain (Reddy, 2004; Rupprecht, Hauser, Trapp, & Holsboer, 1996). THDOC protects against seizures induced by pilocarpine, kindling, and $GABA_A$ receptor antagonists in animal models of epilepsy *in vivo* (Reddy, 2011). Furthermore, *in vitro* studies performed in rat piriform cortex and hippocampus show that inhibitory neurosteroids can produce a concentration-dependent suppression of epileptiform activity in 4AP and picrotoxin-treated slices along with a corresponding reduction in the occurrence of HFOs (Herrington, Lévesque, & Avoli, 2013; Salazar, Tapia, & Rogawski, 2003; Fig. 4.5B). Following pilocarpine treatment in rats, *in vivo* studies have shown that the induction of neurosteroid synthesis can potentially delay the development of spontaneously occurring seizures following SE (Biagini et al., 2006). Neurosteroids play an important role in regulating neural excitability and can potentially contribute to epileptogenesis and ictogenesis, a phenomenon that has also been identified in humans.

Patients with MTLE can experience both sexual and reproductive dysfunction. In men, these changes are associated with low levels of testosterone, which may be attributed to the suppression of the hypothalamic–

pituitary–gonadal axis by limbic seizures (Reddy & Rogawski, 2012). Following surgical removal of the epileptic foci, the levels of serum androgens normalize. Additionally, catamenial seizure exacerbations affect up to 70% of women at a child-bearing age with epilepsy. In these patients, seizures occur in cyclical patterns associated with specific phases of the menstrual cycle, which can be related to the withdrawal of the anticonvulsant effects of neurosteroids (Reddy & Rogawski, 2012).

6. CONCLUSIONS

We have reviewed the neuronal processes leading to epileptiform synchronization under acute pharmacological manipulations as well as in brains made chronically epileptic by experimental procedures such as the pilocarpine-induced SE. These data underscore the importance to distinguish two main patterns of seizure onsets: HYP and LVF seizures. These different types of seizures are associated with specific increases in ripple and fast ripple rates during both preictal period and seizure activity itself. Perhaps and more important, HYP and LVF seizures may mirror and thus result from a decrease and an increase of $GABA_A$ receptor signaling, respectively. In this respect, and with potential implications for the mechanisms underlying epileptogenesis, we anticipate that neurosteroids may play an important modulatory action on limbic networks excitability.

In the context of MTLE, LVF and HYP seizures could reflect distinct patterns of up- or downregulation of specific interneuron types perhaps resulting from the neuronal loss that characterizes mesial temporal sclerosis. We anticipate that further experiments will provide definitive information regarding the mechanisms underlying seizure initiation in MTLE both in animal models and in humans, thus leading to specific pharmacological treatments as well as to the development of seizure prediction tools.

ACKNOWLEDGMENTS

The original work reviewed here was supported by the CIHR (Operating Grants 8109 and 74609), CURE and the Savoy Foundation. We thank Drs. R. Benini, P. de Guzman, J. Gotman, J. Louvel, R. Pumain, and J. Sudbury for participating in some of the original experiments reported in this review.

REFERENCES

Agís-Balboa, R. C., Pinna, G., Zhubi, A., Maloku, E., Veldic, M., Costa, E., et al. (2006). Characterization of brain neurons that express enzymes mediating neurosteroid biosynthesis. *Proceedings of the National Academy of Sciences of the United States of America, 103*, 14602–14607.

Akk, G., Covey, D. F., Evers, A. S., Steinbach, J. H., Zorumski, C. F., & Mennerick, S. (2009). The influence of the membrane on neurosteroid actions at $GABA_A$ receptors. *Psychoneuroendocrinology, 34*(Suppl. 1), S59–S66.
Avoli, M., D'Antuono, M., Louvel, J., Köhling, R., Biagini, G., Pumain, R., et al. (2002). Network and pharmacological mechanisms leading to epileptiform synchronization in the limbic system in vitro. *Progress in Neurobiology, 68*, 167–207.
Avoli, M., & de Curtis, M. (2011). GABAergic synchronization in the limbic system and its role in the generation of epileptiform activity. *Progress in Neurobiology, 95*, 104–132.
Avoli, M., de Curtis, M., & Köhling, R. (2013). Does interictal synchronization influence ictogenesis? *Neuropharmacology, 69*, 37–44.
Avoli, M., Methot, M., & Kawasaki, H. (1998). GABA-dependent generation of ectopic action potentials in the rat hippocampus. *European Journal of Neuroscience, 10*, 2714–2722.
Avoli, M., Panuccio, G., Herrington, R., D'Antuono, M., de Guzman, P., & Lévesque, M. (2013). Two different interictal spike patterns anticipate ictal activity in vitro. *Neurobiology of Disease, 52*, 168–176.
Ayala, G. F., Dichter, M., Gumnit, R. J., Matsumoto, H., & Spencer, W. A. (1973). Genesis of epileptic interictal spikes. New knowledge of cortical feedback systems suggests a neurophysiological explanation of brief paroxysms. *Brain Research, 52*, 1–17.
Barbarosie, M., & Avoli, M. (1997). CA3-driven hippocampal-entorhinal loop controls rather than sustains in vitro limbic seizures. *The Journal of Neuroscience: The Official Journal of the Society for Neuroscience, 17*, 9308–9314.
Barbarosie, M., Louvel, J., Kurcewicz, I., & Avoli, M. (2000). CA3-released entorhinal seizures disclose dentate gyrus epileptogenicity and unmask a temporoammonic pathway. *Journal of Neurophysiology, 83*, 1115–1124.
Ben-Ari, Y. (1985). Limbic seizure and brain damage produced by kainic acid: Mechanisms and relevance to human temporal lobe epilepsy. *Neuroscience, 14*, 375–403.
Berkovic, S. F., Andermann, F., Olivier, A., Ethier, R., Melanson, D., Robitaille, Y., et al. (1991). Hippocampal sclerosis in temporal lobe epilepsy demonstrated by magnetic resonance imaging. *Annals of Neurology, 29*, 175–182.
Biagini, G., Baldelli, E., Longo, D., Pradelli, L., Zini, I., Rogawski, M. A., et al. (2006). Endogenous neurosteroids modulate epileptogenesis in a model of temporal lobe epilepsy. *Experimental Neurology, 201*, 519–524.
Bortel, A., Lévesque, M., Biagini, G., Gotman, J., & Avoli, M. (2010). Convulsive status epilepticus duration as determinant for epileptogenesis and interictal discharge generation in the rat limbic system. *Neurobiology of Disease, 40*, 478–489.
Bragin, A., Azizyan, A., Almajano, J., Wilson, C. L., & Engel, J. (2005). Analysis of chronic seizure onsets after intrahippocampal kainic acid injection in freely moving rats. *Epilepsia, 46*, 1592–1598.
Bragin, A., Benassi, S. K., Kheiri, F., & Engel, J., Jr. (2011). Further evidence that pathologic high-frequency oscillations are bursts of population spikes derived from recordings of identified cells in dentate gyrus. *Epilepsia, 52*, 45–52.
Bragin, A., Wilson, C. L., Fields, T., Fried, I., & Engel, J., Jr. (2005). Analysis of seizure onset on the basis of wideband EEG recordings. *Epilepsia, 46*(Suppl. 5), 59–63.
Buckmaster, P. S. (2012). Mossy fiber sprouting in the dentate gyrus. In J. L. Noebels, M. Avoli, M. A. Rogawski, R. W. Olsen, & A. V. Delgado-Escueta (Eds.), *Jasper's basic mechanisms of the epilepsies* (pp. 416–431). Bethesda, MD: National Center for Biotechnology Information.
Buzsáki, G., & Chrobak, J. J. (1995). Temporal structure in spatially organized neuronal ensembles: A role for interneuronal networks. *Current Opinion in Neurobiology, 5*, 504–510.
Buzsáki, G., & da Silva, F. L. (2012). High frequency oscillations in the intact brain. *Progress in Neurobiology, 98*, 241–249.

Buzsaki, G., Horvath, Z., Urioste, R., Hetke, J., & Wise, K. (1992). High-frequency network oscillation in the hippocampus. *Science*, *256*, 1025–1027.

Chauvière, L., Doublet, T., Ghestem, A., Siyoucef, S. S., Wendling, F., Huys, R., et al. (2012). Changes in interictal spike features precede the onset of temporal lobe epilepsy. *Annals of Neurology*, *71*, 805–814.

Colnat-Coulbois, S., Mok, K., Klein, D., Pénicaud, S., Tanriverdi, T., & Olivier, A. (2010). Tractography of the amygdala and hippocampus: Anatomical study and application to selective amygdalohippocampectomy. *Journal of Neurosurgery*, *113*, 1135–1143.

Crépon, B., Navarro, V., Hasboun, D., Clemenceau, S., Martinerie, J., Baulac, M., et al. (2010). Mapping interictal oscillations greater than 200 Hz recorded with intracranial macroelectrodes in human epilepsy. *Brain: A Journal of Neurology*, *133*, 33–45.

Curia, G., Longo, D., Biagini, G., Jones, R. S. G., & Avoli, M. (2008). The pilocarpine model of temporal lobe epilepsy. *Journal of Neuroscience Methods*, *172*, 143–157.

D'Antuono, M., Benini, R., Biagini, G., D'Arcangelo, G., Barbarosie, M., Tancredi, V., et al. (2002). Limbic network interactions leading to hyperexcitability in a model of temporal lobe epilepsy. *Journal of Neurophysiology*, *87*, 634–639.

De Curtis, M., Jefferys, J. G. R., & Avoli, M. (2012). Interictal epileptiform discharges in partial epilepsy: Complex neurobiological mechanisms based on experimental and clinical evidence. In J. Noebels, M. Avoli, M. A. Rogawski, R. W. Olsen, & A. V. Delgado-Escueta (Eds.), *Jasper's basic mechanisms of the epilepsies* (pp. 213–223). Bethesda: National Center for Biotechnology Information.

De Guzman, P., D'Antuono, M., & Avoli, M. (2004). Initiation of electrographic seizures by neuronal networks in entorhinal and perirhinal cortices in vitro. *Neuroscience*, *123*, 875–886.

Dickson, C. T., & Alonso, A. (1997). Muscarinic induction of synchronous population activity in the entorhinal cortex. *The Journal of Neuroscience: The Official Journal of the Society for Neuroscience*, *17*, 6729–6744.

Diehl, B., & Lüders, H. O. (2000). Temporal lobe epilepsy: When are invasive recordings needed? *Epilepsia*, *41*(Suppl. 3), S61–S74.

Dingledine, R., & Gjerstad, L. (1980). Reduced inhibition during epileptiform activity in the in vitro hippocampal slice. *Journal of Physiology*, *305*, 297–313.

Du, F., Eid, T., Lothman, E. W., Kohler, C., & Schwarcz, R. (1995). Preferential neuronal loss in layer III of the medial entorhinal cortex in rat models of temporal lobe epilepsy. *Journal of Neuroscience*, *15*, 6301–6313.

Dzhala, V. I., & Staley, K. J. (2004). Mechanisms of fast ripples in the hippocampus. *The Journal of Neuroscience: The Official Journal of the Society for Neuroscience*, *24*, 8896–8906.

Engel, J., Jr., Bragin, A., Staba, R., & Mody, I. (2009). High-frequency oscillations: What is normal and what is not? *Epilepsia*, *50*, 598–604.

Engel, J., Dichter, M. A., & Schwartzkroin, P. A. (1997). Basic mechanism of human epilepsy. In J. Engel (Ed.), *Epilepsy: A comprehensive textbook: Vol. 1.* (p. 499). Philadelphia, PA: Lippincott-Raven.

Engel, J., Jr., Henry, T. R., Risinger, M. W., Mazziotta, J. C., Sutherling, W. W., Levesque, M. F., et al. (1990). Presurgical evaluation for partial epilepsy: Relative contributions of chronic depth-electrode recordings versus FDG-PET and scalp-sphenoidal ictal EEG. *Neurology*, *40*, 1670–1677.

Falconer, M. A. (1974). Mesial temporal (Ammon's horn) sclerosis as a common cause of epilepsy. Aetiology, treatment, and prevention. *Lancet*, *2*, 767–770.

Fish, D. R. (1996). Magnetic resonance imaging and supplementary motor area epilepsy. *Advances in Neurology*, *70*, 341–351.

Foffani, G., Uzcategui, Y. G., Gal, B., & Menendez de la Prida, L. (2007). Reduced spike-timing reliability correlates with the emergence of fast ripples in the rat epileptic hippocampus. *Neuron*, *55*, 930–941.

French, J. A., Williamson, P. D., Thadani, V. M., Darcey, T. M., Mattson, R. H., Spencer, S. S., et al. (1993). Characteristics of medial temporal lobe epilepsy: I. Results of history and physical examination. *Annals of Neurology*, *34*, 774–780.

Gloor, P. (1997). *The temporal lobe and limbic system*. USA: Oxford University Press.

Gloor, P., Salanova, V., Olivier, A., & Quesney, L. F. (1993). The human dorsal hippocampal commissure. An anatomically identifiable and functional pathway. *Brain: A Journal of Neurology*, *116*(Pt 5), 1249–1273.

Glykys, J., & Mody, I. (2007). Activation of $GABA_A$ receptors: Views from outside the synaptic cleft. *Neuron*, *56*, 763–770.

Gorter, J. A., van Vliet, E. A., Aronica, E., & Lopes da Silva, F. H. (2001). Progression of spontaneous seizures after status epilepticus is associated with mossy fibre sprouting and extensive bilateral loss of hilar parvalbumin and somatostatin-immunoreactive neurons. *European Journal of Neuroscience*, *13*, 657–669.

Herrington, R., Lévesque, M., & Avoli, M. (2013). Neurosteroids modulate epileptiform activity and associated high-frequency oscillations in the piriform cortex. *Neuroscience*, *256*, 467–477.

Huberfeld, G., Menendez de la Prida, L., Pallud, J., Cohen, I., Le Van Quyen, M., Adam, C., et al. (2011). Glutamatergic pre-ictal discharges emerge at the transition to seizure in human epilepsy. *Nature Neuroscience*, *14*, 627–634.

Ibarz, J. M., Foffani, G., Cid, E., Inostroza, M., & Menendez de la Prida, L. (2010). Emergent dynamics of fast ripples in the epileptic hippocampus. *Journal of Neuroscience*, *30*, 16249–16261.

Jacobs, J., LeVan, P., Chander, R., Hall, J., Dubeau, F., & Gotman, J. (2008). Interictal high-frequency oscillations (80–500 Hz) are an indicator of seizure onset areas independent of spikes in the human epileptic brain. *Epilepsia*, *49*, 1893–1907.

Jefferys, J. G. R., Jiruska, P., de Curtis, M., & Avoli, M. (2012). Limbic network synchronization and temporal lobe epilepsy. In J. Noebels, M. Avoli, M. A. Rogawski, R. W. Olsen, & A. V. Delgado-Escueta (Eds.), *Jasper's basic mechanisms of the epilepsies* (pp. 176–286). Bethesda, MD: National Center for Biotechnology Information.

Jefferys, J. G. R., Menendez de la Prida, L., Wendling, F., Bragin, A., Avoli, M., Timofeev, I., et al. (2012). Mechanisms of physiological and epileptic HFO generation. *Progress in Neurobiology*, *98*, 250–264.

Keller, C. J., Truccolo, W., Gale, J. T., Eskandar, E., Thesen, T., Carlson, C., et al. (2010). Heterogeneous neuronal firing patterns during interictal epileptiform discharges in the human cortex. *Brain: A Journal of Neurology*, *133*, 1668–1681.

Kohling, R., & Staley, K. (2011). Network mechanisms for fast ripple activity in epileptic tissue. *Epilepsy Research*, *97*, 318–323.

Köhr, G., & Heinemann, U. (1989). Effects of NMDA antagonists on picrotoxin-, low Mg^{2+}- and low Ca^{2+}-induced epileptogenesis and on evoked changes in extracellular Na^+ and Ca^{2+} concentrations in rat hippocampal slices. *Epilepsy Research*, *4*, 187–200.

Lévesque, M., & Avoli, M. (2013). The kainic acid model of temporal lobe epilepsy. *Neuroscience & Biobehavioral Reviews*, *37*, 2887–2899.

Lévesque, M., Salami, P., Gotman, J., & Avoli, M. (2012). Two seizure-onset types reveal specific patterns of high-frequency oscillations in a model of temporal lobe epilepsy. *The Journal of Neuroscience: The Official Journal of the Society for Neuroscience*, *32*, 13264–13272.

Lieb, J. P., Engel, J., Jr., Brown, W. J., Gevins, A. S., & Crandall, P. H. (1981). Neuropathological findings following temporal lobectomy related to surface and deep EEG patterns. *Epilepsia*, *22*, 539–549.

Löscher, W., & Hönack, D. (1991). Anticonvulsant and behavioral effects of two novel competitive N-methyl-D-aspartic acid receptor antagonists, CGP 37849 and CGP 39551, in the kindling model of epilepsy. Comparison with MK-801 and carbamazepine. *Journal of Pharmacology and Experimental Therapeutics*, *256*, 432–440.

Majewska, M. D. (1992). Neurosteroids: Endogenous bimodal modulators of the $GABA_A$ receptor. Mechanism of action and physiological significance. *Progress in Neurobiology, 38*, 379–395.
Mantegazza, M., Curia, G., Biagini, G., Ragsdale, D. S., & Avoli, M. (2010). Voltage-gated sodium channels as therapeutic targets in epilepsy and other neurological disorders. *Lancet Neurology, 9*, 413–424.
Margerison, J. H., & Corsellis, J. A. (1966). Epilepsy and the temporal lobes. A clinical, electroencephalographic and neuropathological study of the brain in epilepsy, with particular reference to the temporal lobes. *Brain: A Journal of Neurology, 89*, 499–530.
McIntyre, D., & Schwartzkroin, P. A. (2008). Limbic anatomy and physiology. In J. Engel, & T. Pedley (Eds.), *Epilepsies: A comprehensive textbook* (pp. 337–352). Philadelphia, PA: Lippincott Williams and Wilkins.
Mellon, S. H., & Griffin, L. D. (2002). Neurosteroids: Biochemistry and clinical significance. *Trends in Endocrinology and Metabolism, 13*, 35–43.
Mtchedlishvili, Z., Bertram, E. H., & Kapur, J. (2001). Diminished allopregnanolone enhancement of $GABA_A$ receptor currents in a rat model of chronic temporal lobe epilepsy. *Journal of Physiology, 537*, 453–465.
Niedermeyer, E. (2004). The electrocerebellogram. *Clinical EEG and Neuroscience, 35*, 112–115.
Ogren, J. A., Wilson, C. L., Bragin, A., Lin, J. J., Salamon, N., Dutton, R. A., et al. (2009). Three-dimensional surface maps link local atrophy and fast ripples in human epileptic hippocampus. *Annals of Neurology, 66*, 783–791.
Olivier, A. (2010). Epilepsy surgery at the MNI: From Archibald to the creation of the Shirley and Mark Rayport Fellowship in surgery of epilepsy. *Epilepsia, 51*(Suppl. 1), 97–100.
Panuccio, G., D'Antuono, M., de Guzman, P., De Lannoy, L., Biagini, G., & Avoli, M. (2010). In vitro ictogenesis and parahippocampal networks in a rodent model of temporal lobe epilepsy. *Neurobiology of Disease, 39*, 372–380.
Panuccio, G., Sanchez, G., Lévesque, M., Salami, P., de Curtis, M., & Avoli, M. (2012). On the ictogenic properties of the piriform cortex in vitro. *Epilepsia, 53*, 459–468.
Peng, Z., Huang, C. S., Stell, B. M., Mody, I., & Houser, C. R. (2004). Altered expression of the δ subunit of the $GABA_A$ receptor in a mouse model of temporal lobe epilepsy. *Journal of Neuroscience, 24*, 8629–8639.
Pumain, R., Menini, C., Heinemann, U., Louvel, J., & Silva-Barrat, C. (1985). Chemical synaptic transmission is not necessary for epileptic seizures to persist in the baboon Papio papio. *Experimental Neurology, 89*, 250–258.
Reddy, D. S. (2004). Pharmacology of endogenous neuroactive steroids. *Critical Reviews in Neurobiology, 15*, 197–234.
Reddy, D. S. (2010). Neurosteroids: Endogenous role in the human brain and therapeutic potentials. *Progress in Brain Research, 186*, 113–137.
Reddy, D. S. (2011). Role of anticonvulsant and antiepileptogenic neurosteroids in the pathophysiology and treatment of epilepsy. *Frontiers in Endocrinology, 2*, 38.
Reddy, D. S., & Rogawski, M. A. (2012). Neurosteroids—Endogenous regulators of seizure susceptibility and role in the treatment of epilepsy. In J. Noebels, M. Avoli, M. A. Rogawski, R. W. Olsen, & A. V. Delgado-Escueta (Eds.), *Jasper's basic mechanisms of the epilepsies* (pp. 984–997). Bethesda: National Center for Biotechnology Information.
Rosenow, F., & Lüders, H. (2001). Presurgical evaluation of epilepsy. *Brain: A Journal of Neurology, 124*, 1683–1700.
Rupprecht, R., Hauser, C. A. E., Trapp, T., & Holsboer, F. (1996). Neurosteroids: Molecular mechanisms of action and psychopharmacological significance. *Journal of Steroid Biochemistry and Molecular Biology, 56*, 163–168.
Salanova, V., Markand, O. N., & Worth, R. (1994). Clinical characteristics and predictive factors in 98 patients with complex partial seizures treated with temporal resection. *Archives of Neurology, 51*, 1008–1013.

Salazar, P., Tapia, R., & Rogawski, M. A. (2003). Effects of neurosteroids on epileptiform activity induced by picrotoxin and 4-aminopyridine in the rat hippocampal slice. *Epilepsy Research, 55*, 71–82.

Schwartzkroin, P. A., & Prince, D. A. (1980). Changes in excitatory and inhibitory synaptic potentials leading to epileptogenic activity. *Brain Research, 183*, 61–76.

Smith, S. S., Gong, Q. H., Hsu, F. C., Markowitz, R. S., ffrench-Mullen, J. M., & Li, X. (1998). GABA(A) receptor alpha4 subunit suppression prevents withdrawal properties of an endogenous steroid. *Nature, 392*, 926–930.

Spanedda, F., Cendes, F., & Gotman, J. (1997). Relations between EEG seizure morphology, interhemispheric spread, and mesial temporal atrophy in bitemporal epilepsy. *Epilepsia, 38*, 1300–1314.

Spencer, S. S. (2002). Neural networks in human epilepsy: Evidence of and implications for treatment. *Epilepsia, 43*, 219–227.

Spencer, S. S., Kim, J., & Spencer, D. D. (1992). Ictal spikes: A marker of specific hippocampal cell loss. *Electroencephalography and Clinical Neurophysiology, 83*, 104–111.

Spencer, S. S., & Spencer, D. D. (1994). Entorhinal-hippocampal interactions in medial temporal lobe epilepsy. *Epilepsia, 35*, 721–727.

Spencer, S. S., Williamson, P. D., Bridgers, S. L., Mattson, R. H., Cicchetti, D. V., & Spencer, D. D. (1985). Reliability and accuracy of localization by scalp ictal EEG. *Neurology, 35*, 1567–1575.

Staba, R. J. (2012). Normal and pathologic high-frequency oscillations. In J. Noebels, M. Avoli, M. A. Rogawski, R. W. Olsen, & A. V. Delgado-Escueta (Eds.), *Jasper's basic mechanisms of the epilepsies* (pp. 202–209). Bethesda, MD: National Center for Biotechnology Information.

Staba, R. J., Wilson, C. L., Bragin, A., Fried, I., & Engel, J., Jr. (2002). Quantitative analysis of high-frequency oscillations (80–500 Hz) recorded in human epileptic hippocampus and entorhinal cortex. *Journal of Neurophysiology, 88*, 1743–1752.

Talairach, J., Bancaud, J., Szikla, G., Bonis, A., Geier, S., & Vedrenne, C. (1974). New approach to the neurosurgery of epilepsy. Stereotaxic methodology and therapeutic results. 1. Introduction and history. *Neurochirurgie, 20*(Suppl. 1), 1–240.

Tergau, F., Naumann, U., Paulus, W., & Steinhoff, B. J. (1999). Low-frequency repetitive transcranial magnetic stimulation improves intractable epilepsy. *Lancet, 353*, 2209.

Theodore, W. H., & Fisher, R. S. (2004). Brain stimulation for epilepsy. *Lancet Neurology, 3*, 111–118.

Timofeev, I., & Steriade, M. (2004). Neocortical seizures: Initiation, development and cessation. *Neuroscience, 123*, 299–336.

Townsend, E. (1991). Clinicopathological correlations of low voltage fast and high amplitude spike-and-wave mesial temporal stereoencephalographic ictal onsets. *Epilepsia, 32*, 21.

Traynelis, S. F., & Dingledine, R. (1988). Potassium-induced spontaneous electrographic seizures in the rat hippocampal slice. *Journal of Neurophysiology, 59*, 259–276.

Uusisaari, M., Smirnov, S., Voipio, J., & Kaila, K. (2002). Spontaneous epileptiform activity mediated by GABA(A) receptors and gap junctions in the rat hippocampal slice following long-term exposure to GABA(B) antagonists. *Neuropharmacology, 43*, 563–572.

Velasco, A. L., Wilson, C. L., Babb, T. L., & Engel, J., Jr. (2000). Functional and anatomic correlates of two frequently observed temporal lobe seizure-onset patterns. *Neural Plasticity, 7*, 49–63.

Viitanen, T., Ruusuvuori, E., Kaila, K., & Voipio, J. (2010). The K^+-Cl cotransporter KCC2 promotes GABAergic excitation in the mature rat hippocampus. *Journal of Physiology, 588*, 1527–1540.

Vives, K., Lee, G., Doyle, W., & Spencer, D. (2008). Anterior temporal resection. In J. Engel, & T. Pedley (Eds.), *Epilepsies: A comprehensive textbook* (pp. 1859–1868). Philadelphia, PA: Lippincott Williams and Wilkins.

Vonck, K., Boon, P., Achten, E., De Reuck, J., & Caemaert, J. (2002). Long-term amygdalohippocampal stimulation for refractory temporal lobe epilepsy. *Annals of Neurology*, *52*, 556–565.
Wiebe, S., Blume, W. T., Girvin, J. P., Eliasziw, M., & Effectiveness and Efficiency of Surgery for Temporal Lobe Epilepsy Study Group. (2001). A randomized, controlled trial of surgery for temporal-lobe epilepsy. *New England Journal of Medicine*, *345*, 311–318.
Wieser, H. G., & Yaşargil, M. G. (1982). Selective amygdalohippocampectomy as a surgical treatment of mesiobasal limbic epilepsy. *Surgical Neurology*, *17*, 445–457.
Wohlfarth, K. M., Bianchi, M. T., & Macdonald, R. L. (2002). Enhanced neurosteroid potentiation of ternary GABA(A) receptors containing the delta subunit. *The Journal of Neuroscience: The Official Journal of the Society for Neuroscience*, *22*, 1541–1549.
Ylinen, A., Bragin, A., Nádasdy, Z., Jandó, G., Szabó, I., Sik, A., et al. (1995). Sharp wave-associated high-frequency oscillation (200 Hz) in the intact hippocampus: Network and intracellular mechanisms. *The Journal of Neuroscience: The Official Journal of the Society for Neuroscience*, *15*, 30–46.

CHAPTER FIVE

Limbic Networks: Clinical Perspective

Aylin Y. Reid[1], Richard J. Staba
Department of Neurology, David Geffen School of Medicine, University of California Los Angeles, Los Angeles, California, USA
[1]Corresponding author: e-mail address: areid@mednet.ucla.edu

Contents

1. Introduction 90
2. Clinical Features of Limbic Epilepsy 91
3. Structural Networks in Limbic Epilepsy 95
   3.1 MRI studies 95
   3.2 DTI studies 96
4. Functional Networks in Limbic Epilepsy 97
   4.1 Electro- and magnetoencephalographic studies 97
   4.2 fMRI 105
5. Metabolic Networks in Limbic Epilepsy 109
6. Conclusions 111
References 112

Abstract

Limbic epilepsy refers to a condition that consists of epileptic seizures that originate in or preferentially involve the limbic system. The majority of cases are medically refractory, necessitating surgical resection when possible. However, even resection of structures thought to be responsible for seizure generation may not leave a patient seizure free. While mesial temporal lobe limbic structures are centrally involved, there is growing evidence that the epileptogenic network consists of a broader area, involving structures outside of the temporal lobe and the limbic system. Information on structural, functional, and metabolic connectivity in patients with limbic epilepsy is available from a large body of studies employing methods such as MRI, EEG, MEG, fMRI, PET, and SPECT scanning, implicating the involvement of various brain regions in the epileptogenic network. To date, there are no consistent and conclusive findings to define the exact boundaries of this network, but it is possible that in the future studies of network connectivity in the individual patient may allow more tailored treatment and prognosis in terms of surgical resection.

International Review of Neurobiology, Volume 114
ISSN 0074-7742
http://dx.doi.org/10.1016/B978-0-12-418693-4.00005-4

1. INTRODUCTION

An extraordinary number of detailed experiments have helped to understand the mechanisms underlying the generation of seizures or ictogenesis. In addition, tests such as the electroencephalogram (EEG) identify brain structures where seizures begin (the "seizure onset zone" or SOZ), and collectively data from clinical and basic science studies have been used to develop concepts of the "epileptogenic network," the area of the brain necessary and sufficient to generate spontaneous seizures. Spencer (2002) provides a more elaborate definition of an epileptogenic network as "a functionally and anatomically connected, bilaterally represented, set of cortical and subcortical brain structures and regions in which activity in any one part affects activity in all the others" and that "vulnerability to seizure activity in any one part of the network is influenced by activity everywhere else in the network ... the network as a whole is responsible for the clinical and electrographic phenomena that we associate with human seizures." The concept of networks as the basis for analysis of epilepsy is more widely used and there have been several reviews on this topic in recent years (Bernhardt, Hong, Bernasconi, & Bernasconi, 2013; Bonilha et al., 2013; Engel et al., 2013; Laufs, 2012).

Limbic epilepsy refers to a condition that consists of epileptic seizures that originate in or preferentially involve the limbic system. In many cases, mesial temporal lobe limbic structures are centrally involved which is the basis for a diagnosis of mesial temporal lobe epilepsy (MTLE). MTLE is the most common focal epilepsy (Semah et al., 1998) and is often treated surgically. The resection of mesial temporal structures eliminates seizures in a substantial number, but not all, of patients with MTLE. In those with MTLE who do not benefit from resective surgery, the simplest explanation is the epileptogenic network extends beyond mesial temporal structures and perhaps even the limbic system (Bertram, 2009), and if this is correct, then how far and where does the epileptogenic network extend and how can it be found? For these reasons, limbic epilepsy has been the most studied epilepsy in terms of networks and connections. Networks involved in epilepsy can be conceptualized and studied in a number of ways. One possibility is to consider structural networks, which are comprised of nodes of neurons or brain regions, and the physical connections between them (Kramer & Cash, 2012). Functional networks represent the coupling of dynamic activity in separate areas of the brain (Kramer & Cash, 2012). There is also the concept of metabolic networks in epilepsy, with metabolically coherent regional function and dysfunction (Pan et al., 2012). Data from studies looking at networks within these different frameworks can be complementary, providing different lines of evidence pointing to similar structures being involved in an epileptogenic network. With recent advances in functional and structural imaging, as well as electrophysiological techniques, there is an ever growing body of supporting evidence for a network involving both limbic and extralimbic structures that is important in limbic epilepsy. Understanding limbic epilepsy on a network level may allow further advances into how this pathological network is set up in the first place, and what strategies may be more advantageous for treating, or even preventing, abnormal network activity.

2. CLINICAL FEATURES OF LIMBIC EPILEPSY

The term "limbic seizures" generally refers to epileptic seizures originating in the limbic system but could also include seizures beginning elsewhere that preferentially propagate to limbic structures through direct or indirect pathways (see Fig. 5.1). Clinically, seizures are classified as being limbic based on characteristic ictal manifestations, which often include autonomic, psychic, and motor components, and could also be helpful in determining the anatomical location of ictal onset. Limbic seizures typically begin with an aura, or focal seizure, which can occur in isolation or progress into a focal dyscognitive seizure that usually consists of behavioral arrest and staring. Various clinical features associated with limbic seizures indicate the involvement of different brain regions (see Table 5.1), which is consistent with an epileptogenic network extending beyond limbic areas to extralimbic structures. In addition to the ictal features seen in patients with limbic seizures, there are a number of characteristic interictal disturbances as well. The importance of seizures as well as comorbidities is highlighted by the definition of epilepsy proposed by the ILAE in 2005, which states "Epilepsy is a disorder of the brain characterized by an enduring predisposition to generate epileptic seizures, *and by the neurobiological, cognitive, psychological, and social consequences of this condition*" (Fisher et al., 2005). Psychiatric disorders, such as depression, anxiety, and bipolar disorder, are all seen in higher proportions of patients with epilepsy compared to the general population and even other chronic health conditions (Ettinger, Reed, Cramer, & Epilepsy Impact Project Group, 2004; Ettinger, Reed, Goldberg, & Hirschfeld, 2005; Tellez-Zenteno, Patten, Jette, Williams, & Wiebe, 2007), implicating

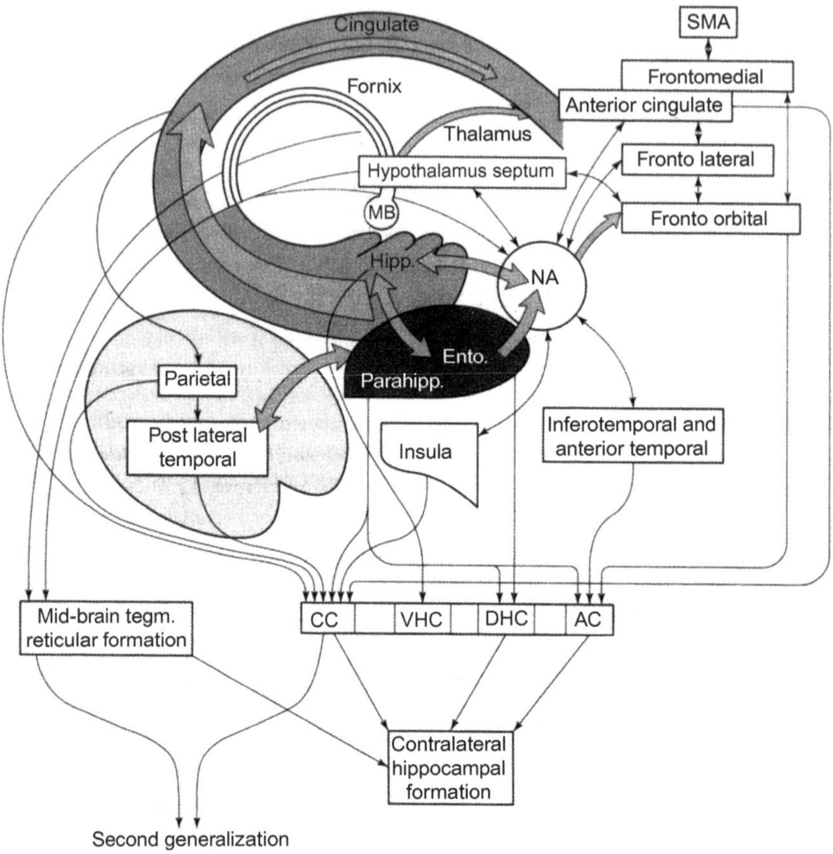

Figure 5.1 Schematic of the human limbic system and its afferent and efferent connections. *Adapted from Wieser (1988) with permission.*

overlap between the networks involved in these disorders and the epileptogenic network. Many patients with limbic epilepsy also have cognitive deficits, especially those that are medically refractory. While memory impairment is common in patients with magnetic resonance imaging (MRI) evidence for hippocampal damage, more diffuse and generalized deficits can also be seen in the areas of intelligence, language, visuospatial function, executive function, and motor function (Marques et al., 2007; Oyegbile et al., 2004), again suggesting a more widespread network is involved.

Hippocampal sclerosis (HS) is often found in patients with MTLE, and MTLE with HS is now recognized as a distinctive constellation within the

Table 5.1 Clinical features of limbic seizures

Clinical feature	Involved brain regions
Olfactory hallucinations	Anterior mesial temporal or orbitofrontal
Gustatory hallucinations	Insula/peri-insular regions, parietal operculum
Fear	Amygdala
Auditory hallucinations	Lateral temporal neocortex
Vestibular symptoms	Lateral temporal neocortex
Visual hallucinations	Occipital lobe
Decreased level of consciousness	Default mode network, bilateral mesial temporal
Oroalimentary automatisms	Amygdala
Head deviation	Frontal lobe

electroclinical syndromes by the International League Against Epilepsy (Berg et al., 2010). The classical pattern of HS consists of cell loss and gliosis that is most prominent in CA1 and CA3 regions, with relative sparing of cells in CA2 (Babb & Brown, 1987; Mathern, Wilson, & Beck, 2008). More recently, other patterns of HS have been described that are associated with different clinical history and response to resective surgery (Blumcke et al., 2007, 2013). Focal cortical dysplasia (FCD) is another common lesion in patients with MTLE. While there are various types of FCDs, FCD Type IIIa comprises cortical lamination abnormalities in the temporal lobe associated with HS (Blumcke et al., 2011). Various tumors have been found in the temporal lobe of patients with limbic seizures, including hamartomas, epidermoids and dermoids, and dysembryoplastic neuroepithelial tumors. Vascular malformations, tubers, and trauma are other pathologies often encountered. However, 20–30% of patients with TLE have no identifiable lesion on MRI (Tellez-Zenteno, Hernandez Ronquillo, Moien-Afshari, & Wiebe, 2010), but obviously have dysfunction within the limbic network.

Treatment of limbic epilepsy initially consists of medical therapy with any one of various antiseizure medications indicated for focal seizures. However, limbic seizures are often refractory to medical treatment, with one study reporting 20% of patients with presumed TLE respond to medical treatment, and only 10% of those with MTLE with HS (Semah et al., 1998). The most common treatment for drug-resistant limbic epilepsy is resective surgery, but newer methods being investigated include implantation of stimulation devices such as the newly approved reactive neurostimulator (Morrell & Group, 2011) and ablative surgery using radiofrequency or radiation (Malikova et al., 2011; Quigg, Rolston, & Barbaro, 2012). A widely used resective approach is anteromesial temporal lobectomy, which extends 3 cm posterior to the temporal pole laterally, and as far back as possible mesially to include the amygdala, pes of the hippocampus and most of the tail of the hippocampus. Another approach is a selective amygdalohippocampectomy that includes some of the parahippocampal gyrus but otherwise spares neocortical tissue. These techniques have been developed and refined based on correlations of surgical outcomes with presurgical results, including electrophysiology and neuroimaging studies of the limbic network (Al-Otaibi, Baeesa, Parrent, Girvin, & Steven, 2012). A recent metaanalysis showed anteromesial temporal resection confers a higher chance of seizure freedom compared to selective amygdalohippocampectomy (Josephson et al., 2013). A separate study found the pooled proportion of seizure-free patients after resective surgery in lesional MTLE was 69% versus 45% in nonlesional MTLE, with the odds of seizure freedom being 2.7 times higher in lesional MTLE at 1-year post-surgery (Tellez-Zenteno et al., 2010). Furthermore, sustained seizure freedom at 5 years post-MTLE surgery is achieved in 66% (Tellez-Zenteno, Dhar, & Wiebe, 2005). Surgical outcome statistics based on retrospective studies is consistent with two randomized controlled trials of resective surgery for TLE that reported seizure-free rates of 64% at 1 year follow-up (Wiebe, Blume, Girvin, & Eliasziw, 2001) and 73% at 2 years after surgery (Engel et al., 2012).

Despite the fact that standard resective surgery for MTLE removes several mesial temporal limbic structures central to the SOZ, as indicated in the preceding paragraph, there are still a large percentage of patients who are not seizure-free or do not remain seizure-free 5 or more years after surgery. Even in patients with electrophysiological-verified epileptogenic lesions, such as HS or other focal pathology, some do not benefit from surgery. Other patients who fail surgery may in fact have temporal "plus" epilepsy, with an epileptogenic network that extends into other cerebral lobes (Barba, Barbati, Minotti, Hoffmann, & Kahane, 2007; Ryvlin, 2005). Results such as these suggest different types of mesial temporal lesions or subtypes of the same pathology (e.g., HS) could be associated with more subtle functional and/or structural abnormalities in extratemporal structures that reflect different entities of MTLE (Thom, Mathern, Cross, & Bertram, 2010). How damage in the mesial temporal lobe affects other components within and outside the limbic system and vice versa is not known. The extent of reorganization that occurs in limbic and extralimbic areas after removal of the epileptogenic lesion is also not known, but in surgical cases with poor outcome, the presence of other epileptogenic sites or possibly the development of new epileptogenic structures could be revealed.

3. STRUCTURAL NETWORKS IN LIMBIC EPILEPSY

Imaging techniques such as MRI are useful for studying various pathologies associated with limbic epilepsy, such as atrophy of gray matter structures, reductions in white matter, and changes in cortical thickness. Differences in white matter tracts in patients compared with controls can help construct structural networks that may be important in limbic epilepsy. Diffusion tensor imaging (DTI) studies the diffusion of water in the brain, allowing fiber tracts to be reconstructed based on water diffusion occurring primarily along a tract. The density of fibers connecting various brain regions, or network nodes, is one way to study structural connectivity in the brain. However, this is limited by the structural resolution of the MRI and small, yet important, tracts may be missed.

3.1. MRI studies

In addition to hippocampal atrophy associated with MTLE, voxel-based morphometry studies have revealed that gray matter pathology also extends to the parahippocampal and isocortical temporal regions, the frontal lobe, parietooccipital regions, cingulate gyrus, thalamus, caudate, midbrain, and cerebellum (Bernasconi et al., 2004; Bonilha et al., 2004). White matter reductions have been reported ipsilateral to the SOZ, including the temporopolar, entorhinal, and perirhinal cortices (Bernasconi et al., 2004). Several of these regions are directly connected to the hippocampal formation and parahippocampal gyrus that could form the structural basis of an epileptogenic network. Some of these changes, such as the degree of thalamic atrophy ipsilateral to the SOZ, correlate with the duration of epilepsy (Natsume, Bernasconi, Andermann, & Bernasconi, 2003). An MRI-based cortical thickness analysis found that TLE patients had cortical thinning in temporal and frontocentral neocortices, and with respect to controls, gray matter thickness in entorhinal cortex was more strongly correlated with thickness in orbitofrontal cortex, but more weakly correlated with thickness of the lateral temporal lobe (Bernhardt et al., 2008). However, there was no significant correlation between mesial temporal thinning and overall cortical surface thinning. Graph theory analysis of MRI-based cortical thickness correlations found that compared to control subjects, patients with TLE had longer path length and increased clustering between anatomical regions, with an altered distribution of network hubs, suggesting reorganization of cortical thickness correlation networks (Bernhardt, Chen, He, Evans, & Bernasconi, 2011).

3.2. DTI studies

Diffusion tensor MRI and probabilistic tractography have shown that MTLE is associated with a regional reduction in fiber density but paradoxical increase in limbic network clustering and increased nodal efficiency in the ipsilateral insula, superior temporal region, and the thalamus (Bonilha, Nesland, et al., 2012). At least 1 year after surgery, those that were not seizure free had higher structural connectivity between the ipsilateral medial and lateral temporal lobe, the ipsilateral medial temporal, and parietal lobe, and the contralateral temporal pole and parietal pole than those that were seizure-free (Bonilha et al., 2013), suggesting some structures not removed during standard resections may be important for seizure control. Hippocampal atrophy also correlates with decreased connectivity density but increased connectivity strength as measured by diffusion anisotropy, showing that even sparse connections can be strong (Ellmore, Pieters, & Tandon, 2011). The fractional anisotropy and apparent diffusion coefficient of language pathways in the uncinate fasciculus and arcuate fasciculus in TLE are abnormal (Kim et al., 2011; Lin, Riley, Juranek, & Cramer, 2008), with bilateral changes in the fornix and cingulum (Concha, Beaulieu, & Gross, 2005) suggesting these white matter tracts are vulnerable to recurrent seizures. There are also DTI changes in the thalami bilaterally (Kim et al., 2010; Wang et al., 2010). While most studies are done in adults, children and adolescents with TLE also have significant WM abnormalities (Meng et al., 2010).

A metaanalysis of 13 studies shows WM structural integrity is affected in the ipsilateral hemisphere more than contralaterally, with tracts closely connected with the affected temporal lobe being the most disturbed (Otte et al., 2012). After anterior temporal lobe resection, there are tract changes compatible with Wallerian degeneration, but changes are also seen in the corona radiata and internal and external capsules that are compatible with plasticity (Winston, Stretton, Sidhu, Symms, & Duncan, 2013). Patients with left TLE show most of these postsurgical changes within a few months after resection, while those with right TLE have changes occurring up to 12 months postoperatively, suggesting slower plasticity (Winston, Stretton, Sidhu, Symms, & Duncan, 2013). While TLE is associated with abnormalities in corpus callosum WM tracts, medical treatment with antiseizure drugs appears to cause additional damage, especially on the splenium (Gunbey et al., 2011). There is also evidence for different patterns of white matter alterations in HS depending on sex (Oguz et al., 2013).

The presence of HS appears to affect the network differently than nonlesional MTLE. While patients with nonlesional MTLE have widespread and bilaterally distributed white matter changes compared to controls (Keller et al., 2013), studies have shown WM changes are more extensive in MTLE with HS compared to nonlesional MTLE (Liacu, Idy-Peretti, Ducreux, Bouilleret, & de Marco, 2012; Liu, Concha, Lebel, Beaulieu, & Gross, 2012). In particular, abnormalities of WM tracts directly related to mesial temporal structures, such as the cingulum and the fornix, may only be seen with MTS (Concha, Beaulieu, Collins, & Gross, 2009). There is conflicting evidence whether patients with right MTS have more extensive changes than those with left MTS (Liu et al., 2012) or vice versa (Ahmadi et al., 2009; Kemmotsu et al., 2011).

4. FUNCTIONAL NETWORKS IN LIMBIC EPILEPSY

There are a variety of tools available to study functional networks in the human brain. Electroencephalography records electrical activity of the brain, either from the scalp or with intracranial electrodes. In particular, intracranial studies have allowed researchers to relate electrical activity in various parts of the brain in patients with MTLE, to study networks of activation, or inactivation, both ictally and interictally. Functional magnetic resonance imaging (fMRI) looks at both temporal and spatial measures of connectivity between different brain regions by assessing changes in blood flow. fMRI can be done both at rest or during different types of tasks, in order to study connectivity during different states. MEG records magnetic fields produced by electric currents in the brain and can better localize these signals compared to EEG.

4.1. Electro- and magnetoencephalographic studies

EEG studies can map functional networks in the epileptic brain on the basis of electric currents. They have the benefit of being easily recorded in both ictal and interictal states, with invasive intracranial recordings being commonly used in presurgical patients leading to a wealth of EEG data. Increased theta-band functional connectivity (FC) can be seen on an EEG after the first seizure in patients who later go on to have a diagnosis of epilepsy versus those with just a single seizure (Douw et al., 2010). A study looking at interictal EEG in TLE showed a trend toward a more regular, and less random, configuration of interictal epileptogenic networks (Bartolomei, Bettus, Stam, & Guye, 2013). Patients with left TLE also have higher power and synchronization than controls in EEG theta bands, but lower synchronization and power in high alpha bands, plus deviations from the small world network architecture seen in health brains. This provides evidence that the epileptic brain differs in its global network characteristics compared to the healthy brain, and not only in focal characteristics (Quraan, McCormick, Cohn, Valiante, & McAndrews, 2013). Computer simulations show EEG discharges representing either focal or generalized seizures arose as a consequence of subtle changes in network structure, without requirement for localized pathologic brain region. Introducing a pathologic region gave rise to either focal, secondary-generalized, or primary-generalized seizures depending on the network structure. Decreased connectivity increased the frequency of seizure-like events (Terry, Benjamin, & Richardson, 2012).

MEG is another useful tool for defining epileptic networks and can extract a significant proportion of the networks found in more invasive studies, such as sEEG (Malinowska et al., 2014). However, there is little work on network connectivity in limbic epilepsy to date. MEG resting state recordings in patients with other types of lesional epilepsy shows a frequency-specific positive correlation between FC in the lower alpha band and seizure frequency. When MEG was repeated 9–15 months after resective surgery, there was an increase in measures of integration of information in the network compared to baseline, which was only seen in the seizure-free patients (van Dellen et al., 2014). In limbic epilepsy, patients with left TLE had decreased ipsilesional medial temporal lobe–inferior temporal cortex backward connectivity and increased bidirectional inferior frontal cortex–medial temporal lobe connectivity contralaterally during verbal working memory tasks. Contralateral frontotemporal interactions were inversely related to performance and level of education (Campo et al., 2012, 2013). Combined MEG/MRI studies have shown that MTLE patients with widely distributed equivalent current dipoles on MEG have widespread cortical thinning on MRI versus patients with vertically oriented anteromedial equivalent current dipoles (Reinsberger et al., 2010). The areas of thinning in these left

MTLE patients included cingulate, supramarginal, occipitotemporal and parahippocampal gyri, precuneus and parietal lobule ipsilaterally, and the frontomedial, frontocentral, and frontobasal gyri and precuneus contralaterally. This is further evidence that some patients with MTLE have alterations to a larger cortical network than others, with MEG being one potential modality to identify this.

Analysis of intracranial EEG often shows seizures that begin independently or nearly simultaneously from multiple mesial temporal lobe structures (e.g., amygdala, hippocampus, entorhinal cortex) and sometimes lateral temporal and even extratemporal structures (insula, orbitofrontal cortex) (Bartolomei et al., 2005; Isnard, Guenot, Ostrowsky, Sindou, & Mauguiere, 2000; Maillard et al., 2004; Ryvlin, 2005; Wennberg, Arruda, Quesney, & Olivier, 2002). Whether the propensity to generate spontaneous seizures (ictogenesis) is equivalent between these brain areas is not known, but results from one study would suggest it is not the same (Bartolomei, Chauvel, & Wendling, 2008). In this latter study, a metric called the "epileptogenicity index" (or EI) was used to quantify the tendency of temporal lobe structures to generate fast ictal activity in patients with MTLE and found higher EI values (i.e., stronger tendency to generate fast ictal activity) were associated with mesial structures (especially anterior hippocampus and entorhinal cortex) compared to lateral temporal areas. Furthermore, patients with normal MRI had higher EI values in lateral temporal sites (e.g., temporopolar and middle temporal gyrus) and a greater number of sites overall with high EI values than patients with HS. The data suggest that compared to patients with HS, patients without obvious hippocampal pathology could have a more widespread epileptogenic network that extends beyond mesial temporal limbic structures. Results such as these have important implications for presurgical localization and prognosis for postsurgical seizure freedom but also indicate the extent of functional alterations that support ictogenesis in limbic epilepsy.

Studies of two common intracranial ictal EEG onset patterns in patients with MTLE—hypersynchronous (HYP) and low voltage fast (LVF)—provide insight into what are likely two different structural and functional mechanisms of ictogenesis. The HYP ictal EEG onset pattern consists of rhythmic low frequency ($\leq$2 Hz), high-amplitude EEG spikes that often appear on a few electrodes in the hippocampus, whereas LVF ictal onsets generally contain high frequency (>13 Hz) ictal activity that can be recorded on several electrodes in hippocampal and extrahippocampal sites (Fig. 5.2; Bartolomei, Gavaret, Badier, Marquis, & Chauvel, 2004;

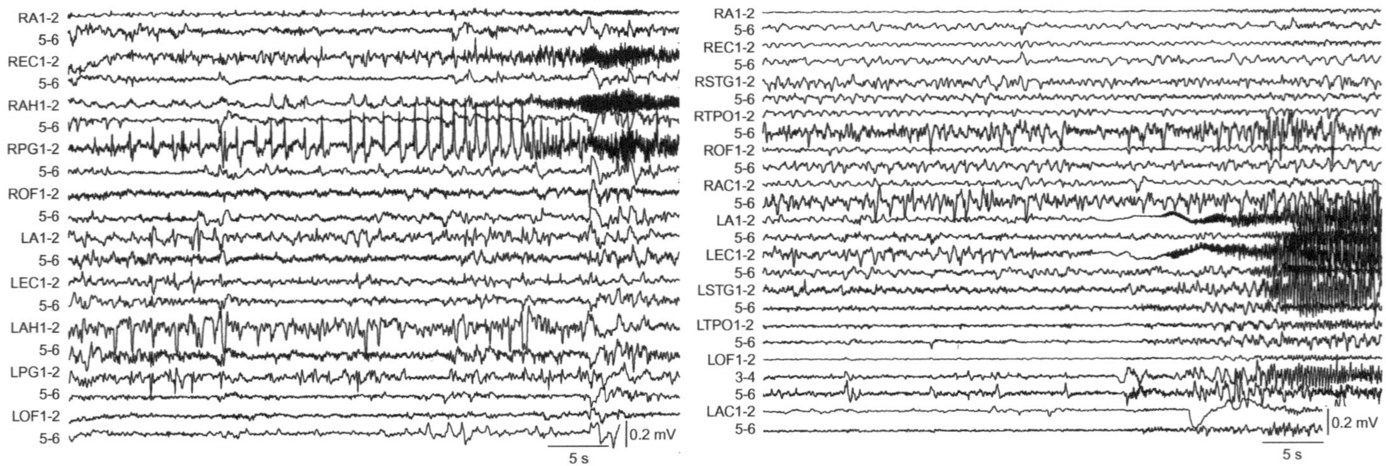

Figure 5.2 Depth electrode recordings of two common ictal EEG onset patterns from two different patients with MTLE. Traces on left illustrate "hypersynchronous" ictal EEG onset consisting of slow (<2 Hz) high-amplitude EEG spikes in the right posterior parahippocampal gyrus (RPG) and persists for $\sim$20 s before spreading to the right anterior hippocampus (RAH) and entorhinal cortex (REC). Traces on right illustrate "low voltage fast" ictal EEG onset consisting of fast (>13 Hz) low-amplitude ictal activity that appears nearly simultaneously in left EC and amygdala (LA) and within seconds in lateral aspects of left orbitofrontal cortex (LOF). Other brain areas sites: anterior cingulate (AC), superior temporal gyrus (STG), temporal–parietal–occipital (TPO). Electrode contact numbers 1–2 indicate mesial, 3–4 middle, and 5–6 lateral locations.

Spencer, Guimaraes, Katz, Kim, & Spencer, 1992; Velasco, Boleag, et al., 2000). HYP ictal discharges can persist in hippocampus for long periods without propagating or spreading to adjacent mesial temporal structures (Perucca, Dubeau, & Gotman, 2014), and eventually contralateral mesial temporal structures, but usually transition to another ictal EEG pattern such as LVF before propagating (Velasco, Wilson, Babb, & Engel, 2000). HYP and LVF onset seizures are frequently recorded in patients with histological or MRI evidence of HS (Perucca et al., 2014; Velasco, Wilson, et al., 2000). However, a detailed neuroimaging analysis of the hippocampal formation found patients with HYP ictal onsets had significant atrophy ipsilateral, but not contralateral, that resembled the classic pattern of HS (i.e., damage in CA3 and CA1, relative sparing in CA2), while those with LVF onset seizures had diffuse ipsilateral hippocampal damage that included CA2 and significant contralateral atrophy (Ogren, Bragin, et al., 2009; Fig. 5.3). Furthermore, more recent work has found in patients with MTLE, cortical GM thinning correlates with longer duration of epilepsy. Patients with HYP onsets had significant thinning per year of disease chiefly in ipsilateral parahippocampal gyrus and bilateral cingulate gyrus, while in patients with LVF onsets, thinning per year of epilepsy was primarily bilateral in temporopolar regions and lateral orbitofrontal gyrus (Memarian et al., 2013). The spatial extent of hippocampal and extrahippocampal damage associated with these two ictal EEG onset patterns provides evidence for different structural alterations that support ictogenesis in two of what are likely several subtypes of MTLE (Bonilha, Martz, Glazier, & Edwards, 2012; Kahane & Bartolomei, 2010).

A major advance in epilepsy research during the last 15 years was the discovery of high-frequency oscillation (HFOs) that were initially recorded from microelectrodes positioned in the hippocampus and entorhinal cortex of patients with MTLE (Bragin, Engel, Wilson, Fried, & Buzsaki, 1999; Bragin, Engel, Wilson, Fried, & Mathern, 1999). In the epileptic brain, interictal HFOs are generally brief in duration (10–100 ms), have variable amplitude (10–1000 μV), and contain spectral power between 100 and 600 Hz (Fig. 5.4A; Bragin, Benassi, Kheiri, & Engel, 2011; Staba, 2012). In patient studies and chronic animal models of epilepsy, there appear to be different types of HFOs with some that have properties that strongly resemble 100–200 Hz ripples in the normal nonprimate and nonhuman primate hippocampal formation (Axmacher, Elger, & Fell, 2008; Bragin, Engel, Wilson, Vizentin, & Mathern, 1999; Buzsaki, 1998; Le Van Quyen et al., 2008; Skaggs et al., 2007; Staba, Wilson, Bragin, Fried, &

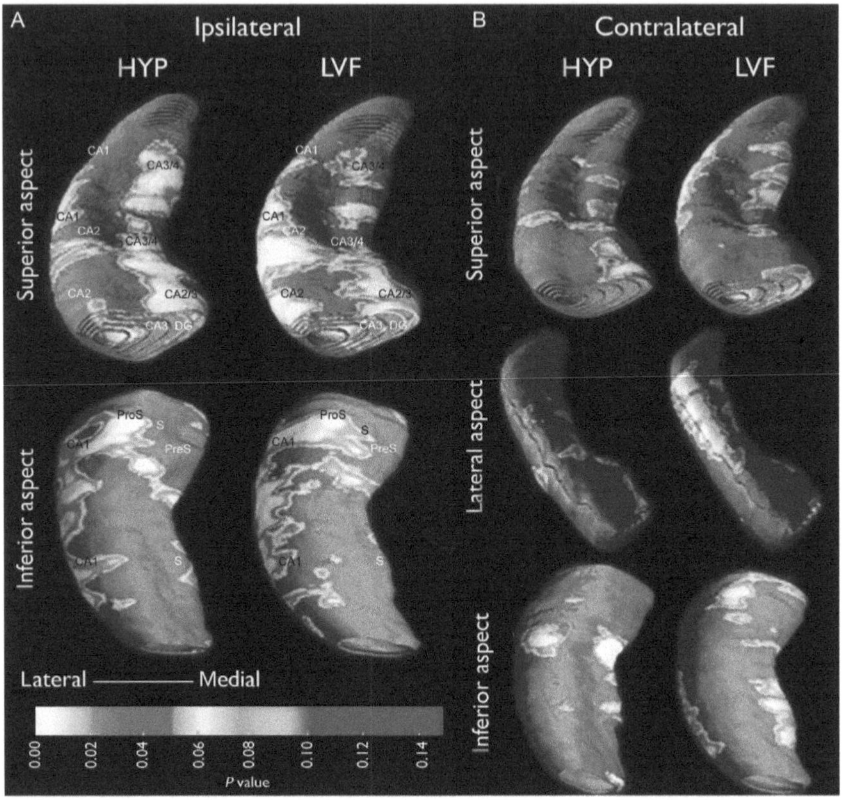

Figure 5.3 Three-dimensional surface contour maps depicting hippocampal atrophy in patients with unilateral hypersynchronous (HYP) or low voltage fast (LVF) ictal EEG onset. (A) In the hippocampus ipsilateral to seizure onset, patients with HYP onsets had significant atrophy (colored white and red (white and light gray in print version)) in several areas that included dentate gyrus (DG), subfields CA3 and CA1, and to lesser extent CA2 (shaded blue (dark gray in print version)). By contrast, patients with LVF onset had atrophy in many of these same areas, but more extensive in damage in CA2. (B) In the contralateral hippocampus, patients with HYP onset had isolated areas of atrophy that overall were not statistically significant compared to matched controls, whereas patients with LVF onset had significant atrophy in CA3 and CA1 and posterior aspects of subicular cortex (S). *Colored illustration modified with permission from Ogren, Bragin, et al. (2009).*

Engel, 2002; Staba et al., 2004). By contrast, other HFOs (200–600 Hz, originally termed "fast ripples") localize to brain regions capable of generating spontaneous seizures (Bragin, Engel, Wilson, Fried, & Mathern, 1999; Jacobs et al., 2008; Staba et al., 2002; Worrell et al., 2008), and rates of their occurrence have been linked with the severity of morphological alterations associated with HS in MTLE (Ogren, Wilson, et al., 2009;

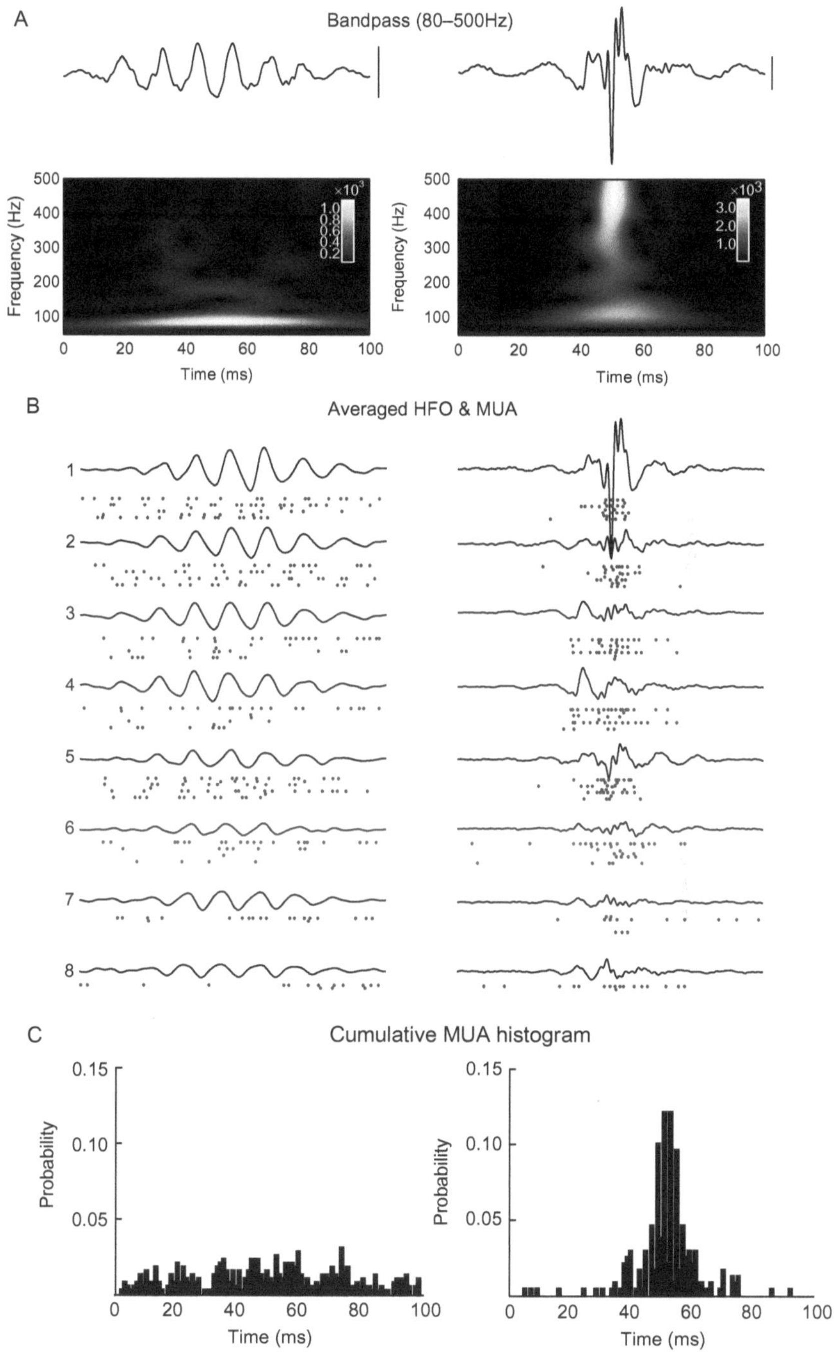

Figure 5.4 Spontaneous HFOs recorded using a spaced microelectrode positioned in entorhinal cortex of patient with MTLE. (A) Bandpass filtered trace of ripple-frequency *(Continued)*

Staba et al., 2002, 2007), but not lesions associated with other epilepsies (Jacobs et al., 2009). Compared to ripple-frequency HFOs, spontaneous fast ripples are associated with an abrupt increase in neuronal spike firing (Fig. 5.4B and C); studies indicate fast ripples reflect abnormal transient bursts of population spikes generated from widely spaced, small groups of pathological interconnected neurons (Bragin et al., 2011; Bragin, Mody, Wilson, & Engel, 2002; Bragin, Wilson, & Engel, 2000, 2007). Studies estimate the volume of tissue generating pathological HFOs could be 1 mm^3 or less (Bragin, Mody, et al., 2002; Bragin, Wilson, et al., 2002; Demont-Guignard et al., 2012; Ibarz, Foffani, Cid, Inostroza, & Menendez de la Prida, 2010). In spite of this small volume, pathological HFOs appear to represent a more robust approach to study synchronous neuronal bursting than microelectrode single neuron studies, which have had difficulty demonstrating neuronal disturbances in epileptogenic tissue (Colder, Frysinger, Wilson, Harper, & Engel, 1996; Colder, Wilson, Frysinger, Chao, et al., 1996), except during sleep or in relation to structural alterations that appear specific to mesial temporal structures capable of generating spontaneous seizures (Bragin, Mody, et al., 2002; Bragin, Wilson, et al., 2002; Staba et al., 2011).

The presence of high-frequency power (>100 Hz) in local field potential recordings indicates proximity to cell bodies and axon terminals that can serve as a proxy for levels of neuronal spike firing (Buzsaki, Anastassiou, & Koch, 2012), thus changes in ictal HFO power could be used to infer possible neuronal mechanisms of ictogenesis (Engel & da Silva, 2012). For example, in chronic models of epilepsy, studies have found fast-ripple frequency HFOs associated with HYP ictal onsets and an increase in ripple-frequency HFOs in sites of onset and regions of propagation during LVF onset seizures (Bragin, Azizyan, Almajano, Wilson, & Engel, 2005; Bragin, Engel, Wilson, Fried, & Buzsaki, 1999; Levesque, Salami, Gotman, & Avoli, 2012). Furthermore, patient studies have found an increase in HFO-frequency power before or during the onset of some focal hippocampal seizures (Jirsch et al., 2006; Khosravani et al., 2008), which appears to coincide with a spatially widespread increase in broadband power at onset of focal seizures (Perucca et al., 2014). Another study showed HYP

Figure 5.4—Cont'd (left column) and fast ripple-frequency (right column) HFO and corresponding time-spectral frequency histogram (below). (B) Averaged ($n=7$) HFO laminar profile recorded from eight microwires with 500 μm inter-tip spacing arranged in a linear array. Red dots beneath each HFO trace depict multiunit activity (MUA) in relation to extracellular HFO field potentials. (C) Cumulative MUA histogram from each of the eight microwires. Calibration bar in panel A is 100 μV.

ictal onset seizures were associated with a progressive increase in ripple- and fast ripple-frequency HFOs during early and late phases of the seizure, while during LVF onset seizures, ripple- and fast ripple-frequency HFOs increased at onset, although in later phases of the seizure, ripple-, but not fast ripple-, frequency HFOs significantly decreased (Perucca et al., 2014). One interpretation of these data is that during HYP onset seizures, the occurrence of ripple- and fast ripple-frequency HFOs likely reflects burst of population spikes from excitatory cells that receives strong inhibitory feedback and rapidly suppresses excitatory cell firing (Bragin et al., 2007; Staba et al., 2007). Recovery from the postinhibitory period could lead to rebound excitation and renewal of the excitatory-inhibitory sequence. By contrast, the initial increase of ripple-frequency HFOs during the LVF onset pattern could reflect interneuron-mediated synchronous IPSPs onto excitatory cells similar to that observed during experimental LVF ictal onset (Gnatkovsky, Librizzi, Trombin, & de Curtis, 2008). Strong inhibitory activity combined with low levels of excitatory cell firing, which is consistent with a relatively low density of fast ripple-frequency HFOs (Perucca et al., 2014), appears to be ideal conditions when inhibitory postsynaptic currents contribute significantly to the extracellular field potential (Trevelyan, 2009). Progression from ictal LVF activity to irregular and eventually regular high-amplitude EEG discharges presumably is associated with a reduction of interneuron-mediated IPSPs (decrease in ripple-frequency HFOs) and strengthening of excitatory cell bursting and synchrony (increase in fast ripple-frequency HFOs). The mechanisms proposed here—enhanced excitation versus enhanced inhibition or synchronization through inhibition (de Curtis & Gnatkovsky, 2009)—likely reflect only two possible neuronal mechanisms of limbic ictogenesis.

4.2. fMRI

fMRI studies have shown increased interictal intrahemispheric FC ipsilateral to the seizure focus (Maccotta et al., 2013), including increased hippocampal FC involving the limbic network (temporal lobe, insula, thalamus), frontal lobes, angular gyrus, basal ganglia, brainstem, and cerebellum (Haneef et al., 2014). In some ways, this seems intuitive, as a disorder of increased excitability and hypersynchrony might be expected to demonstrate increased connectivity between areas involved in seizure generation. However, many studies have actually shown decreased interictal FC in widespread areas (Kucukboyaci et al., 2013), including local and interhemispheric FC

(Maccotta et al., 2013; Morgan, Rogers, Sonmezturk, Gore, & Abou-Khalil, 2011) and with the default mode network (Haneef et al., 2014). Graph theory metric of strength showed reduced connectivity of ipsilateral hippocampus and ipsilateral parahippocampal gyrus compared to controls (James, Tripathi, Ojemann, Gross, & Drane, 2013). These decreases in FC are bilateral, but greater on the epileptogenic side (Doucet, Osipowicz, Sharan, Sperling, & Tracy, 2013a; Pereira et al., 2010; Zhang et al., 2010). It is possible that in the interictal state, decreased connectivity is protective in some ways, helping to prevent propagation of abnormal activity and seizure generation. Negative connectivity across a network including thalamic, brainstem, frontal, and parietal regions supports inhibited function in subcortical and cortical structures during ictal propagation (Morgan, Gore, & Abou-Khalil, 2010). A decreased driving effect from the epileptogenic zone to the thalamus and basal ganglia, with increased feedback, has been demonstrated in left TLE (Ji et al., 2013). A combined approach of both structural and functional connectivity using fMRI and DTI showed decreases in both structural and functional connectivity between the posterior cingulate, precuneus, and bilateral mesial temporal lobes, supporting the fact that decreased connection densities lead to decreased FC (Liao et al., 2011). Cortical atrophy, variations in gray matter density, and microstructural white matter damage assessed by DTI and structural imaging are also accompanied by reduced network integration of the epileptic hippocampus (Voets et al., 2012).

Some changes in connectivity are unique depending on the laterality of the seizure focus. For example, in left TLE, there is reduced FC between the posterior cingulate and the bilateral hippocampi, while patients with right TLE show reduced FC between the posterior cingulate and only the right hippocampus (James et al., 2013). In unilateral TLE, on either side, there is anticorrelated activity in regions outside the epileptogenic temporal lobe, mainly in the contralateral hemisphere and especially the precuneus, which is not seen in bilateral TLE (Tracy et al., 2014). This may be protective in order to constrain epileptiform activity to the affected side in unilateral disease. Many studies have compared changes in FC seen in left versus right TLE and have consistently found reductions may be greatest in left TLE (Kucukboyaci et al., 2013; Pereira et al., 2010). In left TLE there are more extensive default mode networks than with right TLE, which may be relevant to altered cognition and memory (Haneef, Lenartowicz, Yeh, Engel, & Stern, 2012; Haneef et al., 2014). In right MTLE, the FC between the left mesial temporal lobe and medial frontal cortex was positively correlated with the delayed recall score in a non-verbal memory test, while in left MTLE there was a negative correlation between the memory test and the FC of the left mesial temporal lobe and the posterior cingulate (Doucet, Osipowicz, Sharan, Sperling, & Tracy, 2013b). This may be a functional maladaptive change, with left TLE possibly being more impairing than right TLE.

Disease duration plays a role in some changes. The influence from the epileptogenic zone to the right precuneus and brainstem negatively correlates with disease duration, while that from the right hippocampus, fusiform cortex, and lentiform nucleus to the epileptogenic zone shows a positive correlation (Ji et al., 2013). Disrupted interhemispheric hippocampal connectivity actually then increases linearly as epilepsy progresses past 10 years duration, which appears to be due to increased influence of the contralateral onto ipsilateral hippocampus (Morgan et al., 2011), with decreases in FC in the posterior cingulate also seemingly related to disease duration (Zhang et al., 2010).

Apart from its involvement in seizure generation, the limbic network in TLE also appears to be related to cognitive effects, such as impaired memory. There is a negative relationship between working memory performance and FC in the left hippocampus and ipsilateral parahippocampal gyrus. In patients with right TLE there is a positive relationship between performance and FC between the left hippocampus and precuneus, whereas in patients with left TLE there is a negative relationship between performance and FC between both hippocampi and the ipsilateral cerebellum. This suggests patients with right versus left TLE may develop different patterns of FC to implement visuospatial working memory (Doucet et al., 2013b). fMRI during an object-location memory task showed reduced FC to the precuneus in the ipsilateral middle temporal gyrus and hippocampus in left MTLE patients. Patients with left MTLE had lower FC then those with right MTLE in the left middle temporal gyrus, ventral medial prefrontal cortex, and left amygdala (Frings, Schulze-Bonhage, Spreer, & Wagner, 2009). Patients with unilateral MTLE with HS have atypical FC between task-positive and task-negative networks, with co-activation of the task-positive superior parietal lobe and the task-negative epileptic hippocampus. The stronger this abnormal connection, the worse the performance in working memory tasks. This also correlated with abnormal structural connectivity (Stretton et al., 2013). Patients also have activation of bilateral frontoparietal networks during working memory, with reduced activity of the right parietal lobe regardless of side of HS, which is associated with impaired white matter integrity within the frontoparietal network and the contralateral temporal lobe (Winston, Stretton, Sidhu, Symms, Thompson, et al., 2013).

Certain changes in FC are actually able to predict outcome in some studies. In a study using resting state EEG-fMRI pre-operatively, patients who ended up having seizure recurrence after surgery had less lateralized FC than those with a seizure-free outcome (Negishi, Martuzzi, Novotny, Spencer, & Constable, 2011). Patients with stronger FC from the posterior cingulate to the epileptogenic hippocampus showed a greater decline in memory after surgery while those with stronger FC between the posterior cingulate and the contralateral hippocampus showed less (McCormick, Quraan, Cohn, Valiante, & McAndrews, 2013). The resting state FC between the right hippocampus and the ventral lateral nucleus of the right thalamus distinguished seizure-free patients with left versus right TLE, and may be a good marker of laterality (Morgan, Sonmezturk, Gore, & Abou-Khalil, 2012). Changes in FC may also be linked to co-morbidities, as TLE patients with depression had higher activity in the bilateral thalamus, insula, and caudate, and right anterior cingulate, with decreased FC in the prefrontal-limbic system and increased FC within the limbic system and angular gyrus compared to those without depression (Chen et al., 2012). As studies of FC become more widespread, they may one day prove useful in the treatment of individual patients, providing more accurate prognosis in any given patient as to the likelihood of seizure freedom and the risk of cognitive deficits postoperatively.

When interpreting results from fMRI or EEG studies it must be considered that there may be differential effects of epileptic phenomena on electrophysiological and hemodynamic signals. FC measured by iEEG was higher in regions affected by electrical epileptiform abnormalities than non-affected, with opposite pattern for FC using BOLD signal fMRI (Bettus et al., 2011). However, combined studies can be useful. In combined EEG-fMRI studies done in the resting state, it was found that the amygdala and hippocampus on the epileptic side, and to lesser extent the healthy side, are less connected with local networks, and also less connected with dopaminergic mesolimibic networks and the DMN (Pittau, Grova, Moeller, Dubeau, & Gotman, 2012). EEG-fMRI has also shown the hemodynamic response function that peaks 3 s after a spike shows activation involving the ipsilateral mesial temporal structures (likely the hippocampus), putamen/globus pallidus, inferior insula, and superior temporal gyrus. In contrast, the hemodynamic response function peaking at 5 s showed activations involving the ipsilateral and contralateral superior temporal gyrus and inferior insula. Both showed bilateral posterior cingulate deactivations. This is evidence for contralateral temporal involvement later than ipsilateral involvement, and suggests the posterior cingulate deactivation may reflect interconnections between it and limbic structures, and may partially correspond to suspension of DMN, which has been previously described for temporal lobe spikes (Kobayashi, Grova, Tyvaert, Dubeau, & Gotman, 2009).

5. METABOLIC NETWORKS IN LIMBIC EPILEPSY

There is also a somewhat newer concept of metabolic networks in epilepsy, which can be studied using modalities such as positron emission tomography (PET), single-photon emission computed tomography (SPECT), and magnetic resonance spectroscopy (MRS). PET scanning is a functional imaging technique that detects pairs of gamma rays emitted by a positron-emitting radionuclide, or tracer, that is introduced to the body. Fluro-deoxyglucose (FDG) is often used as a tracer to measure metabolic activity. SPECT studies for epilepsy use delivery of a gamma-emitting radionuclide that can be taken up in tissue in a manner proportional to brain blood flow in order to measure perfusion in different areas. MRS can determine the physical and chemical properties of molecules in the brain. For example, increases in choline are associated with membrane breakdown, N-acetyl aspartate (NAA) decreases are seen with neuronal loss or damage, and creatine loss is associated with tissue death.

FDG-PET has shown significant hypometabolism in a broad temporofrontal network ipsilateral to HS, with increased metabolism in mesial and lateral temporal regionals contralaterally. This was positively correlated with episodic memory performance on Wada testing, suggesting a neuroplastic mechanism to preserve these memory processes (Trotta et al., 2011). Naming difficulties in TLE are negatively correlated with resting state brain metabolism in the left posterior and basal temporal regions, implicating changes in the network involving anterior temporal structures responsible for semantic processing and neocortical posterior temporal structures involved in lexical processing (Trebuchon-Da Fonseca et al., 2009). TLE patients with déjà-vu show ipsilateral hypometabolism of the superior temporal gyrus and parahippocampal region compared to TLE patients without déjà-vu or healthy controls, suggesting this phenomenon depends on alterations of the recognition memory system (Guedj, Aubert, McGonigal, Mundler, & Bartolomei, 2010). Another study looking at changes in effective connectivity with the nonepileptic amygdalo–hippocampal complex found decreased effective connectivity with ventral prefrontal cortical areas bilaterally and with the contralateral temporal pole and posterior cingulate cortex, while there was increased connectivity with midline structures such as the anterior cingulate and dorsal medial prefrontal cortex, and the temporoparietal junction bilaterally. Positive connectivity was preserved between the nonepileptic and epileptic amygdalo–hippocampal complex, which may explain how the epileptic side induced these changes contralaterally (Trotta et al., 2013). Similar to other modalities which have shown sex differences in connectivity, PET scans show male patients with TLE more often have ipsilateral frontal lobe hypometabolism and a spread of epileptiform activity to this area versus females who have hypometabolism and ictal spread to the contralateral temporal lobe (Savic & Engel, 1998). A study of focal epilepsy from any location showed EEG–fMRI hemodynamic increases common near the frontal piriform cortex ipsilateral to the focus, and reduced GABA(A) binding with flumazenil PET in same area if the patient had frequent seizures, suggests this area may be a network hub (Laufs et al., 2011). Similar to changes found in FC in the default mode network using fMRI studies, the default mode network also shows lower glucose metabolism and flumazenil-binding ipsilateral to the epileptic focus in TLE (Juhasz et al., 1999). Also, PET abnormalities in metabolism in the thalamus are associated with structural volume loss (Juhasz et al., 1999).

A benefit of SPECT scanning is that it provides valuable information about connectivity during the ictal state. Ictal SPECT scans in MTLE show an increase in tracer uptake in the medial temporal region, cerebellum, thalamus, insula, and putamen (Sequeira et al., 2013), areas also found to have altered FC. The lenticular nucleus and brainstem also appear to be involved in limbic seizures (Grova et al., 2005). Network analyses showed a relatively decreased correlation between the epileptogenic temporal region and the remaining cortex during the interictal state, followed by a surge of cross-correlated perfusion in epileptogenic temporal limbic structures during seizures, corresponding to local network integration. This suggests that MTLE is associated with state-specific perfusion and possibly functional organization and that there is state-specific shift in metabolic networks in MTLE (Sequeira et al., 2013). SPECT studies also provide further evidence for differences in network connectivity in left versus right TLE, as both groups have similar hyperperfusion in the ipsilateral anteromedial temporal–corpus striatum–insula region, but have differences in midbrain reticular formation perfusion, and only right TLE showing contralateral hypoperfusion of the temporoparietal junction (Hogan, Kaiboriboon, Bertrand, Rao, & Acharya, 2006).

While structural studies have shown atrophy in both the hippocampus and the thalamus in MTLE, MRS has shown reduced NAA, NAA/Cr, and NAA/(Cr + Cho) ratios in the epileptic hippocampus and ipsilateral thalamus as well (Fojtikova et al., 2007). Impairment in the ipsilateral hippocampus is correlated with NAA/Cr abnormalities in other subcortical structures, such as the ipsilateral and contralateral anterior and posterior thalami, putamen, and contralateral hippocampus (Hetherington et al., 2007). MRS using the NAA/Cr ratio as a measure of neuronal function and injury in MTLE has demonstrated a set of metabolically covarying regions, including the ipsilateral insula and bilateral basal ganglia, that may be a manifestation of the epileptic network characterizing limbic seizure propagation (Pan et al., 2012).

6. CONCLUSIONS

There is substantial evidence that limbic epilepsy involves a network of structures both within and also external to the limbic system. Certain clinical features of limbic seizures which are known to originate from nonlimbic structures, interictal comorbidities which involve brain regions outside of the limbic system, the range of pathologies causing limbic seizures, and the failure to respond to surgery in some patients with limbic epilepsy has always been suggestive of this. However, the growing body of evidence for altered networks in patients with limbic epilepsy provides evidence for the concept of an epileptogenic network, although it is still not something that can be clearly delineated.

Structural, functional, and metabolic studies all demonstrate alterations in networks in both limbic and nonlimbic areas in patients with limbic epilepsy. Network alterations are related to disease duration, presence of comorbidities such as memory and language deficits, laterality, and even sex. Different network changes in patients with right versus left TLE, and lesional versus nonlesional, are important to consider in the context of response to a surgical treatment that only takes out a limited portion of this epileptogenic network. Surgery failure indicates that in some patients, disruption of the network is not enough to prevent it from generating seizures. Networks can sometimes be useful in predicting outcome after surgery, and plasticity has been demonstrated after resection, however, network studies, outside of localizing language and memory networks, are not routinely used in surgical planning for individual patients. While there is a lack of evidence to date that tailored alterations to the area of resected structures in limbic epilepsy for individual patients with TLE based on their structural, functional, and metabolic connectivity with extralimbic structures can impact outcome, it is possible that one day information from these studies will be useful in guiding individual surgery strategies. Significant gains have been made in recent years about the areas involved in the epileptogenic network in patients with limbic epilepsy, but we are still far from using this information on the individual level and translate it into better patient outcomes.

REFERENCES

Ahmadi, M. E., Hagler, D. J., Jr., McDonald, C. R., Tecoma, E. S., Iragui, V. J., Dale, A. M., et al. (2009). Side matters: Diffusion tensor imaging tractography in left and right temporal lobe epilepsy. *American Journal of Neuroradiology, 30*(9), 1740–1747.

Al-Otaibi, F., Baeesa, S. S., Parrent, A. G., Girvin, J. P., & Steven, D. (2012). Surgical techniques for the treatment of temporal lobe epilepsy. *Epilepsy Research and Treatment, 2012*, 374848.

Axmacher, Nikolai, Elger, Christian E., & Fell, Juergen. (2008). Ripples in the medial temporal lobe are relevant for human memory consolidation. *Brain, 131*(7), 1806–1817.

Babb, T. L., & Brown, W. J. (1987). Pathological findings in epilepsy. In J. Engel, Jr., (Ed.), *Surgical treatment of the epilepsies* (pp. 511–540). New York: Raven Press.

Barba, C., Barbati, G., Minotti, L., Hoffmann, D., & Kahane, P. (2007). Ictal clinical and scalp-EEG findings differentiating temporal lobe epilepsies from temporal 'plus' epilepsies. *Brain, 130*(Pt 7), 1957–1967.

Bartolomei, F., Bettus, G., Stam, C. J., & Guye, M. (2013). Interictal network properties in mesial temporal lobe epilepsy: A graph theoretical study from intracerebral recordings. *Clinical Neurophysiology, 124*(12), 2345–2353.

Bartolomei, F., Chauvel, P., & Wendling, F. (2008). Epileptogenicity of brain structures in human temporal lobe epilepsy: A quantified study from intracerebral EEG. *Brain, 131*(Pt 7), 1818–1830.

Bartolomei, F., Gavaret, M., Badier, J. M., Marquis, P., & Chauvel, P. (2004). EEG and video-EEG explorations in refractory partial epilepsy. *Revue Neurologique (Paris), 160*(Spec. No. 1), 5S81–5S90.

Bartolomei, F., Khalil, M., FWendling, F., Sontheimer, A., Régis, J., Ranjeva, J. P., et al. (2005). Entorhinal cortex involvement in human mesial temporal lobe epilepsy: An electrophysiologic and volumetric study. *Epilepsia, 46*(5), 677–687.

Berg, A. T., Berkovic, S. F., Brodie, M. J., Buchhalter, J., Cross, J. H., van Emde Boas, W., et al. (2010). Revised terminology and concepts for organization of seizures and epilepsies: Report of the ILAE commission on classification and terminology, 2005–2009. *Epilepsia, 51*(4), 676–685.

Bernasconi, N., Duchesne, S., Janke, A., Lerch, J., Collins, D. L., & Bernasconi, A. (2004). Whole-brain voxel-based statistical analysis of gray matter and white matter in temporal lobe epilepsy. *NeuroImage, 23*(2), 717–723.

Bernhardt, B. C., Chen, Z., He, Y., Evans, A. C., & Bernasconi, N. (2011). Graph–theoretical analysis reveals disrupted small-world organization of cortical thickness correlation networks in temporal lobe epilepsy. *Cerebral Cortex, 21*(9), 2147–2157.

Bernhardt, B. C., Hong, S., Bernasconi, A., & Bernasconi, N. (2013). Imaging structural and functional brain networks in temporal lobe epilepsy. *Frontiers in Human Neuroscience*, *7*, 624.

Bernhardt, B. C., Worsley, K. J., Besson, P., Concha, L., Lerch, J. P., Evans, A. C., et al. (2008). Mapping limbic network organization in temporal lobe epilepsy using morphometric correlations: Insights on the relation between mesiotemporal connectivity and cortical atrophy. *NeuroImage*, *42*(2), 515–524.

Bertram, E. H. (2009). Temporal lobe epilepsy: Where do the seizures really begin? *Epilepsy & Behavior*, *14*(Suppl. 1), 32–37.

Bettus, G., Ranjeva, J. P., Wendling, F., Benar, C. G., Confort-Gouny, S., Regis, J., et al. (2011). Interictal functional connectivity of human epileptic networks assessed by intracerebral EEG and bold signal fluctuations. *PLoS ONE*, *6*(5), e20071 (Research Support, Non-U.S. Gov't).

Blumcke, I., Pauli, E., Clusmann, H., Schramm, J., Becker, A., Elger, C., et al. (2007). A new clinico-pathological classification system for mesial temporal sclerosis. *Acta Neuropathologica*, *113*(3), 235–244.

Blumcke, I., Thom, M., Aronica, E., Armstrong, D. D., Bartolomei, F., Bernasconi, A., et al. (2013). International consensus classification of hippocampal sclerosis in temporal lobe epilepsy: A task force report from the ILAE commission on diagnostic methods. *Epilepsia*, *54*(7), 1315–1329.

Blumcke, I., Thom, M., Aronica, E., Armstrong, D. D., Vinters, H. V., Palmini, A., et al. (2011). The clinicopathologic spectrum of focal cortical dysplasias: A consensus classification proposed by an ad hoc task force of the ILAE diagnostic methods commission. *Epilepsia*, *52*(1), 158–174.

Bonilha, L., Helpern, J. A., Sainju, R., Nesland, T., Edwards, J. C., Glazier, S. S., et al. (2013). Presurgical connectome and postsurgical seizure control in temporal lobe epilepsy. *Neurology*, *81*(19), 1704–1710.

Bonilha, L., Martz, G. U., Glazier, S. S., & Edwards, J. C. (2012). Subtypes of medial temporal lobe epilepsy: Influence on temporal lobectomy outcomes? *Epilepsia*, *53*(1), 1–6 (Review).

Bonilha, L., Nesland, T., Martz, G. U., Joseph, J. E., Spampinato, M. V., Edwards, J. C., et al. (2012). Medial temporal lobe epilepsy is associated with neuronal fibre loss and paradoxical increase in structural connectivity of limbic structures. *Journal of Neurology, Neurosurgery & Psychiatry*, *83*(9), 903–909.

Bonilha, L., Rorden, C., Castellano, G., Pereira, F., Rio, P. A., Cendes, F., et al. (2004). Voxel-based morphometry reveals gray matter network atrophy in refractory medial temporal lobe epilepsy. *Archives of Neurology*, *61*(9), 1379–1384.

Bragin, A., Azizyan, A., Almajano, J., Wilson, C. L., & Engel, J., Jr. (2005). Analysis of chronic seizure onsets after intrahippocampal kainic acid injection in freely moving rats. *Epilepsia*, *46*(10), 1592–1598.

Bragin, A., Benassi, S. K., Kheiri, F., & Engel, J., Jr. (2011). Further evidence that pathologic high-frequency oscillations are bursts of population spikes derived from recordings of identified cells in dentate gyrus. *Epilepsia*, *52*(1), 45–52.

Bragin, A., Engel, J., Jr., Wilson, C. L., Fried, I., & Buzsaki, G. (1999). High-frequency oscillations in human brain. *Hippocampus*, *9*, 137–142.

Bragin, A., Engel, J., Jr., Wilson, C. L., Fried, I., & Mathern, G. W. (1999). Hippocampal and entorhinal cortex high-frequency oscillations (100–500 Hz) in human epileptic brain and in kainic acid-treated rats with chronic seizures. *Epilepsia*, *40*(2), 127–137.

Bragin, A., Engel, J., Jr., Wilson, C. L., Vizentin, E., & Mathern, G. W. (1999). Electrophysiologic analysis of a chronic seizure model after unilateral hippocampal ka injection. *Epilepsia*, *40*(9), 1210–1221.

Bragin, A., Mody, I., Wilson, C. L., & Engel, J., Jr. (2002). Local generation of fast ripples in epileptic brain. *The Journal of Neuroscience*, *22*(5), 2012–2021.

Bragin, A., Wilson, C. L., & Engel, J., Jr. (2000). Chronic epileptogenesis requires development of a network of pathologically interconnected neuron clusters: A hypothesis. *Epilepsia*, *41*(Suppl. 6), S144–S152.

Bragin, Anatol, Wilson, Charles L., & Engel, J., Jr. (2007). Voltage depth profiles of high-frequency oscillations after kainic acid-induced status epilepticus. *Epilepsia*, *48*(s5), 35–40.

Bragin, A., Wilson, C. L., Staba, R. J., Reddick, M., Fried, I., & Engel, J., Jr. (2002). Interictal high-frequency oscillations (80–500 Hz) in the human epileptic brain: Entorhinal cortex. *Annals of Neurology*, *52*(4), 407–415.

Buzsaki, G. (1998). Memory consolidation during sleep: A neurophysiological perspective. *Journal of Sleep Research*, 7(Suppl. 1), 17–23.

Buzsaki, G., Anastassiou, C. A., & Koch, C. (2012). The origin of extracellular fields and currents—EEG, ECOG, IFP and spikes. *Nature Reviews. Neuroscience*, *13*(6), 407–420.

Campo, P., Garrido, M. I., Moran, R. J., Garcia-Morales, I., Poch, C., Toledano, R., et al. (2013). Network reconfiguration and working memory impairment in mesial temporal lobe epilepsy. *NeuroImage*, *72*, 48–54.

Campo, P., Garrido, M. I., Moran, R. J., Maestu, F., Garcia-Morales, I., Gil-Nagel, A., et al. (2012). Remote effects of hippocampal sclerosis on effective connectivity during working memory encoding: A case of connectional diaschisis? *Cerebral Cortex*, *22*(6), 1225–1236.

Chen, S., Wu, X., Lui, S., Wu, Q., Yao, Z., Li, Q., et al. (2012). Resting-state fMRI study of treatment-naive temporal lobe epilepsy patients with depressive symptoms. *NeuroImage*, *60*(1), 299–304 (Research Support, Non-U.S. Gov't).

Colder, B. W., Frysinger, R. C., Wilson, C. L., Harper, R. M., & Engel, J., Jr. (1996). Decreased neuronal burst discharge near site of seizure onset in epileptic human temporal lobes. *Epilepsia*, *37*(2), 113–121.

Colder, B. W., Wilson, C. L., Frysinger, R. C., Chao, L. C., Harper, R. M., & Engel, Jerome, Jr. (1996). Neuronal synchrony in relation to burst discharge in epileptic human temporal lobes. *Journal of Neurophysiology*, *75*(6), 2496–2508.

Concha, L., Beaulieu, C., Collins, D. L., & Gross, D. W. (2009). White-matter diffusion abnormalities in temporal-lobe epilepsy with and without mesial temporal sclerosis. *Journal of Neurology, Neurosurgery & Psychiatry*, *80*(3), 312–319.

Concha, L., Beaulieu, C., & Gross, D. W. (2005). Bilateral limbic diffusion abnormalities in unilateral temporal lobe epilepsy. *Annals of Neurology*, *57*(2), 188–196.

de Curtis, M., & Gnatkovsky, V. (2009). Reevaluating the mechanisms of focal ictogenesis: The role of low-voltage fast activity. *Epilepsia*, *50*(12), 2514–2525.

Demont-Guignard, S., Benquet, P., Gerber, U., Biraben, A., Martin, B., & Wendling, F. (2012). Distinct hyperexcitability mechanisms underlie fast ripples and epileptic spikes. *Annals of Neurology*, *71*(3), 342–352.

Doucet, G., Osipowicz, K., Sharan, A., Sperling, M. R., & Tracy, J. I. (2013a). Extratemporal functional connectivity impairments at rest are related to memory performance in mesial temporal epilepsy. *Human Brain Mapping*, *34*(9), 2202–2216.

Doucet, G., Osipowicz, K., Sharan, A., Sperling, M. R., & Tracy, J. I. (2013b). Hippocampal functional connectivity patterns during spatial working memory differ in right versus left temporal lobe epilepsy. *Brain Connectivity*, *3*(4), 398–406.

Douw, L., de Groot, M., van Dellen, E., Heimans, J. J., Ronner, H. E., Stam, C. J., et al. (2010). 'Functional connectivity' is a sensitive predictor of epilepsy diagnosis after the first seizure. *PLoS ONE*, *5*(5), e10839.

Ellmore, T. M., Pieters, T. A., & Tandon, N. (2011). Dissociation between diffusion MR tractography density and strength in epilepsy patients with hippocampal sclerosis. *Epilepsy Research*, *93*(2–3), 197–203.

Engel, J., Jr., & da Silva, F. L. (2012). High-frequency oscillations—Where we are and where we need to go. *Progress in Neurobiology*, *98*(3), 316–318.

Engel, J., Jr., McDermott, M. P., Wiebe, S., Langfitt, J. T., Stern, J. M., Dewar, S., et al. (2012). Early surgical therapy for drug-resistant temporal lobe epilepsy: A randomized trial. *JAMA*, *307*(9), 922–930.

Engel, J., Jr., Thompson, P. M., Stern, J. M., Staba, R. J., Bragin, A., & Mody, I. (2013). Connectomics and epilepsy. *Current Opinion in Neurology*, *26*(2), 186–194.

Ettinger, A., Reed, M., Cramer, J., & Epilepsy Impact Project Group. (2004). Depression and comorbidity in community-based patients with epilepsy or asthma. *Neurology*, *63*(6), 1008–1014.

Ettinger, A. B., Reed, M. L., Goldberg, J. F., & Hirschfeld, R. M. (2005). Prevalence of bipolar symptoms in epilepsy vs other chronic health disorders. *Neurology*, *65*(4), 535–540.

Fisher, R. S., van Emde Boas, W., Blume, W., Elger, C., Genton, P., Lee, P., et al. (2005). Epileptic seizures and epilepsy: Definitions proposed by the international league against epilepsy (ILAE) and the international bureau for epilepsy (IBE). *Epilepsia*, *46*(4), 470–472.

Fojtikova, D., Brazdil, M., Skoch, A., Jiru, F., Horky, J., Marecek, R., et al. (2007). Magnetic resonance spectroscopy of the thalamus in patients with mesial temporal lobe epilepsy and hippocampal sclerosis. *Epileptic Disorders*, *9*(Suppl. 1), S59–S67.

Frings, L., Schulze-Bonhage, A., Spreer, J., & Wagner, K. (2009). Remote effects of hippocampal damage on default network connectivity in the human brain. *Journal of Neurology*, *256*(12), 2021–2029.

Gnatkovsky, V., Librizzi, L., Trombin, F., & de Curtis, M. (2008). Fast activity at seizure onset is mediated by inhibitory circuits in the entorhinal cortex in vitro. *Annals of Neurology*, *64*(6), 674–686.

Grova, C., Jannin, P., Buvat, I., Benali, H., Bansard, J. Y., Biraben, A., et al. (2005). From anatomic standardization analysis of perfusion SPECT data to perfusion pattern modeling: Evidence of functional networks in healthy subjects and temporal lobe epilepsy patients. *Academic Radiology*, *12*(5), 554–565.

Guedj, E., Aubert, S., McGonigal, A., Mundler, O., & Bartolomei, F. (2010). Deja-vu in temporal lobe epilepsy: Metabolic pattern of cortical involvement in patients with normal brain MRI. *Neuropsychologia*, *48*(7), 2174–2181.

Gunbey, H. P., Ercan, K., Findikoglu, A. S., Bilir, E., Karaoglanoglu, M., Komurcu, F., et al. (2011). Secondary corpus callosum abnormalities associated with antiepileptic drugs in temporal lobe epilepsy. A diffusion tensor imaging study. *The Neuroradiology Journal*, *24*(2), 316–323.

Haneef, Z., Lenartowicz, A., Yeh, H. J., Engel, J., Jr., & Stern, J. M. (2012). Effect of lateralized temporal lobe epilepsy on the default mode network. *Epilepsy & Behavior*, *25*(3), 350–357.

Haneef, Z., Lenartowicz, A., Yeh, H. J., Levin, H. S., Engel, J., Jr., & Stern, J. M. (2014). Functional connectivity of hippocampal networks in temporal lobe epilepsy. *Epilepsia*, *55*(1), 137–145.

Hetherington, H. P., Kuzniecky, R. I., Vives, K., Devinsky, O., Pacia, S., Luciano, D., et al. (2007). A subcortical network of dysfunction in TLE measured by magnetic resonance spectroscopy. *Neurology*, *69*(24), 2256–2265.

Hogan, R. E., Kaiboriboon, K., Bertrand, M. E., Rao, V., & Acharya, J. (2006). Composite siscom perfusion patterns in right and left temporal seizures. *Archives of Neurology*, *63*(10), 1419–1426.

Ibarz, J. M., Foffani, G., Cid, E., Inostroza, M., & Menendez de la Prida, L. (2010). Emergent dynamics of fast ripples in the epileptic hippocampus. *The Journal of Neuroscience*, *30*(48), 16249–16261.

Isnard, J., Guenot, M., Ostrowsky, K., Sindou, M., & Mauguiere, F. (2000). The role of the insular cortex in temporal lobe epilepsy. *Annals of Neurology, 48*(4), 614–623.

Jacobs, J., LeVan, P., Chander, R., Hall, J., Dubeau, F., & Gotman, J. (2008). Interictal high-frequency oscillations (80–500 Hz) are an indicator of seizure onset areas independent of spikes in the human epileptic brain. *Epilepsia, 49*(11), 1893–1907.

Jacobs, J., Levan, P., Chatillon, C. E., Olivier, A., Dubeau, F., & Gotman, J. (2009). High frequency oscillations in intracranial EEGs mark epileptogenicity rather than lesion type. *Brain, 132*(Pt 4), 1022–1037.

James, G. A., Tripathi, S. P., Ojemann, J. G., Gross, R. E., & Drane, D. L. (2013). Diminished default mode network recruitment of the hippocampus and parahippocampus in temporal lobe epilepsy. *Journal of Neurosurgery, 119*(2), 288–300.

Ji, G. J., Zhang, Z., Zhang, H., Wang, J., Liu, D. Q., Zang, Y. F., et al. (2013). Disrupted causal connectivity in mesial temporal lobe epilepsy. *PLoS ONE, 8*(5), e63183.

Jirsch, J. D., Urrestarazu, E., LeVan, P., Olivier, A., Dubeau, F., & Gotman, J. (2006). High-frequency oscillations during human focal seizures. *Brain, 129*(6), 1593–1608.

Josephson, C. B., Dykeman, J., Fiest, K. M., Liu, X., Sadler, R. M., Jette, N., et al. (2013). Systematic review and meta-analysis of standard vs selective temporal lobe epilepsy surgery. *Neurology, 80*(18), 1669–1676.

Juhasz, C., Nagy, F., Watson, C., da Silva, E. A., Muzik, O., Chugani, D. C., et al. (1999). Glucose and [^{11}C]flumazenil positron emission tomography abnormalities of thalamic nuclei in temporal lobe epilepsy. *Neurology, 53*(9), 2037–2045.

Kahane, P., & Bartolomei, F. (2010). Temporal lobe epilepsy and hippocampal sclerosis: Lessons from depth EEG recordings. *Epilepsia, 51*(Suppl. 1), 59–62.

Keller, S. S., Ahrens, T., Mohammadi, S., Gerdes, J. S., Moddel, G., Kellinghaus, C., et al. (2013). Voxel-based statistical analysis of fractional anisotropy and mean diffusivity in patients with unilateral temporal lobe epilepsy of unknown cause. *Journal of Neuroimaging, 23*(3), 352–359.

Kemmotsu, N., Girard, H. M., Bernhardt, B. C., Bonilha, L., Lin, J. J., Tecoma, E. S., et al. (2011). MRI analysis in temporal lobe epilepsy: Cortical thinning and white matter disruptions are related to side of seizure onset. *Epilepsia, 52*(12), 2257–2266.

Khosravani, H., Mehrotra, N., Rigby, M., Hader, W. J., Pinnegar, C. R., Pillay, N., et al. (2008). Spatial localization and time-dependant changes of electrographic high frequency oscillations in human temporal lobe epilepsy. *Epilepsia, 50*(4), 605–616.

Kim, C. H., Chung, C. K., Koo, B. B., Lee, J. M., Kim, J. S., & Lee, S. K. (2011). Changes in language pathways in patients with temporal lobe epilepsy: Diffusion tensor imaging analysis of the uncinate and arcuate fasciculi. *World Neurosurgery, 75*(3–4), 509–516.

Kim, C. H., Koo, B. B., Chung, C. K., Lee, J. M., Kim, J. S., & Lee, S. K. (2010). Thalamic changes in temporal lobe epilepsy with and without hippocampal sclerosis: A diffusion tensor imaging study. *Epilepsy Research, 90*(1–2), 21–27.

Kobayashi, E., Grova, C., Tyvaert, L., Dubeau, F., & Gotman, J. (2009). Structures involved at the time of temporal lobe spikes revealed by interindividual group analysis of EEG/fMRI data. *Epilepsia, 50*(12), 2549–2556.

Kramer, M. A., & Cash, S. S. (2012). Epilepsy as a disorder of cortical network organization. *The Neuroscientist, 18*(4), 360–372.

Kucukboyaci, N. E., Kemmotsu, N., Cheng, C. E., Girard, H. M., Tecoma, E. S., Iragui, V. J., et al. (2013). Functional connectivity of the hippocampus in temporal lobe epilepsy: Feasibility of a task-regressed seed-based approach. *Brain Connectivity, 3*(5), 464–474.

Laufs, H. (2012). Functional imaging of seizures and epilepsy: Evolution from zones to networks. *Current Opinion in Neurology, 25*(2), 194–200.

Laufs, H., Richardson, M. P., Salek-Haddadi, A., Vollmar, C., Duncan, J. S., Gale, K., et al. (2011). Converging PET and fMRI evidence for a common area involved in human focal epilepsies. *Neurology*, 77(9), 904–910.

Le Van Quyen, M., Bragin, A., Staba, R., Crepon, B., Wilson, C. L., & Engel, J., Jr. (2008). Cell type-specific firing during ripple oscillations in the hippocampal formation of humans. *Journal of Neuroscience*, *28*(24), 6104–6110.

Levesque, M., Salami, P., Gotman, J., & Avoli, M. (2012). Two seizure-onset types reveal specific patterns of high-frequency oscillations in a model of temporal lobe epilepsy. *The Journal of Neuroscience*, *32*(38), 13264–13272.

Liacu, D., Idy-Peretti, I., Ducreux, D., Bouilleret, V., & de Marco, G. (2012). Diffusion tensor imaging tractography parameters of limbic system bundles in temporal lobe epilepsy patients. *Journal of Magnetic Resonance Imaging*, *36*(3), 561–568.

Liao, W., Zhang, Z., Pan, Z., Mantini, D., Ding, J., Duan, X., et al. (2011). Default mode network abnormalities in mesial temporal lobe epilepsy: A study combining fMRI and DTI. *Human Brain Mapping*, *32*(6), 883–895.

Lin, J. J., Riley, J. D., Juranek, J., & Cramer, S. C. (2008). Vulnerability of the frontal-temporal connections in temporal lobe epilepsy. *Epilepsy Research*, *82*(2–3), 162–170.

Liu, M., Concha, L., Lebel, C., Beaulieu, C., & Gross, D. W. (2012). Mesial temporal sclerosis is linked with more widespread white matter changes in temporal lobe epilepsy. *NeuroImage. Clinical*, *1*(1), 99–105.

Maccotta, L., He, B. J., Snyder, A. Z., Eisenman, L. N., Benzinger, T. L., Ances, B. M., et al. (2013). Impaired and facilitated functional networks in temporal lobe epilepsy. *NeuroImage. Clinical*, *2*, 862–872.

Maillard, L., Vignal, J. P., Gavaret, M., Guye, M., Biraben, A., McGonigal, A., et al. (2004). Semiologic and electrophysiologic correlations in temporal lobe seizure subtypes. *Epilepsia*, *45*(12), 1590–1599.

Malikova, H., Liscak, R., Vojtech, Z., Prochazka, T., Vymazal, J., Vladyka, V., et al. (2011). Stereotactic radiofrequency amygdalohippocampectomy: Does reduction of entorhinal and perirhinal cortices influence good clinical seizure outcome? *Epilepsia*, *52*(5), 932–940.

Malinowska, U., Badier, J. M., Gavaret, M., Bartolomei, F., Chauvel, P., & Benar, C. G. (2014). Interictal networks in magnetoencephalography. *Human Brain Mapping*, *35*, 2789–2805. http://dx.doi.org/10.1002/hbm.22367.

Marques, C. M., Caboclo, L. O., da Silva, T. I., Noffs, M. H., Carrete, H., Jr., Lin, K., et al. (2007). Cognitive decline in temporal lobe epilepsy due to unilateral hippocampal sclerosis. *Epilepsy & Behavior*, *10*(3), 477–485.

Mathern, G. W., Wilson, C. L., & Beck, H. (2008). Hippocampal sclerosis. In J. Engel, & T. A. Pedley, Jr., (Eds.), *Epilepsy: A comprehensive textbook* (pp. 121–136) (2nd ed.). Philadelphia: Lippincott Williams & Wilkins.

McCormick, C., Quraan, M., Cohn, M., Valiante, T. A., & McAndrews, M. P. (2013). Default mode network connectivity indicates episodic memory capacity in mesial temporal lobe epilepsy. *Epilepsia*, *54*(5), 809–818.

Memarian, N., Madsen, S., Kim, S., Nalbandian, K., Macey, P., Dewar, S., et al. (2013). Neocortical gray matter maps linked with two common mesial temporal lobe seizure onset patterns. *Epilepsy Currents*, *14*(Suppl. 1), 374.

Meng, L., Xiang, J., Kotecha, R., Rose, D., Zhao, H., Zhao, D., et al. (2010). White matter abnormalities in children and adolescents with temporal lobe epilepsy. *Magnetic Resonance Imaging*, *28*(9), 1290–1298.

Morgan, V. L., Gore, J. C., & Abou-Khalil, B. (2010). Functional epileptic network in left mesial temporal lobe epilepsy detected using resting fMRI. *Epilepsy Research*, *88*(2–3), 168–178.

Morgan, V. L., Rogers, B. P., Sonmezturk, H. H., Gore, J. C., & Abou-Khalil, B. (2011). Cross hippocampal influence in mesial temporal lobe epilepsy measured with high temporal resolution functional magnetic resonance imaging. *Epilepsia*, *52*(9), 1741–1749.

Morgan, V. L., Sonmezturk, H. H., Gore, J. C., & Abou-Khalil, B. (2012). Lateralization of temporal lobe epilepsy using resting functional magnetic resonance imaging connectivity of hippocampal networks. *Epilepsia*, *53*(9), 1628–1635.

Morrell, M. J., & R. N. S. System in Epilepsy Study Group. (2011). Responsive cortical stimulation for the treatment of medically intractable partial epilepsy. *Neurology*, 77(13), 1295–1304.

Natsume, J., Bernasconi, N., Andermann, F., & Bernasconi, A. (2003). MRI volumetry of the thalamus in temporal, extratemporal, and idiopathic generalized epilepsy. *Neurology*, *60*(8), 1296–1300.

Negishi, M., Martuzzi, R., Novotny, E. J., Spencer, D. D., & Constable, R. T. (2011). Functional MRI connectivity as a predictor of the surgical outcome of epilepsy. *Epilepsia*, *52*(9), 1733–1740.

Ogren, J. A., Bragin, A., Wilson, C. L., Hoftman, G. D., Lin, J. J., Dutton, R. A., et al. (2009). Three-dimensional hippocampal atrophy maps distinguish two common temporal lobe seizure-onset patterns. *Epilepsia*, *50*(6), 1361–1370.

Ogren, J. A., Wilson, C. L., Bragin, A., Lin, J. J., Salamon, N., Dutton, R. A., et al. (2009). Three dimensional surface maps link local atrophy and fast ripples in human epileptic hippocampus. *Annals of Neurology*, *66*(6), 783–791.

Oguz, K. K., Tezer, I., Sanverdi, E., Has, A. C., Bilginer, B., Dolgun, A., et al. (2013). Effect of patient sex on white matter alterations in unilateral medial temporal lobe epilepsy with hippocampal sclerosis assessed by diffusion tensor imaging. *American Journal of Neuroradiology*, *34*(5), 1010–1015.

Otte, W. M., van Eijsden, P., Sander, J. W., Duncan, J. S., Dijkhuizen, R. M., & Braun, K. P. (2012). A meta-analysis of white matter changes in temporal lobe epilepsy as studied with diffusion tensor imaging. *Epilepsia*, *53*(4), 659–667.

Oyegbile, T. O., Dow, C., Jones, J., Bell, B., Rutecki, P., Sheth, R., et al. (2004). The nature and course of neuropsychological morbidity in chronic temporal lobe epilepsy. *Neurology*, *62*(10), 1736–1742.

Pan, J. W., Spencer, D. D., Kuzniecky, R., Duckrow, R. B., Hetherington, H., & Spencer, S. S. (2012). Metabolic networks in epilepsy by MR spectroscopic imaging. *Acta Neurologica Scandinavica*, *126*(6), 411–420.

Pereira, F. R., Alessio, A., Sercheli, M. S., Pedro, T., Bilevicius, E., Rondina, J. M., et al. (2010). Asymmetrical hippocampal connectivity in mesial temporal lobe epilepsy: Evidence from resting state fMRI. *BMC Neuroscience*, *11*, 66.

Perucca, P., Dubeau, F., & Gotman, J. (2014). Intracranial electroencephalographic seizure-onset patterns: Effect of underlying pathology. *Brain*, *137*, 183–196. http://dx.doi.org/10.1093/brain/awt299.

Pittau, F., Grova, C., Moeller, F., Dubeau, F., & Gotman, J. (2012). Patterns of altered functional connectivity in mesial temporal lobe epilepsy. *Epilepsia*, *53*(6), 1013–1023.

Quigg, M., Rolston, J., & Barbaro, N. M. (2012). Radiosurgery for epilepsy: Clinical experience and potential antiepileptic mechanisms. *Epilepsia*, *53*(1), 7–15.

Quraan, M. A., McCormick, C., Cohn, M., Valiante, T. A., & McAndrews, M. P. (2013). Altered resting state brain dynamics in temporal lobe epilepsy can be observed in spectral power, functional connectivity and graph theory metrics. *PLoS ONE*, *8*(7), e68609.

Reinsberger, C., Tanaka, N., Cole, A. J., Lee, J. W., Dworetzky, B. A., Bromfield, E. B., et al. (2010). Current dipole orientation and distribution of epileptiform activity correlates with cortical thinning in left mesiotemporal epilepsy. *NeuroImage*, *52*(4), 1238–1242.

Ryvlin, P. (2005). The modern challenges of drug resistant epilepsy. Introduction. *Epileptic Disorders*, 7(Suppl. 1), S1–S2 (Introductory).

Savic, I., & Engel, J., Jr. (1998). Sex differences in patients with mesial temporal lobe epilepsy. *Journal of Neurology, Neurosurgery & Psychiatry*, *65*(6), 910–912.

Semah, F., Picot, M. C., Adam, C., Broglin, D., Arzimanoglou, A., Bazin, B., et al. (1998). Is the underlying cause of epilepsy a major prognostic factor for recurrence? *Neurology*, *51*(5), 1256–1262.

Sequeira, K. M., Tabesh, A., Sainju, R. K., DeSantis, S. M., Naselaris, T., Joseph, J. E., et al. (2013). Perfusion network shift during seizures in medial temporal lobe epilepsy. *PLoS ONE*, *8*(1), e53204.

Skaggs, W. E., McNaughton, B. L., Permenter, M., Archibeque, M., Vogt, J., Amaral, D., et al. (2007). EEG sharp wave and sparse ensemble unit activity in the macaque hippocampus. *Journal of Neurophysiology*, *98*, 898–910.

Spencer, S. S. (2002). Neural networks in human epilepsy: Evidence of and implications for treatment. *Epilepsia*, *43*(3), 219–227.

Spencer, S. S., Guimaraes, P., Katz, A., Kim, J., & Spencer, D. (1992). Morphological patterns of seizures recorded intracranially. *Epilepsia*, *33*(3), 537–545 (Conference Paper).

Staba, R. J. (2012). Normal and pathologic high-frequency oscillations. In J. L. Noebels, M. Avoli, M. A. Rogawski, R. W. Olsen, & A. V. Delgado-Escueta (Eds.), *Jasper's basic mechanisms of the epilepsies* (4th ed.). Bethesda, MD: National Center for Biotechnology Information (US).

Staba, R. J., Ekstrom, A. D., Suthana, N. A., Burggren, A., Fried, I., Engel, J., Jr., et al. (2011). Gray matter loss correlates with mesial temporal lobe neuronal hyperexcitability inside the human seizure-onset zone. *Epilepsia*, *53*(1), 25–34.

Staba, R. J., Frighetto, L., Behnke, E. J., Mathern, G. W., Fields, T. A., Bragin, A., et al. (2007). Increased fast ripple to ripple ratios correlate with reduced hippocampal volumes and neuron loss in temporal lobe epilepsy patients. *Epilepsia*, *48*(11), 2130–2138.

Staba, R. J., Wilson, C. L., Bragin, A., Fried, I., & Engel, J., Jr. (2002). Quantitative analysis of high-frequency oscillations (80–500 Hz) recorded in human epileptic hippocampus and entorhinal cortex. *Journal of Neurophysiology*, *88*(4), 1743–1752 (Research Support, U.S. Gov't, P.H.S.).

Staba, R. J., Wilson, C. L., Bragin, A., Jhung, D., Fried, I., & Engel, J. (2004). High-frequency oscillations recorded in human medial temporal lobe during sleep. *Annals of Neurology*, *56*, 108–115.

Stretton, J., Winston, G. P., Sidhu, M., Bonelli, S., Centeno, M., Vollmar, C., et al. (2013). Disrupted segregation of working memory networks in temporal lobe epilepsy. *NeuroImage Clinical*, *2*, 273–281.

Tellez-Zenteno, J. F., Dhar, R., & Wiebe, S. (2005). Long-term seizure outcomes following epilepsy surgery: A systematic review and meta-analysis. *Brain*, *128*(5), 1188–1198 (Review).

Tellez-Zenteno, J. F., Hernandez Ronquillo, L., Moien-Afshari, F., & Wiebe, S. (2010). Surgical outcomes in lesional and non-lesional epilepsy: A systematic review and meta-analysis. *Epilepsy Research*, *89*(2–3), 310–318.

Tellez-Zenteno, J. F., Patten, S. B., Jette, N., Williams, J., & Wiebe, S. (2007). Psychiatric comorbidity in epilepsy: A population-based analysis. *Epilepsia*, *48*(12), 2336–2344.

Terry, J. R., Benjamin, O., & Richardson, M. P. (2012). Seizure generation: The role of nodes and networks. *Epilepsia*, *53*(9), e166–e169.

Thom, M., Mathern, G. W., Cross, J. H., & Bertram, E. H. (2010). Mesial temporal lobe epilepsy: How do we improve surgical outcome? *Annals of Neurology*, *68*(4), 424–434.

Tracy, J. I., Osipowicz, K., Spechler, P., Sharan, A., Skidmore, C., Doucet, G., et al. (2014). Functional connectivity evidence of cortico–cortico inhibition in temporal lobe epilepsy. *Human Brain Mapping, 35*(1), 353–366.

Trebuchon-Da Fonseca, A., Guedj, E., Alario, F. X., Laguitton, V., Mundler, O., Chauvel, P., et al. (2009). Brain regions underlying word finding difficulties in temporal lobe epilepsy. *Brain, 132*(Pt 10), 2772–2784.

Trevelyan, A. J. (2009). The direct relationship between inhibitory currents and local field potentials. *The Journal of Neuroscience, 29*(48), 15299–15307.

Trotta, N., Goldman, S., Legros, B., Baete, K., Van Laere, K., Van Bogaert, P., et al. (2013). Changes in functional integration with the non-epileptic temporal lobe of patients with unilateral mesiotemporal epilepsy. *PLoS ONE, 8*(6), e67053.

Trotta, N., Goldman, S., Legros, B., Ligot, N., Guerry, N., Baete, K., et al. (2011). Metabolic evidence for episodic memory plasticity in the nonepileptic temporal lobe of patients with mesial temporal epilepsy. *Epilepsia, 52*(11), 2003–2012.

van Dellen, E., Douw, L., Hillebrand, A., de Witt Hamer, P. C., Baayen, J. C., Heimans, J. J., et al. (2014). Epilepsy surgery outcome and functional network alterations in longitudinal MEG: A minimum spanning tree analysis. *NeuroImage, 86*, 354–363.

Velasco, A. L., Boleaga, B., Brito, F., Jimenez, F., Gordillo, J. L., Velasco, F., et al. (2000). Absolute and relative predictor values of some non-invasive and invasive studies for the outcome of anterior temporal lobectomy. *Archives of Medical Research, 31*(1), 62–74.

Velasco, A. L., Wilson, C. L., Babb, T. L., & Engel, J., Jr. (2000). Functional and anatomic correlates of two frequently observed temporal lobe seizure-onset patterns. *Neural Plasticity*, 7(1–2), 49–63.

Voets, N. L., Beckmann, C. F., Cole, D. M., Hong, S., Bernasconi, A., & Bernasconi, N. (2012). Structural substrates for resting network disruption in temporal lobe epilepsy. *Brain, 135*(Pt 8), 2350–2357.

Wang, X. Q., Lang, S. Y., Hong, L. U., Lin, M. A., Yan-ling, M. A., & Yang, F. (2010). Changes in extratemporal integrity and cognition in temporal lobe epilepsy: A diffusion tensor imaging study. *Neurology India, 58*(6), 891–899.

Wennberg, R., Arruda, F., Quesney, L. F., & Olivier, A. (2002). Preeminence of extrahippocampal structures in the generation of mesial temporal seizures: Evidence from human depth electrode recordings. *Epilepsia, 43*(7), 716–726.

Wiebe, S., Blume, W. T., Girvin, J. P., & Eliasziw, M. (2001). A randomized, controlled trial of surgery for temporal-lobe epilepsy. *New England Journal of Medicine, 345*(5), 311–318.

Wieser, H. G. (1988). Human limbic seizures: EEG studies, origin, and patterns of spread. In B. S. Meldrum, J. A. Ferrendelli, & H. G. Wieser (Eds.), *Anatomy of epileptogenesis* (pp. 127–138). London: John Libbey.

Winston, G. P., Stretton, J., Sidhu, M. K., Symms, M. R., & Duncan, J. S. (2013). Progressive white matter changes following anterior temporal lobe resection for epilepsy. *NeuroImage. Clinical, 4*, 190–200.

Winston, G. P., Stretton, J., Sidhu, M. K., Symms, M. R., Thompson, P. J., & Duncan, J. S. (2013). Structural correlates of impaired working memory in hippocampal sclerosis. *Epilepsia, 54*(7), 1143–1153.

Worrell, Greg A., Gardner, Andrew B., Stead, S. Matt, Hu, Sanqing, Goerss, Steve, Cascino, Gregory J., et al. (2008). High-frequency oscillations in human temporal lobe: Simultaneous microwire and clinical macroelectrode recordings. *Brain, 131*(4), 928–937.

Zhang, Z., Lu, G., Zhong, Y., Tan, Q., Liao, W., Wang, Z., et al. (2010). Altered spontaneous neuronal activity of the default-mode network in mesial temporal lobe epilepsy. *Brain Research, 1323*, 152–160.

CHAPTER SIX

Modern Concepts of Seizure Modeling

Christophe Bernard*,†,[1], Sebastien Naze*,†, Timothée Proix*,†, Viktor K. Jirsa*,†
*Institut de Neurosciences des Systèmes, Aix Marseille Université, Marseille, France
†Inserm UMR_S 1106, Aix Marseille Universite, Marseille, France
[1]Corresponding author: e-mail address: christophe.bernard@univ-amu.fr

Contents

1. Introduction 122
2. Detailed Biophysical Models 122
   2.1 Role of potassium and other ion channels 123
   2.2 Gap junctions and VFOs 128
   2.3 Posttraumatic epilepsy and homeostatic synaptic plasticity 129
   2.4 Network topology 130
   2.5 Thalamocortical loop and absence seizures 135
3. Lumped Models 137
   3.1 Abstract models 138
   3.2 Neural mass and neural field models 140
   3.3 Epileptor model 143
   3.4 Large-scale models 143
4. Analogies 146
   4.1 Stochastic models 146
   4.2 Power-law behavior 146
5. Discussion 148
Acknowledgments 149
References 149

Abstract

Seizures are complex phenomena spanning multiple spatial and temporal scales, from ion dynamics to communication between brain regions, from milliseconds (spikes) to days (interseizure intervals). Because of the existence of such multiple scales, the experimental evaluation of the mechanisms underlying the initiation, propagation, and termination of epileptic seizures is a difficult problem. Theoretical models and numerical simulations provide new tools to investigate seizure mechanisms at multiple scales. In this chapter, we review different theoretical approaches and their contributions to our understanding of seizure mechanisms.

International Review of Neurobiology, Volume 114
ISSN 0074-7742
http://dx.doi.org/10.1016/B978-0-12-418693-4.00006-6

1. INTRODUCTION

Seizure mechanisms can be investigated at different spatial scales, ranging from molecular to emergent properties of whole-brain network dynamics. Such mechanisms are difficult to assess experimentally as it would require recording simultaneously a large number of variables at various scales. Theoretical approaches provide functional insights, since it is possible to investigate all variables simultaneously. We review here different modeling approaches; from the single-cell level, networks of cells, to the whole-brain scale. The first section focuses on single neurons and their dynamics emerging from the coupling among them. Ion channel mechanisms and key features of the extracellular environment are introduced to reproduce empirical complex behaviors observed *in vitro*. The role of network microcircuitry topology, in particular synaptic plasticity, on seizure dynamics is addressed. The involvement of other mechanisms, such as gap junction communication, in the emergence of very-fast oscillations (VFOs) and the thalamocortical loop for generalized spike–wave discharges is also considered.

The second section is dedicated to larger spatial scale descriptions of the disorder, focusing on macroscopic dynamical features. Neural field/neural mass models suggest alternative analysis of seizure dynamics on a more macroscopic level of organization. The use of dynamical system theory and bifurcation theory allows development of a novel theoretical framework leading to taxonomies of seizures and the nature of transitions between ictal and interictal state. Recent developments introduce fully autonomous epileptogenic neural masses that do not require manual parameter change. We further expand on the development of large-scale epileptic networks which rely on diffusion MRI imaging and tractography to measure long-range white matter connections and show their potential applications in clinical studies.

Finally, a third section presents concepts brought by analogy with other fields such as machine learning and statistics. Models of spontaneous transitions from interictal to ictal stages are introduced alongside with relevant statistic on seizures and interictal periods. These analogies have notably interesting application in seizures predictions.

2. DETAILED BIOPHYSICAL MODELS

Epilepsy is characterized by complex reorganizations of the underlying neuronal circuitry. These modifications can occur at different levels, including cell death, axon sprouting, synaptic changes, and alterations of the intrinsic properties of the neuronal membrane. It is very difficult to assess experimentally the functional consequences of such reorganizations. Detailed biophysical models are useful to gain some insight into the role of a given parameter.

Biophysical modeling studies on single neuron dynamics mostly rely on the conductance-based modeling approach as formulated by Hodgkin and Huxley (1952). At least two currents are needed to generate action potentials, those carried by specific classes of potassium and sodium channels. Numerous other ionic channels can be added according to the properties of the neuronal membrane (e.g., to take into account the fact that the neuron is a fast-spiking, bursting, regular firing, stuttering, etc., cell).

A detailed representation of the neuron can be obtained with a multicompartmental model, which includes the soma, the axon, and the dendritic tree (for more details, see Koch & Segev, 1998). Each compartment contains ionic channels and postsynaptic receptors. It is thus possible to model the interplay between the different ionic conductances, which ultimately determines whether or not the neuron will emit an action potential. We can then vary biophysical parameters (e.g., the kinetic properties of a given ionic channel) to mimic experimental observations in pathological conditions. Conversely, a parametric study can be performed to determine which parameter values lead to the investigated behavior (e.g., high-frequency bursting), leading to testable predictions.

We review findings and predictions from such approaches in this section, starting with the role of extracellular ionic concentrations on membrane voltage dynamics, further identifying the role of synaptic and electrotonic coupling on the network behavior, and then extending on topological architectures of the hippocampus and thalamocortical loop.

2.1. Role of potassium and other ion channels

The current carried by ionic channels primarily depends upon the electrochemical gradient for the corresponding ionic species. Modeling neural activity with Hodgkin–Huxley equations usually assumes constant extracellular ionic concentrations. In nonpathological conditions, extracellular ionic concentration is kept mostly constant by different mechanisms such as ionic pumps, spatial diffusion across extracellular environment, and glial buffering. However, under pathological conditions such as epileptic discharges, the concentration of ions in the extracellular space can be dramatically altered, leading to shifts in the reversal potential for a given ion species, hence changing the amount of ions flowing through the channel and thus the current carried by the channel. As there exist a large number of channels carrying ions between the intra- and extracellular space, changes in ion concentration may alter cellular function, possibly favoring the occurrence of seizures. We review here recent computational studies investigating the effects of changes on the extracellular environment.

2.1.1 Extracellular potassium

Kager, Wadman, and Somjen (2000) were among the first to add extracellular environment in the context of seizure discharges. They implemented an anatomically realistic hippocampal pyramidal cell made of 201 compartments with five ionic currents, sodium–potassium exchange pumps, and glial uptake (Fig. 6.1A), based on a model of a CA3 pyramidal cell (Traub, 1994). They investigated the role of elevated extracellular potassium concentration $[K^+]_o$ in seizure genesis and spreading depression (SD). Their results show that impairing glial uptake leads to $[K^+]_o$ accumulation and to clonic epileptiform discharges.

Building on this approach, Bazhenov, Timofeev, Steriade, and Sejnowski (2004) used a reduced two-compartmental neuron model and included extracellular potassium concentration dynamics via Na/K pumps and glial uptake. Their model reproduced slow spike–wave and poly spike–wave behaviors. They simulated networks of such interconnected cells to study spatial propagation of epileptiform discharges. In their simulated network, they extended the model of Kager et al. (2000) with short-term depression and lateral potassium diffusion to avoid spontaneous continuous paroxysmal oscillations to happen without external stimulation. Typical 2–3 Hz paroxysmal oscillations observed in clonic seizures were obtained by injecting extracellular potassium manually or by blocking K^+ pumps (Fig. 6.1C).

Adding intracellular calcium ($[Ca^{2+}]_i$) as a control parameter, Fröhlich et al. (2006) analyzed more systematically the different dynamics of a pyramidal cell as a function of $[K^+]_o$ using bifurcation analysis to unravel the precise mechanisms underlying the transitions between states. They identified four stable states in single neuron dynamics: rest, tonic spiking, slow bursting, and depolarized state (Fig. 6.1B). In particular, there was bistability between tonic firing and slow bursting (i.e., coexistence of the two regimes with possible transitions from one to another). Using a network of 60 pyramidal cells and 15 inhibitory cells, they reproduced a sustained regime of oscillations between slow bursting and tonic spiking, which has been described *in vivo* during transitions from tonic to clonic seizure phases.

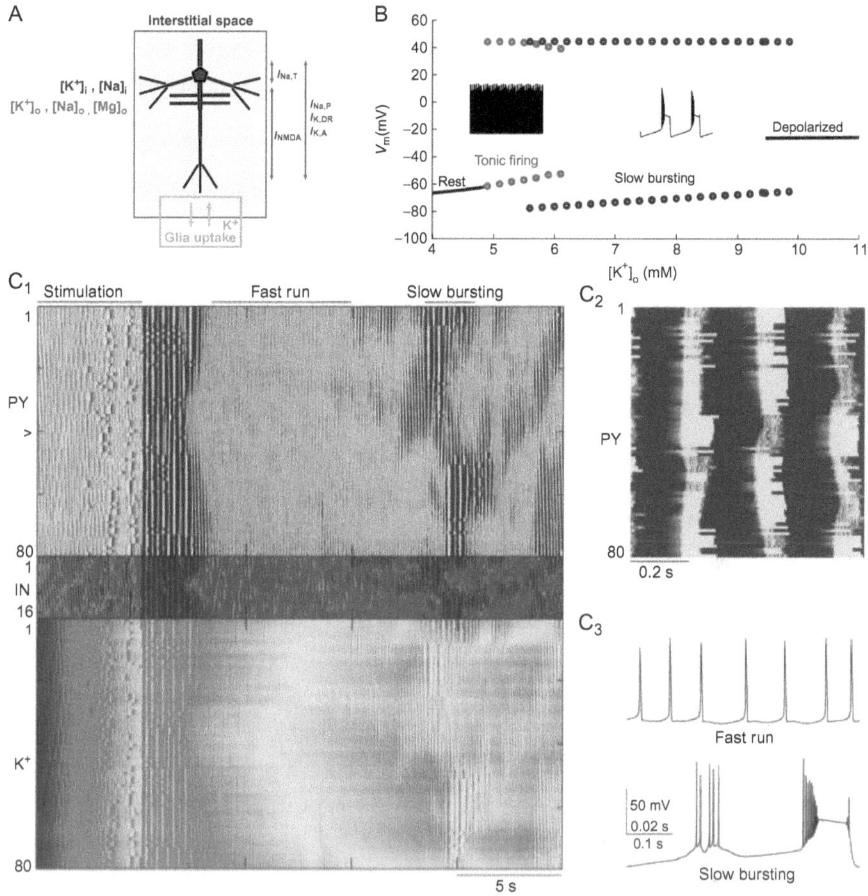

Figure 6.1 Effect of $[K^+]_o$ on neurons and network dynamics. (A) Schematic view of simulated single neuron model with extracellular environment including interactions between neuron, interstitial space, and glial cells. (B) Bifurcation diagram showing transitions from rest to tonic firing to slow bursting to depolarized state as a function of $[K^+]_o$. Inner plots display time series of neuron membrane potential for tonic firing (left) and slow bursting (right). (C_1) Population activity of 80 pyramidal cells (top), 16 interneurons (middle), and $[K^+]_o$ (bottom) are shown following a transient stimulation of PYs. (C_2) Zoom on pyramidal cell activity during slow bursting. (C_3) Time series of an individual neuron membrane potentials at the fast run regime (top) and slow bursting (bottom). *Panel (B) is adapted with permission from Fröhlich, Bazhenov, Timofeev, Steriade, and Sejnowski (2006). Copyright (2006) Society for Neuroscience. Panel (C) is adapted with permission from Fröhlich et al. (2007). Copyright (2007) Cambridge University Press.*

They further tested the coexistence of healthy and pathological neural activity in the same network without parameter variations (Fröhlich, Sejnowski, & Bazhenov, 2010). In this study, they investigated precise perturbations (number of neurons and stimulation length) that triggered transition to pathological activity. They concluded that factors such as synaptic depression, fast $[K^+]_o$ decrease due to Na/K pumps, activity-dependent outward currents and progressive depolarization of Cl^- reversal potential may contribute to the termination of a seizure and promote postictal depression. However, they put forward that these factors are less likely to be significant near seizure onset. Impairment of potassium regulation has long been known as an epileptogenic factor in epilepsy research. Such computational investigations, using detailed single neuron modeling (Kager et al., 2000), or reduced forms (Bazhenov et al., 2004; Fröhlich et al., 2006, 2010) provide insights about the role of extracellular K^+ for the genesis of paroxysmal discharges. Other ion species can also play a key role. They are considered hereafter.

2.1.2 Sodium concentrations

Cressman, Ullah, Ziburkus, Schiff, and Barreto (2009) and Ullah, Cressman, Barreto, and Schiff (2009) studied a single-compartment neuron model and a network of such neurons, respectively, incorporating both intra and extracellular potassium and sodium concentration dynamics. The single neuron model included sodium, potassium, calcium-gated potassium, and leak current channels. Intra and extracellular concentrations of sodium and potassium ($[Na^+]_i$, $[Na^+]_o$, $[K^+]_i$, and $[K^+]_o$) were computed according to the Na/K pump, glial uptake, and K^+ diffusion. They found an oscillatory regime when increasing $[K^+]_o$ and when long periods of rapid firing and rest were alternating autonomously under the control of $[Na^+]_i$ and $[K^+]_o$ (Fig. 6.2A).

The network model included 100 excitatory and 100 inhibitory neurons. A Gaussian excitatory stimulation was applied for 20 ms to simulate a working memory task, in which the network is expected to return to a resting state after a persistent activity of several seconds. Using this model, they demonstrated that if glial reuptake is impaired, the activity induced by the perturbations lead to a seizure with depolarization block observed on neuron membrane potentials (Fig. 6.2B).

In conclusion, the analysis of the dynamics of a system including intra and extracellular sodium and potassium interactions revealed a new oscillatory regime, which switches autonomously between ictal and interictal state on a slow timescale, being particularly critical for the stability of the network.

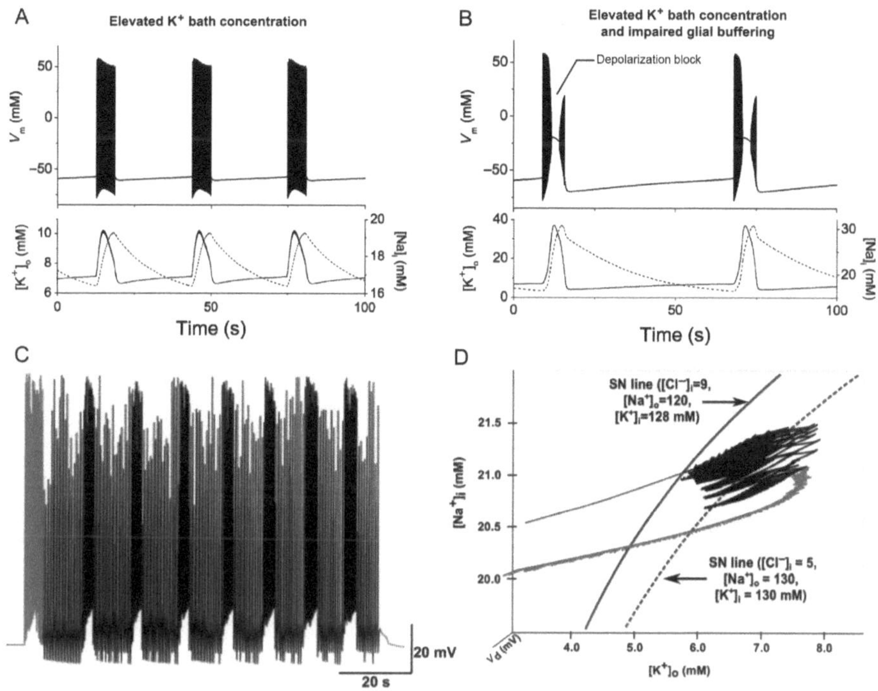

Figure 6.2 Effects of ionic concentrations on neural dynamics. (A) Traces of membrane potential (V_m), $[K^+]_o$ (solid trace) and $[Na^+]_i$ (dashed trace) in elevated $[K^+]_o$. (B) Same as (A) with additional impaired glial buffering. (C) Simulated membrane time series of a neuron dynamics at onset (red, first 5 s), during paroxysmal activity (black, middle 2 min), and offset (green, last 5 s) of a seizure. (D) Phase space of (C) spanned by dendritic membrane potential V_d, $[K^+]_o$, and $[Na^+]_i$. Dashed and solid blue lines indicate bifurcation boundaries between resting and seizure, respectively, at the beginning of the seizure and at the end of the seizure. *Panels (A) and (B) are adapted with permission from Cressman et al. (2009). Copyright (2009) Springer. Panels (C) and (D) are adapted with permission from Krishnan and Bazhenov (2011). Copyright (2011) Society for Neuroscience.*

2.1.3 Chloride

Krishnan and Bazhenov (2011) introduced chloride and calcium intracellular concentrations $[Cl^-]_i$ and $[Ca^{2+}]_i$ dynamics. As in previous studies, potassium and sodium concentrations were mediated by leak and intrinsic currents, Na/K pump, glial uptake, and K^+ diffusion. The model also included short-term synaptic depression and an additional potassium chloride cotransporter (KCC2) was introduced to take into account intracellular chloride concentration. After direct current (DC) stimulation, the network exhibited epileptic activity for more than 100 s, then spontaneously returned to rest (Fig. 6.2C). Using bifurcation diagrams, they showed that $[Na^+]_i$,

$[Cl^-]_i$, and $[K^+]_o$ critically affect neural activity during the seizure. In particular, the KCC2 pump became less effective when $[K^+]_o$ increased, allowing $[Cl^-]_i$ to rise higher than normal with a tendency to increase the length of the seizure. In parallel, $[Na^+]_i$ increased and was responsible for seizure termination by modifying the excitation/inhibition balance (Fig. 6.2D).

Thus, not only glutamate and GABA transmission play a role in the excitability of the system but also intracellular and extracellular ionic concentrations dynamics. These depend nonlinearly on intrinsic currents of individual neurons and suggest different possible scenarios for paroxysmal behaviors to emerge. In conclusion, these studies demonstrated the importance of intracellular and extracellular ion dynamics in seizure initiation, propagation, and termination. Until now, we only considered chemical synaptic connections; however, other types of synaptic coupling such as gap junctions are also critical for seizure dynamics, as we see in the Section 2.2.

2.2. Gap junctions and VFOs

VFOs are oscillatory patterns of electrographic activity above 70 Hz. They have been shown to occur *in vivo* when recorded with electrocorticography (ECoG) or intracranial EEG and may be of importance for localizing seizure onset. A proposed mechanism for such oscillations is based on electrotonic coupling through gap junctions. Previous studies highlighted the role of axonal electrical coupling in the origin of VFOs *in vitro* (Draguhn, Traub, Schmitz, & Jefferys, 1998). Simulations predicted that gap junctions facilitate seizure genesis in disinhibited networks (Traub, Contreras, & Whittington, 2005). Gap junctions also contributed to the conversion from single to multiple network bursts and fast runs in neocortical circuits, which are decreased under the action of gap junction blockers (e.g., carbenoxolone) or $GABA_A$ agonists (e.g., barbiturates).

Numerous studies further affirm the occurrence of VFO superimposed on epileptiform field transients (i.e., interictal bursts) in networks deprived of chemical synaptic transmission, both in computational simulations and experimental models, in which interictal spike discharges were eliminated by gap junction blockade (for review, see Traub, Whittington, & Cunningham, 2012). However, considering the unreliable selectivity of gap junction blockers, this change of interictal activity may be due to unspecific binding to different other targets (Juszczak & Swiergiel, 2009). Although recent findings argue that gap junctions are not necessarily required to generate HFOs (Quilichini et al., 2012).

Another novel study (Volman, Perc, & Bazhenov, 2011) questions this theory, as under some circumstances, neuronal electrical coupling increases the stability of a network and suppresses the collective electrical activity associated with seizure. Their model predicted the existence of critical gap junction strength (in terms of number of connections and conductance value) above which such coupling would no longer suppress but facilitate escalation toward epileptic seizure. In this new perspective, the observation of increased number of neuronal gap junctions after seizure-like stimulation was seen as an adaptive response of the network rather than an anatomical defect.

Given the difficulty of measuring gap junction communications, the subject is still open to discussion.

We see in the following section how structural changes of neural networks alone (i.e., without changing dynamics of single neurons or the nature of the coupling) can lead to pathological behaviors.

2.3. Posttraumatic epilepsy and homeostatic synaptic plasticity

Homeostatic synaptic plasticity (HSP) refers to the essential adaptation mechanisms of neurons to avoid excessive excitation or inhibition by controlling synaptic strength, and is critical for recovery of normal brain function after brain insults, such as traumatic injury. In terms of computation, it is usually assumed that HSP targets at maintaining the pretraumatic average network-firing rate. Houweling, Bazhenov, Timofeev, Steriade, and Sejnowski (2005) and Fröhlich, Bazhenov, and Sejnowski (2008) investigated the role of HSP in the context of brain injuries involving deafferentation, i.e., removing presynaptic cortical structures resulting in loss of external input in a network of neurons. They demonstrated that severe deafferentation led to paroxysmal activity such as network bursts in simulated large networks of pyramidal cells and interneurons. The same phenomenon has been observed after activity deprivation in hippocampal slices (Trasande & Ramirez, 2007). Deafferentation was mimicked by reducing the extrinsic input frequency from both pyramidal cells and interneurons (Fröhlich et al., 2008), or by removing extrinsic inputs completely (Houweling et al., 2005). In the latter, they showed that paroxysmal activity was induced by homeostatic upregulation of excitatory synapses (Fig. 6.3A). Such activity included burst discharges that could only emerge from network interactions since single cells did not have intrinsic bursting mechanisms. The burst frequency increased proportionally to propagation velocity of pathological activity as HSP increased throughout the network (Fig. 6.3B).

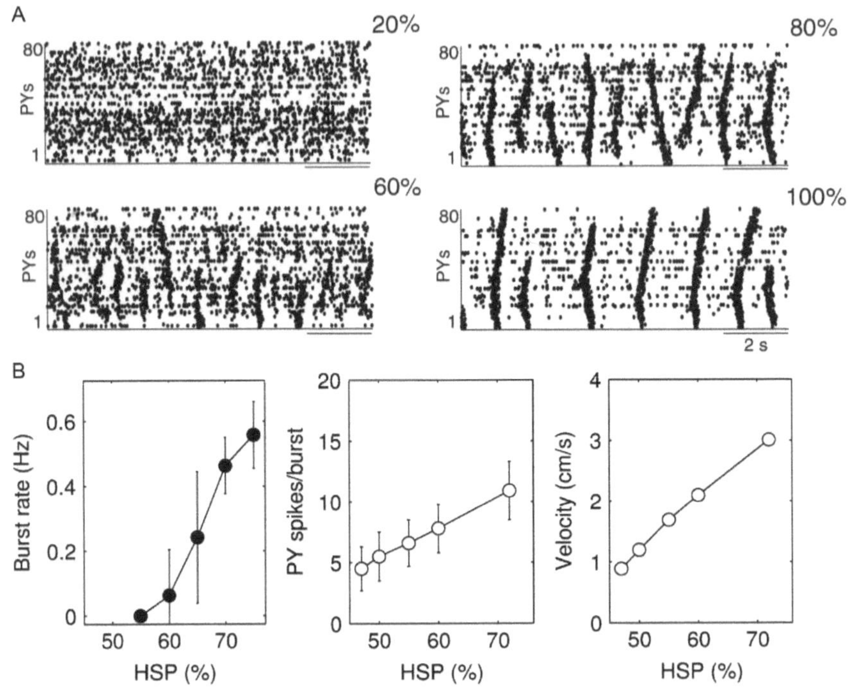

Figure 6.3 Effect of homeostatic synaptic plasticity (HSP) on network activity. (A) Raster plots showing network activity for different ratios of maximum HSP (20%, 60%, 80%, and 100%). (B) Single neuron burst rate, spike per burst, and network propagation velocity as function of HSP ratio. *Panel (A) is adapted with permission from Fröhlich et al. (2008). Copyright (2008) Society for Neuroscience. Panel (B) is adapted with permission from Houweling et al. (2005). Copyright (2005) Oxford University Press.*

They proposed that axonal sprouting and the formation of new synapses as observed in posttraumatic epileptogenic tissue are induced by paroxysmal activity, the prime factor being the upregulation of excitatory synapses via HSP processes within the existing network.

We next address the consequences of axonal sprouting and cell loss in the context of topological variations.

2.4. Network topology

2.4.1 Topological changes in the hippocampus

Influence of network topology has mainly been studied in the well-known case of temporal lobe epilepsy and hippocampal sclerosis. Hippocampal sclerosis involves major cell loss and it is usually associated with axon fiber sprouting. As these phenomena have been extensively studied in the dentate gyrus of the hippocampus, we will focus on this region.

Previous computational studies considered mossy fiber sprouting in the dentate gyrus and cell death in the hilus (Lytton, Hellman, & Sutula, 1998; Santhakumar, Aradi, & Soltesz, 2005) as typically encountered in animal models and human epilepsies. Experimental observations suggested that hyperexcitability would be due to reduction of inhibition caused by either neuronal loss or reduced excitatory drive onto interneurons. Lytton et al. (1998) tested this hypothesis in a computational model involving a scaled-down version of the dentate gyrus (50 excitatory granule cells, 2 excitatory mossy cells, and 2 inhibitory aspiny interneurons) stimulated via the perforant path. Mossy fiber sprouting was modeled by increasing collateral connections between granule cells, and network disinhibition was produced by removing GABAergic synapses. Together, these changes resulted in increased firing activity of aspiny neurons, depolarization block of mossy cells, but they did not affect granule cell activity. Such robust behavior from granule cells is believed to come from their high firing threshold. The high concentration of voltage and calcium-sensitive potassium channels repolarizing the cell after the spike may maintain the cell in a prolonged refractory period and would explain such stability. Furthermore, granule cells are the primary drivers for the whole network behavior. Their hypersynchronization during interictal spikes (as observed before and between seizure-like activities) could be thus a protective mechanism avoiding the cascade toward continuous discharge.

This pioneering study was followed by more detailed studies of mossy fiber sprouting in the dentate gyrus. Santhakumar et al. (2005) showed in larger simulations (500 granule cells, 15 mossy cells, 12 interneurons) the importance of using a realistic network topology in the model. Indeed, mossy fiber sprouting had little effect on a nontopographic network as found previously by Lytton et al. (1998). However, in a topographic network taking realistic axon distribution into account, even low mossy fiber sprouting increases dentate excitability. Dyhrfjeld-Johnsen et al. (2007) used very large networks, more accurate single neurons compartmentalization and dendritic arborization, to perform 1:1 scale analysis with up to eight types of identified dentate cells. They further investigated effects of structural topological network changes in the generation and spread of paroxysmal activity. They used two graph theoretical measures to assess the topology of a biological dentate gyrus: the average path length (L), i.e., the average minimal number of synapses between two neurons, and the clustering coefficient (C), i.e., the tendency of neurons to cluster in small interconnected groups (Fig. 6.4A).

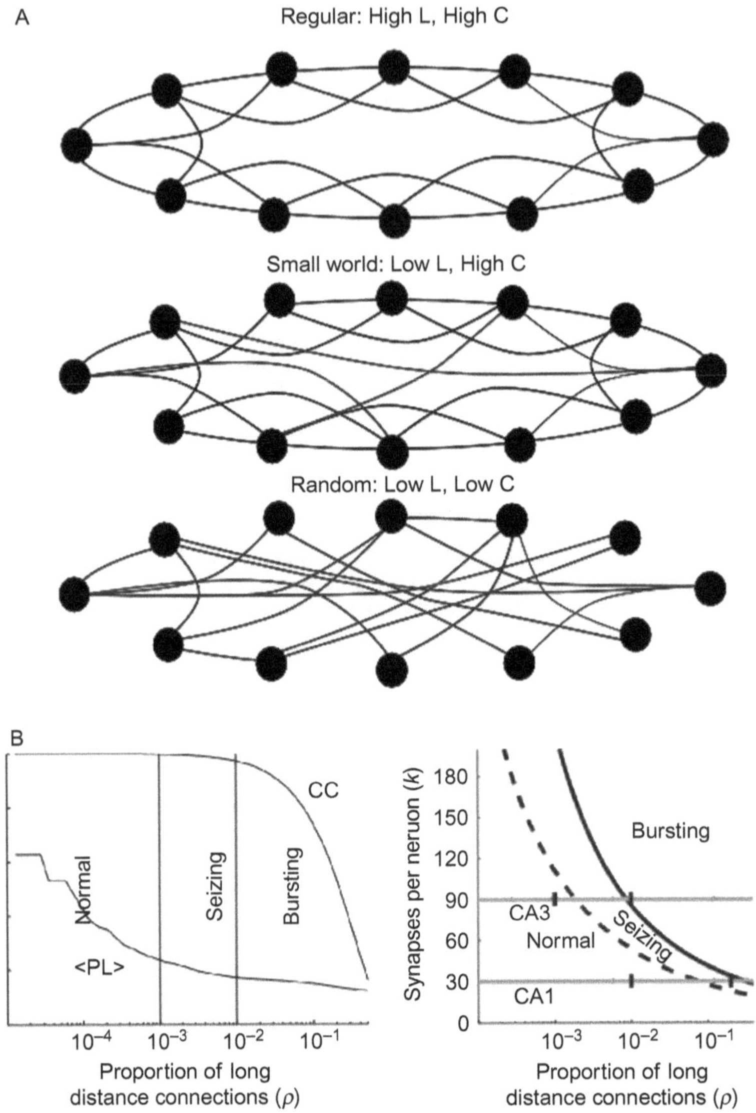

Figure 6.4 Effect of network topology on resting, seizing, and bursting transitions. (A) Topological organization of regular, small world, and random networks. (B) Left: network regimes (Normal, Seizing, Bursting), average path length (<PL>), and clustering coefficient (CC) as function of long distance connections ratio. Right: number of synapses per neuron as function of long distance connection proportion. Horizontal gray lines are empirical number of synapses per neuron observed in CA3 (at k=90) and CA1 (k=30). Regimes are transposed with solid and dashed curved lines. *Panel (B) is a dapted with permission from Netoff, Netoff, Clewley, Arno, Keck, and White (2004). Copyright (2004) Society for Neuroscience.*

They found that the dentate gyrus has a small world topology and investigated the consequences of sclerosis on the network dynamics. Sclerosis is biologically characterized by hilar cell loss and mossy fiber sprouting, which decrease L and increase C, respectively. During the first sclerotic phase, mossy fiber sprouting compensated for hilar cell loss, so C increased whereas L stayed almost constant, resulting paradoxically in enhanced excitability. However, when the number of dead hilar cells reached a critical threshold (around 90% of cell loss), L dramatically increased, decreasing hyperexcatibility. Interestingly, small-worldness was enhanced during the majority of the sclerotic process, with maximal epileptiform activity observed at around 80% sclerosis, a ratio that coincides with human patients' data.

Using a slightly different approach, Netoff et al. (2004) investigated the impact of increasing the number of long-range connections (thus increasing L only) in a one-dimensional ring network of spontaneously firing neurons. The study aimed to investigate the functional relevance of the different topographical maps of the CA1 and the CA3 circuits, from which seizure and paroxysmal network bursts can originate, using a set of three different neuron models, which ensured that results were robust across various modalities. Large networks (24,000 neurons) and reduced ones (3000 neurons) were simulated to study scaling properties of the findings. Overall, this study confirmed that irregular short (~100 ms) network bursts were more likely to originate from CA3 while sustainable seizures would rather be generated by CA1, in which less recurrent connections are observed experimentally (Fig. 6.4B).

These studies clearly show that the topology of the network plays a key role in seizure genesis. Another clinically critical process to understand is how seizures propagate.

2.4.2 Spatial propagation of seizure activity

The local spread of seizures was studied by Ursino and La Cara (2006) and Hall and Kuhlmann (2013) in two-dimensional leaky integrate-and-fire networks, varying connection extent and strength. In both studies, a depolarizing current was injected at the center of a lattice and the propagation of activity via traveling waves was examined for a variety of parameters involving strengths and spatial extensions of both excitatory and inhibitory synapses (Fig. 6.5). Epileptiform discharges such as small-amplitude high-frequency signals or isolated spikes could be observed and mapped to their corresponding 2D patterns of ECoG. For example, an isolated spike corresponded to the abrupt birth of one or multiple circular waves or their

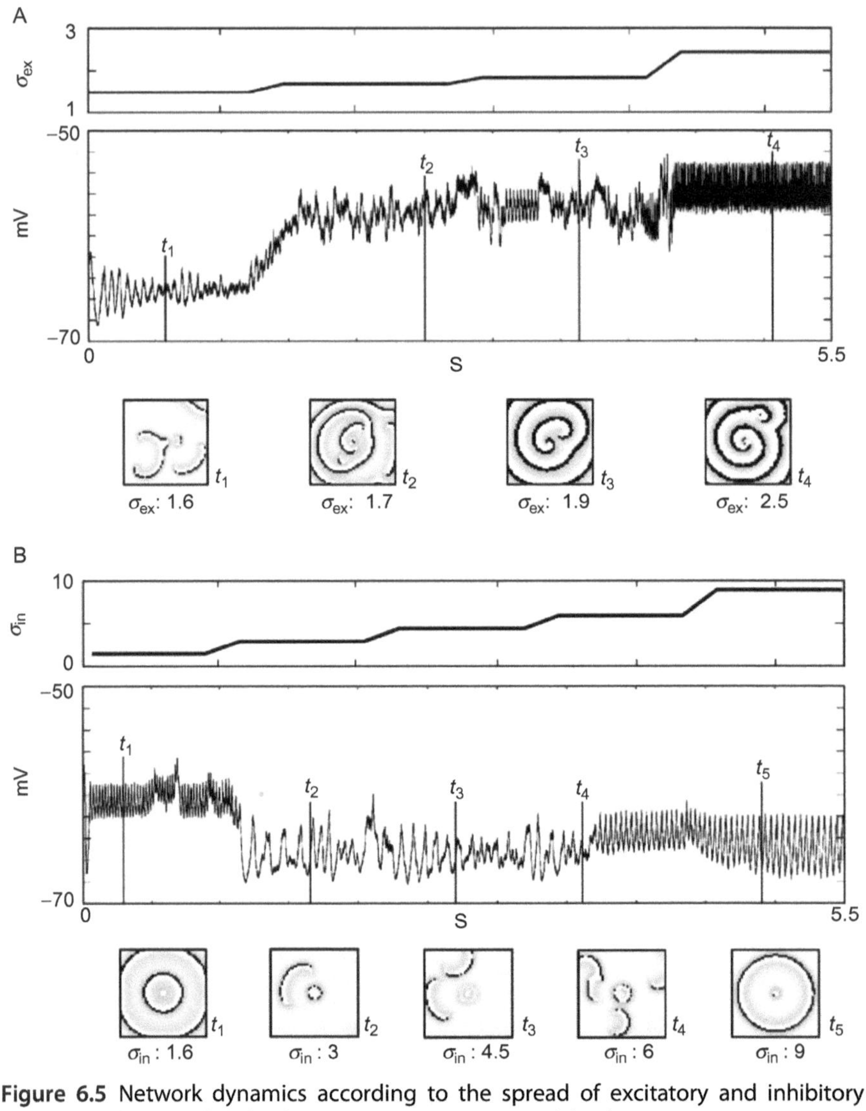

Figure 6.5 Network dynamics according to the spread of excitatory and inhibitory synapses. (A) Manual shift of excitatory synaptic spread (top), average neurons membrane potential evolution in time (middle), and respective 2D spatial network activity patterns at time corresponding to vertical lines drawn in the middle panel (bottom). (B) Same as (A) but the manual shift is operated on the inhibitory synaptic spread. *Adapted with permission from Ursino and La Cara (2006). Copyright (2006) Elsevier.*

sudden halt. Coexistence of segregated rotating waves or spiral patterns were associated to irregular rhythms with large amplitude, wide frequency content, and trends to display rapid low-amplitude discharges as excitation strength was increased.

Propagation speeds in simulated slice models were found to be either slow (0.1–20 mm/s) in low extracellular magnesium concentration but fast (10–100 mm/s) when GABAergic transmission was blocked (Hall & Kuhlmann, 2013). Although such values in animal and simulated slices differ significantly with experimental setups (i.e., Traub, Miles, & Jefferys, 1993 found 100–200 mm/s in disinhibited guinea pig slices) and are somewhat far from the observed propagation speed on human cortical recording (ECoG) (Kramer, Kirsch, & Szeri, 2005; Kramer, Szeri, Sleigh, & Kirsch, 2007), it is clear that propagation speed is strongly linked to the amount of presynaptic depression. In particular, in the low-magnesium model, propagation speed increased when the inhibitory synapses were more depressed.

So far, we discussed the influence of internal structure topologies and now consider networks of networks on a larger scale, addressed in Section 2.5 about corticothalamic dynamics.

2.5. Thalamocortical loop and absence seizures

Thalamocortical circuitry plays an important role in the generation and spread of spike–wave oscillations typically observed during absence seizures. In a first study, Destexhe (1998) described how spindle oscillations usually observed during slow-wave sleep can be gradually transformed into spike–wave discharges by blockade of $GABA_A$ receptors while preserving the slower $GABA_B$ synaptic transmission. The neural architecture involving corticothalamic loop and its variety of neuronal firing properties is shown in Fig. 6.6A. The thalamus model consisted of excitatory thalamocortical cells (TC) and inhibitory thalamoreticular cells (RE) with intrinsic bursting dynamics generated via the presence of T-type Ca^{2+} currents (with slower kinetics in RE cells). The cortical circuit was modeled with excitatory pyramidal (PY) cells and inhibitory interneurons (IN) with regular and fast intrinsic spiking, respectively, and with a slow K^+ current responsible for the spike–frequency adaptation of PY cells (see Fig. 6.6B and C). They showed that the short spike component of the observed spike–wave discharge was generated by strong excitatory postsynaptic potential, while the slower wave component was due to the belated G-protein activation mechanism of the inhibitory postsynaptic potential (IPSP). Such nonlinear

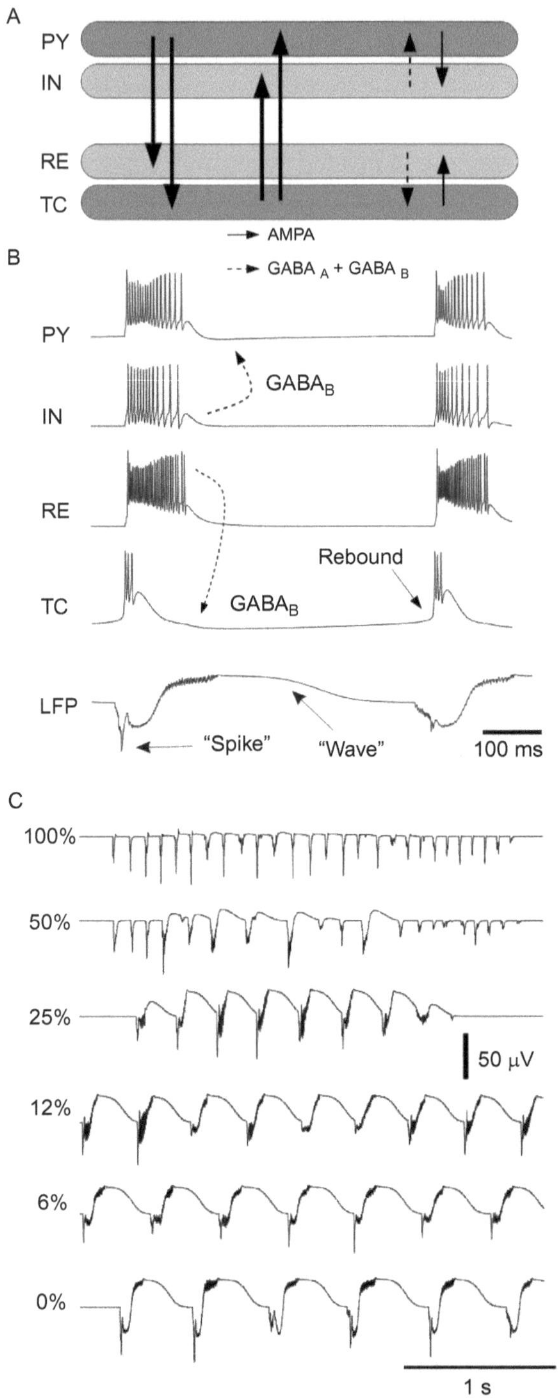
A
PY
IN
RE
TC
AMPA
GABA A + GABA B
B
PY
GABAB
IN
RE
Rebound
TC
GABAB
LFP
"Spike"
"Wave"
100 ms
C
100%
50%
25%
50 μV
12%
6%
0%
1 s

activation property from the $GABA_B$ receptor seems responsible for the silence recorded in all cell types during the wave, until TC cells fire again, provoking the rebound effect that drives the network oscillation. Thus, coexistence of two oscillation types was observed in the intact circuit: mild corticothalamic feedback recruited the system in an approximate 10-Hz spindle oscillatory regime, whereas strong feedback from the cortex at 3 Hz could force the thalamus to operate at the same frequency.

However, experimental evidence later unraveled that cortical circuits alone are sufficient to generate spike–wave discharges (Steriade & Contreras, 1998). This issue is addressed in another computational modeling study (Destexhe, Contreras, & Steriade, 2001) including low-threshold spike cells able alone to stand for the rebound mechanism in the isolated cortical circuit.

Together with the large-scale network simulations of intra- and intercortical spread of paroxysmal neural activity mentioned earlier (Van Drongelen, Lee, Stevens, & Hereld, 2007), those results account for the complex and various biophysical mechanisms involved in the generation and spread of epileptiform discharges. To understand the emergence of such patterns of neural behavior at the whole-brain scale, more abstract models have been developed to perform simulations at reasonable computational cost; such models are introduced in Section **3**.

3. LUMPED MODELS

Microscopic models have several drawbacks. First, each neuron model includes one or several variables. The number of variables scales linearly with the number of neurons simulated, which requires high-performance computing capabilities. Second, the number of free parameters increases with the number of neurons simulated. Extracting valuable information thus becomes more difficult as the size of the parameter space increases.

Figure 6.6 Mechanisms of generation of spike–wave discharges. (A) Neural coupling between areas. PY and IN are pyramidal cells and inhibitory interneurons in the cortex, TC and RE are excitatory thalamocortical cells and reticular inhibitory neurons, respectively. (B) Single neuron time series for each type of cells and the resulting local field potential (LFP) where the spike component and the wave component are clearly identified. $GABA_B$ action from inhibitory cells on excitatory neurons are shown with dashed arrows, and thalamocortical cells rebound effect (i.e., first cells to reignite after the strong inhibition) is pointed. (C) Evolution from sleep spindle to spike–wave discharge as $GABA_B$ conductance is reduced (from top to bottom). *Adapted with permission from Destexhe (1998). Copyright (1998) Society for Neuroscience.*

Third, microscopic models spatial scales are difficult to link with clinically relevant scales of brain regions that are usually recorded in patients with EEG, MEG, and iEEG. These drawbacks led neuroscientists to develop lumped models, collapsing the complex dynamics of millions of neurons in only a few macroscopic variables.

Lumped models have a long tradition in neuroscience, going back to the seminal work of Amari, Wilson and Cowan (Amari, 1977; Wilson & Cowan, 1972, 1973). The main hypothesis of lumped models is based on the highly redundant structure of the cortex and assumes that collective dynamics of neurons is well-captured by spatial and temporal averages. So-called mean-field approaches allow casting the many degrees of freedom into a few macroscopic population variables. This assumption applies particularly well for epilepsy where EEG, MEG, and iEEG are particularly low-dimensional during the seizure, i.e., most of the signal can be explained with the help of only a few variables.

Lumped models are also called neural mass or neural population models. Although they are unable to probe molecular mechanisms, they can help to elucidate the nature of the transitions between different stages of the seizure (interictal state, initiation, propagation and/or recruitment of other brain regions, and finally termination). In particular, transitions between states can be explained by nonlinear dynamical systems and bifurcation theory. Understanding the nature of such transitions is especially important for clinical applications because unraveling mechanisms leading to the ictal state would help finding solutions to avoid it.

3.1. Abstract models

One of the critical questions to address is how seizures start. Lopes da Silva et al. (2003) considered some very simple and generic models to illustrate three different routes to epileptic seizures, which do not result in the same electrographic signals. The dynamics of a chosen model depends upon specific control parameters, which affects the attractors of the system. In dynamical system theory, an attractor is a state, toward which the system is attracted to. In the first scenario, the control parameter changes the stability of the attractor from a normal healthy state (that we represent here by a fixed point, i.e., the system stays in a static state) to an ictal state (e.g., a limit cycle, in which the variables of interest oscillate, as in seizures). Here, the limit cycle progressively grows or decays in amplitude. Such dynamical deformation of the state space presumably accounts for mechanisms of partial epilepsy (Fig. 6.7A). In such a situation, only one stable state exists and is

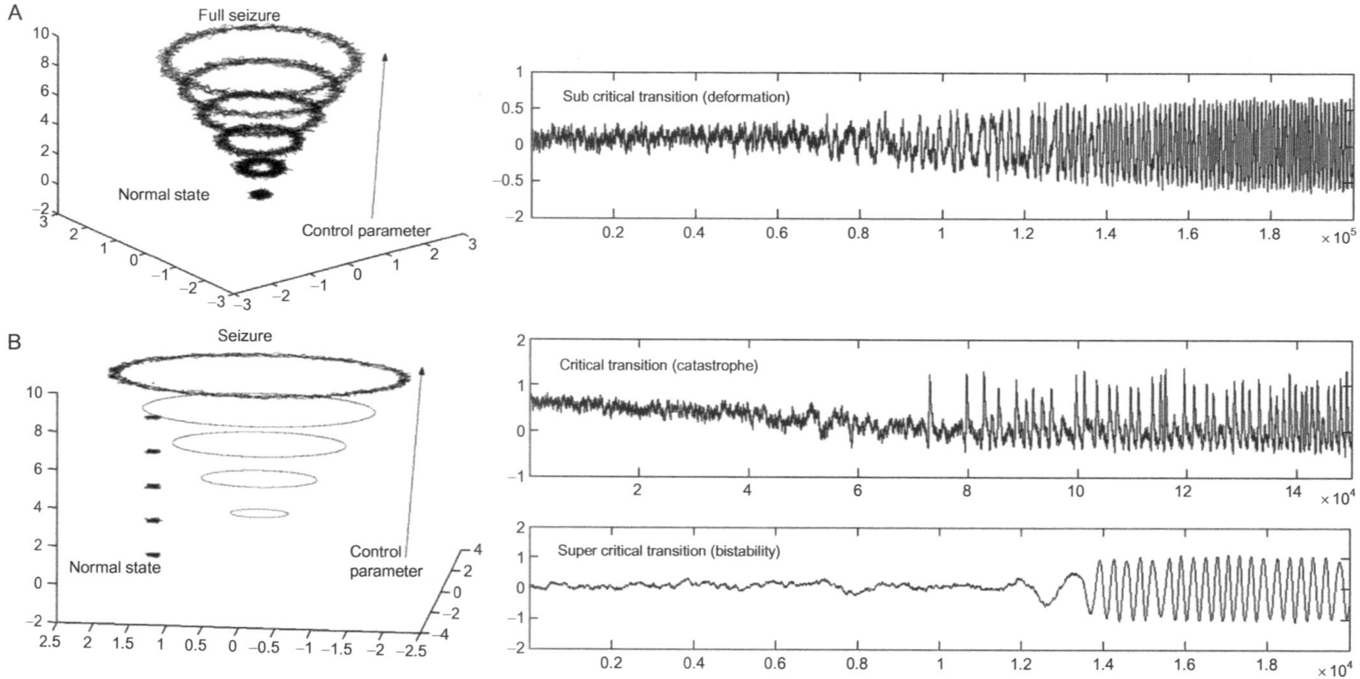

Figure 6.7 Different routes toward seizure. (A) Left: 2D time series plotted for different values of a control parameter on the vertical axis. The control parameter deforms the attractor of the system and leads it toward a full seizure. Right: for the same scenario, example of a time series in which the control parameter changes in time, going from a normal state (stable fix point) toward a full seizure (limit cycle). (B) Left: same space as in (A). When the control parameter changes, the seizure state (limit cycle) appears and coexists with the normal state (bistability). Right: for the same scenario, times series showing transition from a normal state to a seizure state. In this case, two routes are possible. On top, the time series switches back and forth between interictal and ictal states before entering into the seizure. On bottom, the time series stays on the stable fix point and then abruptly switches to the seizure. *Adapted with permission from Kalitzin, Velis, and Lopes Da Silva (2010). Copyright (2010) Elsevier.*

progressively distorted. In the second scenario, the change of the parameter leads to a bistable state, i.e., two stable states coexist in the same phase space: the healthy and the seizure state. In such a bistable scenario, transition from one state to the other occurs randomly in a fixed parameter setting (Fig. 6.7B, top right). On the other hand, a parameter change alters the distance between both states (and therefore the probability of transition) or even removes one of them (Fig. 6.7B, bottom right). Importantly, transitions in a bistable scenario predict a terse change in the time series of one or several variables and can be recorded experimentally.

Although very abstract, these models lay the ground for clinical applications, e.g., in seizure prediction. In such a context, prediction of random transitions remains difficult. Kalitzin et al. (2010) developed a method based on electrical stimulation where they measured the phase clustering index of a given stimulation, i.e., the propensity of a system to have an epileptic response to stimulation. Because their bistable system exhibits stronger phase clustering when approaching seizure threshold, a measure based on phase clustering differences can be computed and predict the distance from seizure onset. Then, a simple stimulation at the appropriate time can prevent seizure.

3.2. Neural mass and neural field models

We now discuss more realistic neural mass models, where parameters represent biophysical variables and time series closely resemble real time series of EEG, ECoG, and iEEG recordings. Still, they can be understood in the light of the preceding discussion. We start with neural mass models and continue with neural field models.

Neural mass models collapse in a single point several thousands of neurons by averaging their dynamics. Wendling, Bartolomei, Bellanger, and Chauvel (2002) modified a neural mass model previously introduced by Freeman (1975), Lopes da Silva, Hoeks, Smits, and Zetterberg (1974), and Lopes da Silva, van Rotterdam, Barts, van Heusden, and Burr (1976). The model consists of four interconnected populations: excitatory pyramidal cells, excitatory interneurons, inhibitory interneurons with slow kinetics, and inhibitory interneurons with fast kinetics. The two main parameters are the slow and the fast inhibitory synaptic gains. Their systematic variations demonstrate different regimes corresponding to different stages of the seizure (Fig. 6.8A). The choice of proper paths in the parameter space can reproduce time series close to real seizures, or be fitted to human data in order to find a realistic path through the parameter space heading

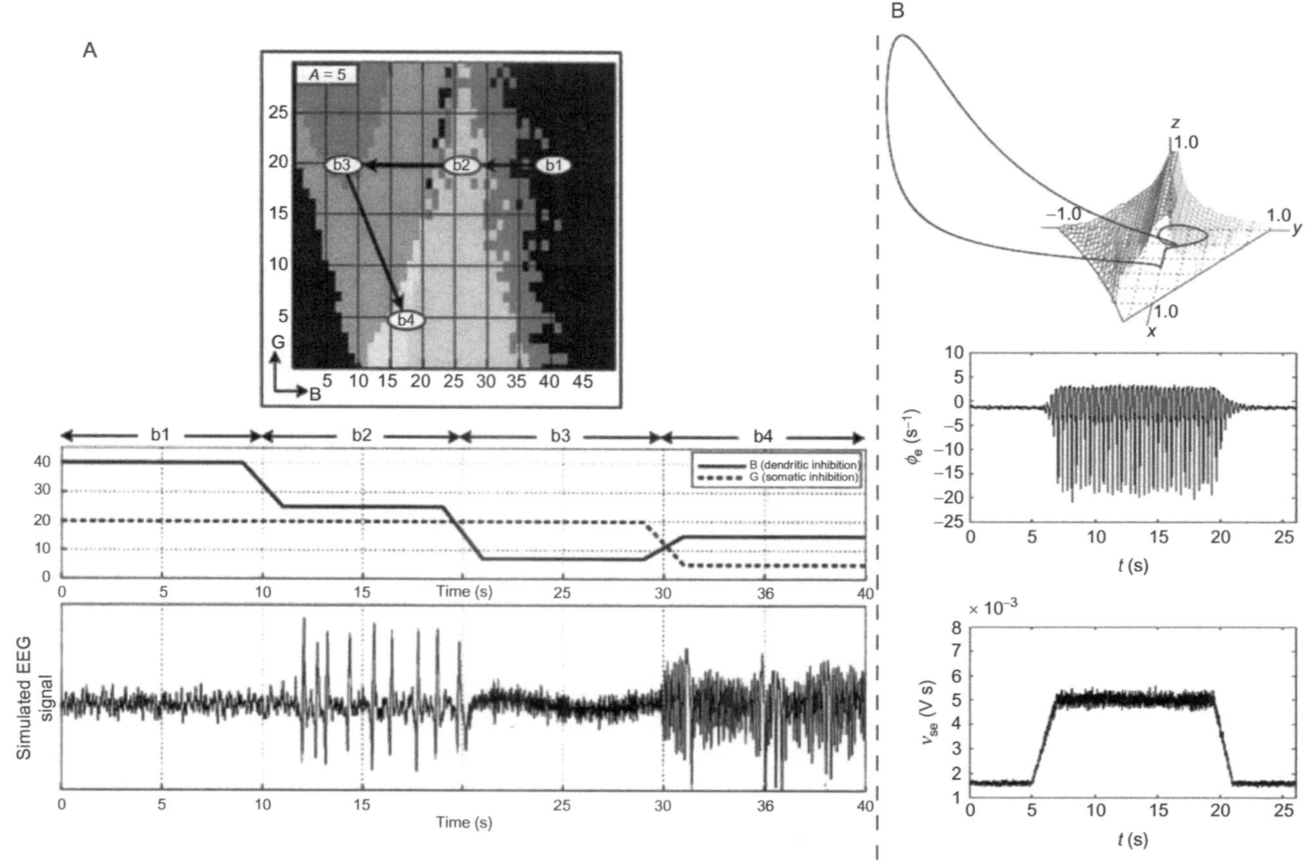

Figure 6.8—See legend on next page.

toward seizure (Fabrice Wendling, Hernandez, Bellanger, Chauvel, & Bartolomei, 2005). From this study, they concluded that, although excitation increases from interictal to ictal activity as expected, it is also compensated by an increase in slow dendritic GABAergic inhibition, which suddenly drops at the onset of the seizure and is replaced during low-voltage fast activity (15–40 Hz) by an increase of fast somatic inhibition. The understanding of the Wendling model rests on the analyses of the Jansen–Rit model (Jansen & Rit, 1995; Jansen, Zouridakis, & Brandt, 1993), and Touboul, Wendling, Chauvel, and Faugeras (2011) showed that they indeed have the same bifurcation structure.

Neural field models, in contrast to neural mass models, are spatially extended. Authors used different versions of these models, but they are all based on the neural field formulation of the cortex in which the cortical sheet is represented as a continuous medium (Freeman, 1975; Jirsa & Haken, 1996; Nunez, 1974; Robinson, Rennie, & Wright, 1997; Wilson & Cowan, 1973). Breakspear et al. (2006) studied a corticothalamic neural field model, as proposed by Robinson, Rennie, and Rowe (2002) based on the Jirsa–Haken wave equation (Jirsa & Haken, 1996) with a corticothalamic loop; they reproduced both tonic–clonic and absence seizures by only changing the value of three parameters, which are combinations of physiological parameters (Fig. 6.8B). They analyzed extensively the nature of the bifurcations at the transitions between ictal and interictal states, explaining why absence seizure are sudden events rapidly returning to the normal state after seizure offset, whereas tonic–clonic seizures are longer to recover from, with the existence of a cognitively deficient postictal state. Wilson, Sleigh, Steyn-Ross, and Steyn-Ross (2006), using cortical neural

---

Figure 6.8—Cont'd Two examples of path toward a seizure through bifurcations. (A) A neural mass model of partial epilepsy. Top: parameter space depending of slow inhibitory synaptic gain (*x* axis) and fast inhibitory synaptic gain (*y* axis). The colors identify regimes corresponding to the different time series patterns. The path b1–b2–b3–b4 is an example of a route toward and through the seizure. Middle: change in the values of the parameters in time. Bottom: resulting time series. (B) A neural field model of absence seizure. Top: inside the tent shape, the dynamics is stable. Outside it is considered epileptic. For well-chosen values of the parameters *x*, *y*, and *z*, one can be close to the alpha instability (tonic–clonic seizures) or 3 Hz instability (absence seizures). Then change of a single physiological parameter (combination of *x*, *y*, and *z* parameters) will lead to the route drawn and a seizure. Middle: values in time of a control parameter. Bottom: resulting time series. *Panel (A) is adapted with permission from Wendling et al. (2002). Copyright (2002) John Wiley and Sons. Panel (B) is adapted with permission from Breakspear et al. (2006). Copyright (2006) Oxford University Press.*

field model without subthalamic structures, explained anesthetic-induced seizures by introducing a disproportionate prolongation of the inhibitory postsynaptic potentials.

Most researches previously focused on seizure onset and only recently researchers started to investigate mechanisms of seizure termination, of which very little is known. For instance, how can seizures stop so abruptly in all regions of the cortex? Kramer et al. (2012) analyzed human seizure data to identify markers of the bifurcation occurring at the end of the seizure such as flickering, slowing down of oscillations, increasing autocorrelation and spatial correlation, and reproducing them with a model of secondary generalized seizures with a localized mean-field model.

3.3. Epileptor model

Jirsa, Stacey, Ivanov, and Bernard (2014) proposed a taxonomy of seizure-like events based on first principles in the bifurcation theory, which clearly classify onset and offset bifurcations. The predominant seizure class is characterized by a neural mass model of partial seizures, the so-called Epileptor. One major difference introduced in the Epileptor compared to the above models is the absence of control parameters: it performs autonomously the transition from ictal to interictal states. The Epileptor is indeed composed of a slow and a fast population mutually coupled and further linked to a very slow variable called the permittivity variable, acting as a control parameter but constrained by the activity of both populations (Fig. 6.9A-D). Moreover, the nature of the bifurcations occurring at onset and offset of the seizure is clearly identified and leads to strong predictions; e.g., interspike intervals at the end of the seizure grow logarithmically because of the homoclinic bifurcation, causing the flickering phenomena (Fig. 6.9E).

3.4. Large-scale models

Very little is known about seizure propagation and recruitment at a large scale, i.e., the scale of brain regions. Neural field models are useful to investigate propagation of partial seizures throughout the brain. Kramer et al. (2005, 2007) demonstrated in a neural field model that entering into the seizure is performed by an increase of the subcortical excitation together with a decrease of excitatory input on the mean soma membrane potentials of both excitatory and inhibitory cells. They compared their simulations with electrocorticography (ECoG) recordings, in particular in the maximum

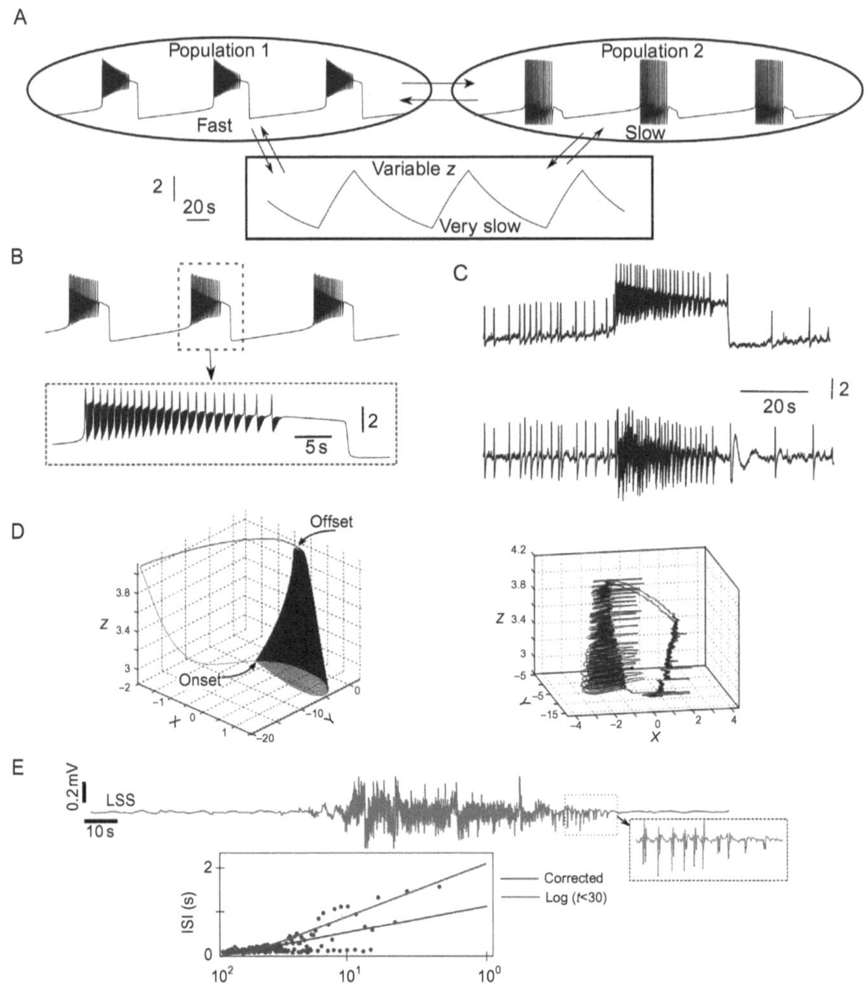

Figure 6.9 A neural mass model generates seizures autonomously. (A) The model is made of three different populations, one fast, responsible for both the switch from interictal to ictal state and the fast oscillations during the seizures (beta–gamma range), one slow, responsible for the spike–wave events during the seizures, and a third population, a slow permittivity variable, encompassing intra and extracellular impaired properties detailed above in biophysical models. These three populations are coupled together (B) A complete time series of the full deterministic model, and a zoom into the seizure temporal dynamics. (C) Top: same model with noise; bottom: same model after bandpass filtering 0.1–96 Hz as performed on intracranial EEG recordings. (D) Three-dimensional phase space spanned by three variables of the model without (left) and with noise (right). (E) The distribution of interictal spikes toward the end of the seizure follows a logarithmic scale, resulting from the homoclinic bifurcation.

frequency power and the speed of voltage propagation across the cortex. Kim, Roberts, and Robinson (2009) used the Robinson corticothalamic model to reproduce partial seizures. By tuning the range of corticocortical axons, they can keep the epileptiform discharges confined or make it spread as in the case of secondarily generalized seizures, therefore controlling spatial synchrony.

Nowadays, with the advent of large-scale brain connectivity, knowledge brought about by diffusion-weighted imaging and tractography algorithms (Sporns, Tononi, & Kötter, 2005), the human large-scale connectivity is being unscrambled. Neural fields do not in general introduce spatial heterogeneities in the network or very few when subthalamic structures are included. An alternative model for seizure spatial propagation is to approximate neural field models with neural masses coupled via short-range connectivity using a kernel function (Mexican hat for instance) and long-range connections (based on heterogeneous structural connectivity matrix). Terry, Benjamin, and Richardson (2012) studied a simple phenomenological seizure model with a probability of transitions between interictal and ictal states. They showed that network topology strongly influences the probability of transition into ictal state, as well as the nature of the seizure (focal, secondary generalized, or primary generalized). Interestingly, reducing the number of connections increased the probability of seizure appearance. Several authors have searched to spatially spread out neural mass models. Goodfellow, Schindler, and Baier (2011) locally extended a Jansen–Rit model to a set of 25 compartments and reproduced spike-and-wave discharges. They showed the existence of a zone of bistability between interictal and ictal states; thus, a transition to a full ictal state could be driven by noise. They further demonstrated the propagation of epileptic activity from one neural mass to others in such a local model. However, they did not include the thalamus, whereas the thalamocortical loop presumably plays an important role in absence seizures.

Taylor, Goodfellow, Wang, and Baier (2013) explored a spatially extended version of a modified Amari neural field to model spike-and-wave discharges, adding a second slow inhibitory timescale resulting from $GABA_B$ receptor activation. Introducing a slow timescale allows an autonomous transition between a stable point representing the slow wave and a fast oscillation accounting for the spike (Marten, Rodrigues, Benjamin, Richardson, & Terry, 2009; Taylor & Baier, 2011). They grouped together four of these models via short-range connectivity, and connected these groups via long-range connectivity measured with diffusion-weighted magnetic resonance imaging. The probabilistic tractography reconstruction was performed with realistic time delays, introducing spatial heterogeneities in the model as opposed to previous studies. Altogether, they accounted for a wide range of dynamical behaviors, such as polyspike–wave, spike-and-wave, and simple waves, as observed across patients, but could also create patient-specific fingerprint of their stereotypical seizures.

4. ANALOGIES

Analogies are models based on statistically relevant features extracted by training on samples. It has brought new perspectives in seizure prediction, based on other models parameter tuning.

4.1. Stochastic models

An important issue in seizure management would be to detect seizures as early as possible and preferably before the recruitment of other brain areas, giving the possibility to abort the seizure, e.g., via electrical stimulation (seizure termination system). Guirgis, Serletis, Zhang, and Florez (2014) trained a hidden Markov model on features (amplitude, phase, and their derivatives of Hilbert transform) extracted from hippocampal preparations. The model segregated three distinct states by testing time series: preictal, ictal, and postictal. A hidden Markov model was also used by Direito et al. (2012), but based on the spatiotemporal configuration of the spectral power of EEG for interictal, preictal, ictal, and postictal periods in 10 patients. Interestingly, preictal period was detectable several minutes before the seizure with a very good accuracy, which suggests potential application in real-time seizure detection and prevention.

A different approach was brought by Sunderam, Osorio, Frei And, and Watkins (2001) who tried to predict the next interictal period and seizure duration, considering prior ones using Markov chain models. Using recordings in rodents, they showed that seizures and interictal duration followed a bimodal distribution and predicted the duration of interictal and ictal periods with great accuracy for each distribution.

4.2. Power-law behavior

Osorio, Frei, Sornette, Milton, and Lai (2010) made the analogy between epileptic seizures and earthquakes regarding their power-law behavior: the probability density functions of earthquakes moments and seizure energy

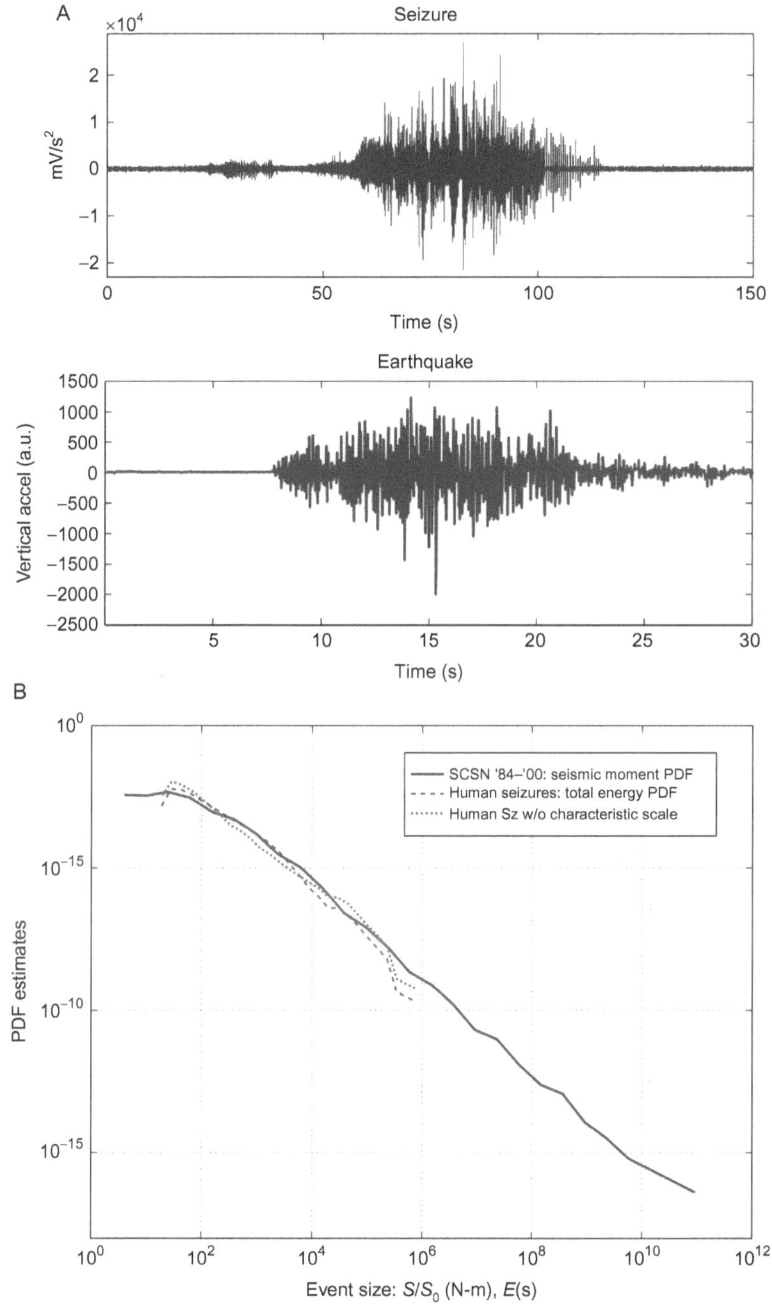

Figure 6.10 Analogy between seizures and earthquakes (A) Top: time series (acceleration, i.e., second derivative) of an epileptic seizure; human electrocorticogram recording. Bottom: time series (acceleration) of an earthquake. (B) Power spectral density estimate of seizures energy (in red, dashed lines) and earthquakes moments (in blue, solid lines) in log–log scale. *Adapted with permission from Osorio et al. (2010). Copyright (2010) American Physical Society.*

both follow a power-law statistics with an exponent of 5/3 (Fig. 6.10). Indeed, both systems are networks of coupled heterogeneous threshold oscillators, with a multiscale organization. This analogy is interesting for epilepsy since it would benefit from the statistical knowledge of earthquake research. For instance, the role of small earthquakes was recently employed for prediction of future larger earthquakes (Helmstetter, 2003). An analogy is drawn with interictal spikes, restating their importance in predictive models.

5. DISCUSSION

Through this review, we have presented novel developments in the computational modeling of epilepsy and highlighted their significant contributions in understanding this disorder. Computational modeling assists experimental paradigms in such a context, as it offers a malleable conceptual framework hardly accessible with biological setups. Reproduction of experimental findings and further systematic analysis allow challenging previous conclusions and investigating novel theories.

Single neuron models inform about the role of ion dynamics in the autonomous emergence of paroxysmal activity. Additional impairment of the regulation of extracellular ionic concentrations caused the entrainment of larger neuronal populations into abnormal discharges. The effects of network reorganization, such as observed after a head injury or sclerotic process, have also been investigated by systematically changing the topological organization of the simulated neural network. The effort of the system to maintain its activity at a given operating firing rate via HSP can induce hyperexcitability when strong reorganization happens after traumatic events.

The speed at which the seizure-like discharges propagate seems to be also greatly affected by changes in the organization of the network, with peculiar emphasis when inhibitory connection strengths decrease. Single neuron models also aid in evaluating other debated theories such as the role of electronic neural coupling via gap junctions in high-frequency oscillations or the implication of the thalamocortical loop in spike–wave discharges such as observed during absence seizures.

Large-scale simulations use simplifications in modeling neural dynamics where individual neuron activities are spatially averaged. In such lumped models, transitions between ictal and interictal states occur and can be characterized via their underlying dynamical mechanisms (bifurcations) underlying the transition. Since the number of such mechanisms is finite, seizure-like events can be classed in a well-defined taxonomy as assessed by the bifurcation theory. The nature of these transitions can be further constrained when the leading parameter is itself driven by the system dynamics, providing strong prediction on temporal properties of the electrographic signal. Whereas still in infancy, lumped models also open the way to whole-brain simulations and spatial study of seizure propagation; recent advances in the field of connectomics open large avenues to potentially clinically relevant applications through biologically realistic brain network simulations (Sanz Leon et al., 2013).

New ideas also arise from advances in seizure predictions, by taking advantages of machine-learning algorithms or statistical laws in analyzing patient-specific or experimental recordings. Seizure detection algorithms could benefit from such analogy, for instance by reevaluating the role of interictal spikes in seizure prediction.

We here reviewed different modeling approaches that are able to reproduce seizures in different ways over very different spatial and temporal scales. Though different in spirit and techniques, all approaches are integrated by the guiding idea of understanding how paroxysmal activity can be understood as a reorganization (state transition) process. Many different mechanisms argue for existence of multiple paths to seizure onset, the mechanisms at play being patient-, epilepsy type- or time-dependent. This underlines the relevance of general mathematical models such as the Epileptor, which unraveled the macroscopic paths of seizure transitions in dynamic terms without precisely identifying the underlying biophysical mechanisms. Thus, those different approaches are complementary.

We believe that integrative approaches combining biological experiments, brain imaging methods, and computational modeling at multiple spatial and temporal scales will lead, in the forthcoming decades, to a deeper understanding of epilepsy.

ACKNOWLEDGMENTS

Supported by INSERM and the James S. McDonnell Foundation.

REFERENCES

Amari, S. (1977). Dynamics of pattern formation in lateral-inhibition type neural fields. *Biological Cybernetics*, *87*, 77–87.

Bazhenov, M., Timofeev, I., Steriade, M., & Sejnowski, T. J. (2004). Potassium model for slow (2-3 Hz) in vivo neocortical paroxysmal oscillations. *Journal of Neurophysiology*, *92*(2), 1116–1132. http://dx.doi.org/10.1152/jn.00529.2003.

Breakspear, M., Roberts, J. A., Terry, J. R., Rodrigues, S., Mahant, N., & Robinson, P. A. (2006). A unifying explanation of primary generalized seizures through nonlinear brain modeling and bifurcation analysis. *Cerebral Cortex*, *16*(9), 1296–1313. http://dx.doi.org/10.1093/cercor/bhj072.

Cressman, J., Ullah, G., Ziburkus, J., Schiff, S. J., & Barreto, E. (2009). The influence of sodium and potassium dynamics on excitability, seizures, and the stability of persistent states: I. Single neuron dynamics. *Journal of Computational Neuroscience*, *26*(2), 159–170. http://dx.doi.org/10.1007/s10827-008-0132-4.The.

Destexhe, A. (1998). Spike-and-wave oscillations based on the properties of GABAB receptors. *The Journal of Neuroscience*, *18*(21), 9099–9111.

Destexhe, A., Contreras, D., & Steriade, M. (2001). LTS cells in cerebral cortex and their role in generating spike-and-wave oscillations. *Neurocomputing*, *38–40*, 555–563. http://dx.doi.org/10.1016/S0925-2312(01)00348-4.

Direito, B., Teixeira, C., Ribeiro, B., Castelo-Branco, M., Sales, F., & Dourado, A. (2012). Modeling epileptic brain states using EEG spectral analysis and topographic mapping. *Journal of Neuroscience Methods*, *210*(2), 220–229. http://dx.doi.org/10.1016/j.jneumeth.2012.07.006.

Draguhn, A., Traub, R. D., Schmitz, D., & Jefferys, J. G. R. (1998). Electrical coupling underlies high-frequency oscillations in the hippocampus in vitro. *Nature*, *394*(6689), 189–192. http://dx.doi.org/10.1038/28184.

Dyhrfjeld-Johnsen, J., Santhakumar, V., Morgan, R. J., Huerta, R., Tsimring, L., & Soltesz, I. (2007). Topological determinants of epileptogenesis in large-scale structural and functional models of the dentate gyrus derived from experimental data. *Journal of Neurophysiology*, *97*(2), 1566–1587. http://dx.doi.org/10.1152/jn.00950.2006.

Freeman, W. J. (1975). *Mass action in the nervous system*. New York: Academic Press.

Fröhlich, F., Bazhenov, M., Timofeev, I., Steriade, M., & Sejnowski, T. J. (2006). Slow state transitions of sustained neural oscillations by activity-dependent modulation of intrinsic excitability. *The Journal of Neuroscience*, *26*(23), 6153–6162. http://dx.doi.org/10.1523/JNEUROSCI.5509-05.2006.

Fröhlich, F., Bazhenov, M., Timofeev, I., & Sejnowski, T. J. (2007). Maintenance and termination of neocortical oscillations by dynamic modulation of intrinsic and synaptic excitability. *Thalamus and related systems*, *3*(2), 147–156. http://dx.doi.org/10.1017/S1472928807000155.

Fröhlich, F., Bazhenov, M., & Sejnowski, T. J. (2008). Pathological effect of homeostatic synaptic scaling on network dynamics in diseases of the cortex. *The Journal of Neuroscience*, *28*(7), 1709–1720. http://dx.doi.org/10.1523/JNEUROSCI.4263-07.2008.

Fröhlich, F., Sejnowski, T. J., & Bazhenov, M. (2010). Network bistability mediates spontaneous transitions between normal and pathological brain states. *The Journal of Neuroscience*, *30*(32), 10734–10743. http://dx.doi.org/10.1523/JNEUROSCI.1239-10.2010.

Goodfellow, M., Schindler, K., & Baier, G. (2011). Intermittent spike-wave dynamics in a heterogeneous, spatially extended neural mass model. *NeuroImage*, *55*(3), 920–932. http://dx.doi.org/10.1016/j.neuroimage.2010.12.074.

Guirgis, M., Serletis, D., Zhang, J., & Florez, C. (2014). Classification of multiple seizure-like states in three different rodent models of epileptogenesis. *IEEE Transactions on Neural Systems and Rehabilitation Engineering*, *22*(1), 21–32.

Hall, D., & Kuhlmann, L. (2013). Mechanisms of seizure propagation in 2-dimensional centre-surround recurrent networks. *PloS One*, *8*(8), e71369. http://dx.doi.org/10.1371/journal.pone.0071369.

Helmstetter, A. (2003). Foreshocks explained by cascades of triggered seismicity. *Journal of Geophysical Research*, *108*(B10), 2457. http://dx.doi.org/10.1029/2003JB002409.

Hodgkin, A., & Huxley, A. (1952). A quantitative description of membrane current and its application to conduction and excitation in nerve. *The Journal of Physiology*, *117*(4), 500–544.

Houweling, A. R., Bazhenov, M., Timofeev, I., Steriade, M., & Sejnowski, T. J. (2005). Homeostatic synaptic plasticity can explain post-traumatic epileptogenesis in chronically isolated neocortex. *Cerebral Cortex*, *15*(6), 834–845. http://dx.doi.org/10.1093/cercor/bhh184.

Jansen, B., & Rit, V. (1995). Electroencephalogram and visual evoked potential generation in a mathematical model of coupled cortical columns. *Biological Cybernetics*, *366*, 357–366.

Jansen, B., Zouridakis, G., & Brandt, M. (1993). A neurophysiologically-based mathematical model of flash visual evoked potentials. *Biological Cybernetics*, *283*, 275–283.

Jirsa, V., & Haken, H. (1996). Field theory of electromagnetic brain activity. *Physical Review Letters*, 77(5), 960–963.

Jirsa, V. K., Stacey, W. C., Ivanov, A., & Bernard, C. (2014). On the nature of seizure dynamics. *Brain* (in press). http://dx.doi.org/10,1093/brain/awu133.

Juszczak, G. R., & Swiergiel, A. H. (2009). Properties of gap junction blockers and their behavioural, cognitive and electrophysiological effects: Animal and human studies. *Progress in Neuro-Psychopharmacology & Biological Psychiatry*, *33*(2), 181–198. http://dx.doi.org/10.1016/j.pnpbp.2008.12.014.

Kager, H., Wadman, W. J., & Somjen, G. G. (2000). Simulated seizures and spreading depression in a neuron model incorporating interstitial space and ion concentrations. *Journal of Neurophysiology*, *84*(1), 495–512.

Kalitzin, S. N., Velis, D. N., & Lopes Da Silva, F. H. (2010). Stimulation-based anticipation and control of state transitions in the epileptic brain. *Epilepsy & Behavior*, *17*(3), 310–323. http://dx.doi.org/10.1016/j.yebeh.2009.12.023.

Kim, J. W., Roberts, J. A., & Robinson, P. A. (2009). Dynamics of epileptic seizures: Evolution, spreading, and suppression. *Journal of Theoretical Biology*, *257*(4), 527–532. http://dx.doi.org/10.1016/j.jtbi.2008.12.009.

Koch, C., & Segev, I. (Eds.), (1998). *Methods in neuronal modeling: From ions to networks*. Cambridge, MA: MIT Press.

Kramer, M. A., Kirsch, H. E., & Szeri, A. J. (2005). Pathological pattern formation and cortical propagation of epileptic seizures. *Journal of the Royal Society Interface*, *2*(2), 113–127. http://dx.doi.org/10.1098/rsif.2004.0028.

Kramer, M. A., Szeri, A. J., Sleigh, J. W., & Kirsch, H. E. (2007). Mechanisms of seizure propagation in a cortical model. *Journal of Computational Neuroscience*, *22*(1), 63–80. http://dx.doi.org/10.1007/s10827-006-9508-5.

Kramer, M., Truccolo, W., Eden, U. T., Lepage, K., Hochberg, L. R., Eskandar, E. N., et al. (2012). Human seizures self-terminate across spatial scales via a critical transition. *Proceedings of the National Academy of Sciences of the United States of America*, *109*(51), 21116–21121. http://dx.doi.org/10.1073/pnas.1210047110/-/DCSupplemental. www.pnas.org/cgi/doi/10.1073/pnas.1210047110.

Krishnan, G. P., & Bazhenov, M. (2011). Ionic dynamics mediate spontaneous termination of seizures and postictal depression state. *The Journal of Neuroscience*, *31*(24), 8870–8882. http://dx.doi.org/10.1523/JNEUROSCI.6200-10.2011.

Lopes da Silva, F. H., Blanes, W., Kalitzin, S. N., Parra, J., Suffczynski, P., & Velis, D. N. (2003). Dynamical diseases of brain systems: Different routes to epileptic seizures. *IEEE Transactions on Biomedical Engineering*, *50*(5), 540–548. http://dx.doi.org/10.1109/TBME.2003.810703.

Lopes da Silva, F. H., Hoeks, A., Smits, H., & Zetterberg, L. H. (1974). Model of brain rhythmic activity. The alpha-rhythm of the thalamus. *Kybernetik*, *15*(1), 27–37.

Lopes da Silva, F. H., van Rotterdam, A., Barts, P., van Heusden, E., & Burr, W. (1976). Models of neuronal populations: The basic mechanisms of rhythmicity. *Progress in Brain Research, 45*, 281–308.

Lytton, W. W., Hellman, K. M., & Sutula, T. P. (1998). Computer models of hippocampal circuit changes of the kindling model of epilepsy. *Artificial Intelligence in Medicine, 13*(1–2), 81–97.

Marten, F., Rodrigues, S., Benjamin, O., Richardson, M. P., & Terry, J. R. (2009). Onset of polyspike complexes in a mean-field model of human electroencephalography and its application to absence epilepsy. *Philosophical Transactions of the Royal Society of London A, 367*(1891), 1145–1161. http://dx.doi.org/10.1098/rsta.2008.0255.

Netoff, T. I., Clewley, R., Arno, S., Keck, T., & White, J. A. (2004). Epilepsy in small-world networks. *The Journal of Neuroscience, 24*(37), 8075–8083. http://dx.doi.org/10.1523/JNEUROSCI.1509-04.2004.

Nunez, P. (1974). The brain wave equation: A model for the EEG. *Mathematical Biosciences, 291*, 279–297.

Osorio, I., Frei, M. G., Sornette, D., Milton, J., & Lai, Y.-C. (2010). Epileptic seizures: Quakes of the brain? *Physical Review E, 82*(2), 021919. http://dx.doi.org/10.1103/PhysRevE.82.021919.

Quilichini, P. P., Le Van Quyen, M., Ivanov, A., Turner, D. A., Carabalona, A., Gozlan, H., et al. (2012). Hub GABA neurons mediate gamma-frequency oscillations at ictal-like event onset in the immature hippocampus. *Neuron, 74*(1), 57–64. http://dx.doi.org/10.1016/j.neuron.2012.01.026.

Robinson, P., Rennie, C., & Rowe, D. (2002). Dynamics of large-scale brain activity in normal arousal states and epileptic seizures. *Physical Review E, 65*(4), 1–9. http://dx.doi.org/10.1103/PhysRevE.65.041924.

Robinson, P. A., Rennie, C. J., & Wright, J. J. (1997). Propagation and stability of waves of electrical activity in the cerebral cortex. *Physical Review E, 56*(1), 826.

Santhakumar, V., Aradi, I., & Soltesz, I. (2005). Role of mossy fiber sprouting and mossy cell loss in hyperexcitability: A network model of the dentate gyrus incorporating cell types and axonal topography. *Journal of Neurophysiology, 93*(1), 437–453. http://dx.doi.org/10.1152/jn.00777.2004.

Sanz Leon, P., Knock, S. A., Woodman, M. M., Domide, L., Mersmann, J., McIntosh, A. R., et al. (2013). The Virtual Brain: A simulator of primate brain network dynamics. *Frontiers in Neuroinformatics*, 7, 1–23. http://dx.doi.org/10.3389/fninf.2013.00010.

Sporns, O., Tononi, G., & Kötter, R. (2005). The human connectome: A structural description of the human brain. *PLoS Computational Biology, 1*(4), e42. http://dx.doi.org/10.1371/journal.pcbi.0010042.

Steriade, M., & Contreras, D. (1998). Spike-wave complexes and fast components of cortically generated seizures. I. Role of neocortex and thalamus. *Journal of Neurophysiology, 80*(3), 1439–1455.

Sunderam, S., Osorio, I., Frei And, M. G., & Watkins, J. F. (2001). Stochastic modeling and prediction of experimental seizures in Sprague-Dawley rats. *Journal of Clinical Neurophysiology, 18*(3), 275–282.

Taylor, P. N., & Baier, G. (2011). A spatially extended model for macroscopic spike-wave discharges. *Journal of Computational Neuroscience, 31*(3), 679–684. http://dx.doi.org/10.1007/s10827-011-0332-1.

Taylor, P. N., Goodfellow, M., Wang, Y., & Baier, G. (2013). Towards a large-scale model of patient-specific epileptic spike-wave discharges. *Biological Cybernetics, 107*(1), 83–94. http://dx.doi.org/10.1007/s00422-012-0534-2.

Terry, J. R., Benjamin, O., & Richardson, M. P. (2012). Seizure generation: The role of nodes and networks. *Epilepsia, 53*(9), e166–e169. http://dx.doi.org/10.1111/j.1528-1167.2012.03560.x.

Touboul, J., Wendling, F., Chauvel, P., & Faugeras, O. (2011). Neural mass activity, bifurcations, and epilepsy. *Neural Computation*, *23*(12), 3232–3286. http://dx.doi.org/10.1162/NECO_a_00206.

Trasande, C. A., & Ramirez, J.-M. (2007). Activity deprivation leads to seizures in hippocampal slice cultures: Is epilepsy the consequence of homeostatic plasticity? *Journal of Clinical Neurophysiology*, *24*(2), 154–164. http://dx.doi.org/10.1097/WNP.0b013e318033787f.

Traub, R. D. (1994). Enhanced NMDA conductance can account for epileptiform activity induced by low Mg^{2+} in the rat hippocampal slice. *The Journal of Physiology*, *478*(3), 379–393.

Traub, R. D., Contreras, D., & Whittington, M. A. (2005). Combined experimental/simulation studies of cellular and network mechanisms of epileptogenesis in vitro and in vivo. *Journal of Clinical Neurophysiology*, *22*(5), 330–342.

Traub, R. D., Miles, R., & Jefferys, J. (1993). Synaptic and intrinsic conductances shape picrotoxin-induced synchronized after-discharges in the guinea-pig hippocampal slice. *The Journal of Physiology*, *461*, 525–547.

Traub, R. D., Whittington, M. A., & Cunningham, M. O. (2012). Fast oscillations and synchronization examined with in vitro models of epileptogenesis. In J. L. Noebels, M. Avoli, M. A. Rogawski, R. W. Olsen, & A. V. Delgado-Escueta (Eds.), *Japser's basic mechanisms of the epilepsies* (4th ed., pp. 286–297). Bethesda, MD: National Center for Biotechnology.

Ullah, G., Cressman, J. R., Barreto, E., & Schiff, S. J. (2009). The influence of sodium and potassium dynamics on excitability, seizures, and the stability of persistent states: II. Network and glial dynamics. *Journal of Computational Neuroscience*, *26*(2), 171–183. http://dx.doi.org/10.1007/s10827-008-0130-6.

Ursino, M., & La Cara, G.-E. (2006). Travelling waves and EEG patterns during epileptic seizure: Analysis with an integrate-and-fire neural network. *Journal of Theoretical Biology*, *242*(1), 171–187. http://dx.doi.org/10.1016/j.jtbi.2006.02.012.

Van Drongelen, W., Lee, H. C., Stevens, R. L., & Hereld, M. (2007). Propagation of seizure-like activity in a model of neocortex. *Journal of Clinical Neurophysiology*, *24*(2), 182–188. http://dx.doi.org/10.1097/WNP.0b013e318039b4de.

Volman, V., Perc, M., & Bazhenov, M. (2011). Gap junctions and epileptic seizures—Two sides of the same coin? *PLoS One*, *6*(5), e20572. http://dx.doi.org/10.1371/journal.pone.0020572.

Wendling, F., Bartolomei, F., Bellanger, J. J., & Chauvel, P. (2002). Epileptic fast activity can be explained by a model of impaired GABAergic dendritic inhibition. *European Journal of Neuroscience*, *15*(9), 1499–1508. http://dx.doi.org/10.1046/j.1460-9568.2002.01985.x.

Wendling, F., Hernandez, A., Bellanger, J.-J., Chauvel, P., & Bartolomei, F. (2005). Interictal to ictal transition in human temporal lobe epilepsy: Insights from a computational model of intracerebral EEG. *Journal of Clinical Neurophysiology*, *22*(5), 343–356.

Wilson, H. R., & Cowan, J. D. (1972). Excitatory and inhibitory interactions in localized populations of model neurons. *Biophysical Journal*, *12*(1), 1–24. http://dx.doi.org/10.1016/S0006-3495(72)86068-5.

Wilson, H. R., & Cowan, J. D. (1973). A mathematical theory of the functional dynamics of cortical and thalamic nervous tissue. *Kybernetik*, *13*(2), 55–80.

Wilson, M. T., Sleigh, J. W., Steyn-Ross, D. A., & Steyn-Ross, M. L. (2006). General anesthetic-induced seizures can be explained by a mean-field model of cortical dynamics. *Anesthesiology*, *104*(3), 588–593.

CHAPTER SEVEN

Mechanisms of Ictogenesis

Thomas Blauwblomme[*,†,‡,§,¶], **Premysl Jiruska**[||,#], **Gilles Huberfeld**[‡,§,¶,**,††,1]

[*]Neurosurgery Unit, Hopital Necker—Enfants Malades, APHP, Paris, France
[†]Université Paris Descartes, Paris, France
[‡]INSERM U1129—Infantile Epilepsies and Brain Plasticity, Paris, France
[§]University Paris Descartes, PRES Sorbonne Paris Cité, Paris, France
[¶]CEA, Gif sur Yvette, France
[||]Department of Developmental Epileptology, Institute of Physiology, Academy of Sciences of Czech Republic, Prague, Czech Republic
[#]Department of Neurology, 2nd Faculty of Medicine, Charles University in Prague, Motol University Hospital, Prague, Czech Republic
[**]Clinical Neurophysiology Department, CHU Pitié-Salpêtrière, APHP, Paris, France
[††]Université Pierre et Marie Curie, Paris, France
[1]Corresponding author: e-mail address: gilles.huberfeld@upmc.fr

Contents

1. Introduction 156
2. Recording the Brain and its Networks During the Transition to Seizures 159
   2.1 Preictal dynamics 159
   2.2 Transition to seizure 162
3. Cellular Substrates of Ictogenesis: Neurons and Neurotransmission 162
   3.1 Neuronal behavior during ictogenesis 162
   3.2 Pyramidal cells 164
   3.3 Glutamatergic neurotransmission 165
   3.4 Interneurons 165
   3.5 GABAergic neurotransmission 167
4. The Environment of Neurons: Ions, Astrocytes, and the Extracellular Space 170
   4.1 Potassium 170
   4.2 Calcium 172
   4.3 Extracellular space 172
   4.4 Astrocytes 173
5. Conclusion 174
Acknowledgments 175
References 175

Abstract

Epilepsy is a paroxysmal condition characterized by repeated transient seizures separated by longer interictal periods. Ictogenesis describes the processes of transition from the interictal state to a seizure. The processes include a preictal state, with specific clinical signs and a distinct electrophysiology which may provide opportunities to anticipate, or even prevent, seizures. Biological mechanisms of ictogenesis remain poorly

International Review of Neurobiology, Volume 114
ISSN 0074-7742
http://dx.doi.org/10.1016/B978-0-12-418693-4.00007-8

understood and may vary between conditions/syndromes. We review here ictogenic processes including the involvement of pyramidal cells, interneurons and astrocytes, GABAergic and glutamatergic signaling, and ionic perturbations. Our review suggests that specific excitatory influences at the transition to an ictal event include (1) GABA receptor activation with a neuronal Cl^- load and (2) a transient increase in external K^+.

1. INTRODUCTION

In epilepsy disorders, the brain has an enduring predisposition to generate seizures (Fisher et al., 2005). Seizures result from transient episodes of abnormally excessive or synchronous neuronal activity which generate specific signs and symptoms. While it remains unclear how seizures start, the recent data and new concepts reviewed here may improve our understanding of the mechanisms contributing to seizure initiation.

Epilepsy is a paroxysmal condition in which transient seizures are repeated, separated by longer interictal periods. Seizures are complex, variable, and prolonged electrical events that last for 10s of seconds to several minutes. Between seizures an "epileptic" brain may generate synchronous discharges of duration 10s or 100 s of milliseconds, which may be too short for an evident behavioral correlate. These interictal discharges are usually monophasic (spike) or combine a few stereotyped electroencephalographic (EEG) patterns such as spike-wave or polyspike (de Curtis & Avanzini, 2001). Ictal discharges are more complex, organized patterns of electrical discharges (Pacia & Ebersole, 1999). They may be initiated by preictal discharges or by a single slow deflection, continue with irregular low-voltage fast activity period of variable duration and then conclude with a period of oscillating bursts of activity. Other seizures begin with recurring oscillating bursts. Epilepsy surgery evaluation allowed to define distinct types of zones involved in the epileptic process (Rosenow & Lüders, 2001). The epileptic zone is the area responsible for the genesis of seizures, which may differ or only partially overlap with that of the eventual epileptic lesion. The seizure onset zone is the area in which the seizure is initiated and will spread to other areas that will produce ictal symptoms (symptomatogenic zone). Finally, the area producing interictal discharges, often wider than the epileptogenic zone is called irritative zone. Therefore, interictal and ictal activities may be generated by distinct brain regions, where they presumably depend on different neuronal populations and possibly distinct signaling systems

(Avoli, de Curtis, & Köhling, 2013). Some evidence suggests interictal events tend to impede the genesis of ictal discharges (de Curtis & Avanzini, 2001).

A brain becomes epileptic, able to generate repeated seizures, due to molecular, structural and functional changes occurring during a process of epileptogenesis (Fig. 7.1; Pitkänen et al., 2007). Current theories on acquired epilepsy suggest that during a latent phase of epileptogenesis after an insult or injury or due to a genetic defect the brain is reconfigured in multiple ways before seizures occur. This brain reorganization can include neuronal loss, changes in expression of voltage-gated channels (Bernard et al., 2004; Su et al., 2002) receptors (Epsztein, Represa, Jorquera, Ben-Ari, & Crepel, 2005), metabolic alterations, immunological processes, and gliosis (Vezzani, Friedman, & Dingledine, 2013). Epileptogenic processes may aggravate once seizures begin. Williams, Hellier, White, Staley, and Dudek (2007) and Williams et al. (2009) examined the long-term dynamics of epilepsy following an initial injury in experimental epilepsies induced by kainic acid or hypoxic-ischemic injury. They showed that seizure frequency increases in a sigmoid rather than a step fashion after a first seizure. Clinical studies also document a progressive nature of certain epilepsies especially those of the temporal lobe (Engel, Van Ness Paul, Rasmussen, & Ojemann, 1996). Epilepsy progression includes both an increase in seizure frequency and an aggravation in structural abnormalities. Seizures are also influenced by multiple endogenous and exogenous factors (Huberfeld

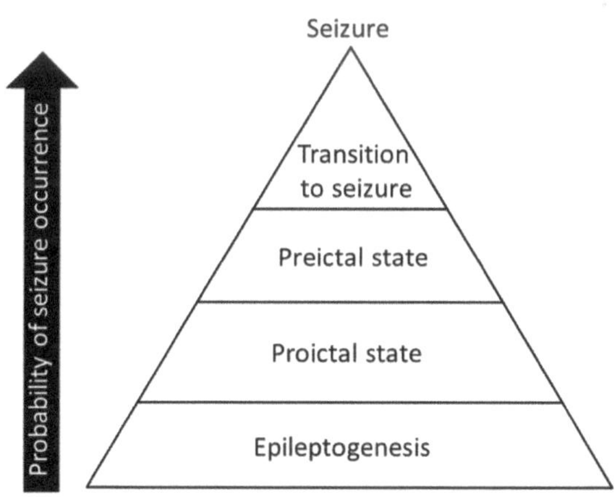

Figure 7.1 Seizures are the result of several processes, which act at different timescales and involve different pathophysiological mechanisms. The preictal state indicates that probability of seizure occurrence is high. The preictal state marks the period when the brain moves deterministically toward the seizure.

et al., 2013). The most common is hippocampal sclerosis, which may progress from mild forms only detectable with magnetic resonance (MR) spectroscopy, to a reduced volume and hippocampal sclerosis visible in MR images (Fuerst, Shah, Shah, & Watson, 2003). Changes during epileptogenesis result in a persistently increased neuronal excitability, which is permissive for seizures and for epileptiform phenomena including interictal discharges and high-frequency oscillations and also for augmented responses to stimuli (Sloviter, 1991). Even so, once established epileptogenic states do not represent a uniform but rather a fluctuating increase in excitability, together with a susceptibility to generate seizures.

Ictogenesis describes processes of seizure genesis by an epileptogenic brain. These processes are faster, occurring over periods of seconds to minutes, and less easy to detect from clinical signs or electrical activity. Prolonged EEG records over several months are needed for a precise understanding of long-term dynamics of human seizure activity. As this technically demanding goal becomes possible, it appears that seizures in some patients may be predicted (Cook et al., 2013). Analyses of the long-term dynamics of seizures in animal models of epilepsy suggest that the interval between ictal events is not random (Williams et al., 2009). Seizures tend to occur in temporal clusters rather than with a Poisson interval distribution. Presumably, periods of increased susceptibility to seizure are responsible. The susceptibility to generate other epileptiform activities also fluctuates. While interictal discharges have often been used as a marker of disease activity, even if they may act as a protection of the emergence of ictal discharges (de Curtis & Avanzini, 2001), their spatiotemporal dynamics with respect to seizure susceptibility are unclear. However, high-frequency EEG oscillations (200–500 Hz) are well correlated in space and time with seizure initiation in patients and in animal models of epilepsy. Zijlmans, Jacobs, Zelmann, Dubeau, and Gotman (2009) showed that as antiepileptic drugs were withdrawn from patients during presurgical records with implanted electrodes, the incidence of high-frequency oscillations was increased in parallel with the increase in seizure probability. In the same way, an increased frequency of fast ripples generated from specific regions follows the temporal profile of seizure occurrence in the pilocarpine model of epilepsy (Lévesque, Bortel, Gotman, & Avoli, 2011). Seizure clustering presumably results from proconvulsive effects induced by a seizure. These effects may be associated with long-term endogenously initiated changes in tissue excitability (Casillas Espinosa, Powell, & O'Brien, 2012; Reddy, 2013). Seizure probability should then be enhanced indirect during proictal periods by long-term fluctuations in excitability (Richardson & Jefferys, 2011).

Ictogenic processes should be prime targets in attempt to suppress seizure initiation. And yet, most antiepileptic dugs act less specifically to reduce neuronal excitability and burst firing or to reduce neurotransmitter release. Thus, they affect a large spectrum of brain activities other than seizures, resulting in unwanted side effects (Rogawski & Löscher, 2004). To develop drugs acting at the transition to seizure, necessitates careful studies on ictogenesis.

In vivo, ictal discharges are less frequent than interictal ones. *In vitro*, interictal activity is often generated spontaneously by slices of epileptic tissue, but proconvulsant stimuli are needed to generate ictal events (Huberfeld et al., 2011). Such stimuli, whether electrical or pharmacological, could bias mechanisms of ictogenesis. Alternatively, convulsant stimuli have been applied to healthy brain or tissue, a situation distant from that in a patient where a seizure is generated "spontaneously" by an epileptic network/focus. Probably, the most useful information may emerge from studies on epileptic and even human tissue. Even so, seizure onset patterns *in vivo* are variable suggesting that diverse preictal mechanisms may operate in a convergent fashion.

This review aims to describe the activity of the brain and its networks at the transition to seizures. It focuses first on clinical and electrophysiological data associated with this transition period and on synchrony dynamics and high-frequency oscillations at seizure onset. The cellular substrates of ictogenesis both at the neuronal and neurotransmission level are reviewed, for principal neurons that signal via glutamatergic synapses, and for interneurons and GABAergic transmission. Finally, the influence of the neuronal environment including ionic homeostasis and glial cells must be explored.

2. RECORDING THE BRAIN AND ITS NETWORKS DURING THE TRANSITION TO SEIZURES

2.1. Preictal dynamics

The preictal state moves in deterministic fashion toward a seizure (Richardson & Jefferys, 2011) presumably according to the dynamics of brain states between seizures. It thus discriminates preictal states from proictal states, which occur on much longer timescale. However, definition of the preictal state may vary. New analytical tools from nonlinear systems and chaos dynamics were introduced into epilepsy research in the 1990s. Applied to scalp and intracranial EEG records from epileptic patients, they revealed detectable changes in brain activity preceding seizure onset by minutes or even hours (Lehnertz et al., 2007; Litt & Lehnertz, 2002; Mormann,

Andrzejak, Elger, & Lehnertz, 2007). Several measures derived from these approaches, including correlation dimension, Lyapunov exponent, and Kolmogorov entropy, display preictal dynamics during the period before a seizure. Preictal changes have also been detected in cerebral hemodynamics, characterized by increased regional cerebral blood flow (Baumgartner et al., 1998; Weinand et al., 1997) before a seizure. The most successful indices are those of brain synchrony which demonstrate preictal changes more than 20 min before some seizures (Mormann et al., 2003; Rudrauf et al., 2006). However, most data were obtained from retrospective studies and prospective use of the same indices has not always predicted seizures reliably (see Cook et al., 2013). In general, these algorithmic approaches have not been linked to underlying pathological processes and so they possess little power of biological explanation (Mormann et al., 2007).

Records of neuronal and circuit dynamics are starting to reveal stereotyped changes preceding ictal events. Technological advances now permit multineuronal and multisite records of activity from epileptic animals and from patients (Bower, Stead, Meyer, Marsh, & Worrell, 2012; Buzsáki, 2004; Truccolo et al., 2011). Slice records show that seizures are preceded by changes in unit activity and high-frequency field potential oscillations. Epileptic seizures induced by low calcium or high potassium are preceded by progressive increases in neuronal firing, organized into small functional clusters with size that increases as each seizure approaches (Jiruska et al., 2010). In extracellular recordings, this process results in a progressive build up in high-frequency activity (~200 Hz) as in slice models of seizures induced by low magnesium or high potassium (Dzhala & Staley, 2003; Khosravani et al., 2005). In neocortical slices, Trevelyan has shown a progressive increase in inhibition preceding seizures, possibly reflecting enhanced activity in inhibitory circuits as an ictal event approaches (Trevelyan, Sussillo, & Yuste, 2007). Fast inhibitory currents seem likely to contribute significantly to high-frequency extracellular oscillations. Huberfeld et al. (2011) identified glutamatergic preictal discharges, distinct from interictal discharges, which precede ictal events in slices of human tissue. In chronic epileptic animals, high-frequency oscillations increase, most notably in terms of their duration, before seizures (Bragin, Azizyan, Almajano, & Engel, 2009; Bragin, Sanderson, Peterson, Connor, & Müller, 2009) and in humans, the energy of high-frequency oscillations (>80 Hz) increases up to 20 min before seizure onset (Worrell et al., 2004). It must be noted that some studies have not detected preictal changes in high-frequency activity (Jacobs et al., 2009). Overall however, pre-seizure phenomena in patients seem to reflect slow, progressive changes in neuronal population dynamics possibly corresponding to an increased excitability, eventually determined by seizure-triggering stimuli (Jiruska et al., 2010). Furthermore, changes in dynamic multineuronal behaviors have recently been suggested in regions of seizure spread, outside the onset zone, at several minutes before seizure onset in human patients with neocortical epilepsy (Truccolo et al., 2011).

Seizures are a dynamically distinct brain state; in many physical systems, transitions between different dynamic states follow similar rules. In epilepsy, the transition from an interictal state to a seizure may follow different pathways (Lopes da Silva et al., 2003). Data suggest that the transition in dynamic state at seizure onset is preceded by discrete system dynamic changes corresponding to a progressive reduction of system stability (Dakos et al., 2008; Scheffer et al., 2012). These early-warning signals of a critical transition are manifest as an increased autocorrelation and variance, a slowing oscillation frequency, increased skewness and spatial expansion (Scheffer et al., 2009). Seizures represent an extreme transition to an alternative brain dynamic state with preceding neurobiological changes that are similar to such early-warning signals. In epilepsy, these dynamics may play important roles in the transition to seizure (Jiruska et al., 2010, 2013a, 2013b) and also in seizure termination (Kramer et al., 2012). In addition to early signal changes, systems in transition to a different state, also show progressive decreases in resilience to small stochastic perturbations, manifest as a slowing of recovery from internal or external perturbations (Fig. 7.2; Thrush et al., 2009; van Nes & Scheffer, 2007). Active perturbations applied to probe brain dynamics during the transition to seizure have identified changes consistent with early changes in system dynamics (Kalitzin, Velis, Suffczynski,

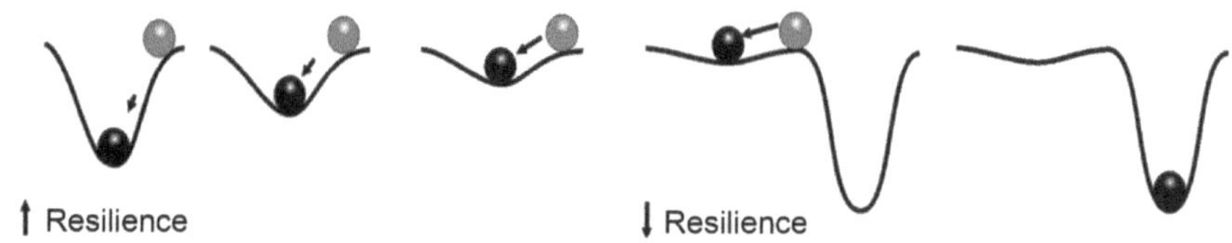

Figure 7.2 Schema showing decreased system resilience during the critical transition to an alternate dynamic regime. The system becomes less stable, less resilient to weak perturbations (shallow basin of attraction) and recovery from perturbations is slowed (critical slowing down). When system stability and resilience are both low, even small perturbations can shift the system to an alternate state (seizure). The tipping point is at the edge of the basin of attraction. In epilepsy, the tipping point represents events near seizure onset.

Parra, & da Silva, 2005; Suffczynski et al., 2008; Wright, Orth, Patsalos, Smith, & Richardson, 2006). Preictal changes in neuronal and brain dynamics exist and contribute to drive the neuronal population system toward a tipping point where seizure initiation is inevitable. At such a tipping point, or threshold, the probability that small perturbations drive the system to seizure is greatly enhanced.

2.2. Transition to seizure

Transition to seizure involves proictal and preictal mechanisms together with seizure initiation. Here, we consider the transition to seizure as a process which involves cellular, synaptic, network, and homeostatic components.

Neuronal networks within epileptic foci appear to be organized as clusters of pathologically interconnected epileptic neurons. Abnormal activity of these neuronal domains may generate both interictal events associated with high-frequency field oscillations and seizures. Such neuronal clusters have been inferred from *in vitro* (Bikson, Fox, & Jefferys, 2003; Jefferys et al., 2012; Jiruska et al., 2010), *in vivo* (Bragin, Wilson, & Engel, 2000), and human studies (Stead et al., 2010). At transition to seizure, these clusters coalesce into larger neuronal aggregates that generate ictal activities. Both local and larger scale mechanisms may contribute to this transition (Barbarosie & Avoli, 1997; Jensen & Yaari, 1988). Precise mechanisms of the transition to seizure may vary and so govern the electrographic pattern of seizure onset (Bragin, Azizyan, et al., 2009).

3. CELLULAR SUBSTRATES OF ICTOGENESIS: NEURONS AND NEUROTRANSMISSION

Epileptic seizures depend on neurons firing excessively and synchronously. Here, we review data on pyramidal cell and interneuron firing just before and during seizures to ask how the resulting glutamatergic and GABAergic signals may contribute to seizure initiation.

3.1. Neuronal behavior during ictogenesis

Traditionally, the human EEG has been recorded with surface or intracranial electrodes placed on or inserted into the parenchyma. Large electrode contacts and a limited sampling frequency of classical EEG records have limited data to extracellular signals generated by very large neuronal assemblies (Buzsáki, 2004). They are generated by synchronous synaptic events (local field potential at the site of origin) over large cortical areas (Tao, Baldwin, Hawes-Ebersole, & Ebersole, 2007) and modulated by the orientation of cortical neurons and the extracellular conductance. Signal resolution has been improved for studies on epileptic patients (Bragin et al., 2002) by supplementing conventional macroelectrodes with microelectrodes as flexible bundles, linear microarrays or 2D electrode arrays (Worrell et al., 2012). These electrodes permit records of synaptic fields and multiunit neuronal firing from smaller regions, on the order of 200 μm^2 (Schevon et al., 2008).

Comparing micro- and macroelectrode records reveals mismatches in ictal onset time detected from low-frequency field potentials and from neuronal firing (Schevon et al., 2012). Neuronal firing before seizure onset could reflect both *microseizure* behavior and/or reflect events preceding the transition to seizure. Microseizures reflect an increase in neuronal firing without the coherence needed to generate a detectable field (Stead et al., 2010). They are detected in seizure onset zones and surrounding regions at times close to seizure onset and also between seizures. In noncontiguous local microdomains, these temporally clustered firing events show a degree of stability between events. Their timing before seizure initiation in restricted areas suggests they represent a precursor event at a submillimeter scale, which may need to recruit neurons of surrounding zones before conventional EEG records can detect a field. Microseizure activity appears to spread by direct neuronal propagation rather than volume conduction (Schevon, Goodman, McKhann, & Emerson, 2010).

Microelectrode records provide precise data on neuronal spiking so that changes in single-cell activity can be detected *in situ* in patients. Neuronal spiking during seizures has been shown to be much more variable than the monolithic synchronous field potential records (Truccolo et al., 2011). This heterogeneity was maximal at seizure onset before decreasing and was reproducible between consecutive seizures. 20–30% of recorded units increased or decreased firing before seizure onset, with greater increases in firing for interneuron than pyramidal cell units. However, another study with multielectrode arrays in humans, closer to the seizure initiation sites (Schevon et al., 2012) presented contradictory results. This work distinguished two periods of seizure initiation: a prerecruitment period with a small increase in firing, low coherence with no phase relationship of LFP, and a postrecruitment period with much higher firing rates, phase locked to the LFP, suggesting recruited cells fire homogeneously. The preictal period included heterogeneous spikes. Possibly, a focal group of intensely firing neurons initiates an ictal event against the restraint of a "penumbra" of feedforward inhibition (Weiss et al., 2013). It therefore appears that the increase in neuronal firing, eventually not synchronous may better localize the seizure onset zone that the field potentials which depend on synaptically driven synchronization processes that may be recorded in more distant and widespread areas (Morman & Jefferys, 2013).

3.2. Pyramidal cells

Pyramidal cells certainly fire strongly during epileptic seizures. It is less clear how they contribute at the transition to seizure. Slices of human postoperative tissue preserve sufficient elements of the networks that underlie temporal lobe epilepsies for interictal and ictal-like events to be studied at the cellular level. Spontaneous interictal discharges are generated by an output area of the hippocampus, the subiculum. They are initiated by interneurons which depolarize 20% of subicular pyramidal cells but hyperpolarize the others (Cohen, Navarro, Clemenceau, Baulac, & Miles, 2002). Depolarizing effects of GABA are related to a perturbed chloride regulation, resulting from a downregulation, in some neurons, of the Cl^- extruding cotransporter KCC2 and an upregulation of the Cl^- loading cotransporter NKCC1 (Huberfeld et al., 2007). Pyramidal cells with disturbed Cl^- homeostasis fire during interictal events but do not initiate them. Ictal events are induced by increasing cellular excitability in human hippocampal slices. They emerge after a transition period of several 10s of minutes (Huberfeld et al., 2011). During this transition period, specific preictal discharges emerge progressively. Mature preictal discharges trigger seizures. They involve distinct networks from those that generate interictal events: glutamatergic signals are generated by pyramidal cell firing and most subicular pyramidal cells fire during preictal bursts. Recurring preictal events in the seconds before seizure onset involve intense pyramidal cell firing. Preictal discharges become more complex at the transition to seizure presumably via interactions mediated by recurrent excitation between pyramidal cells. Slice work in tissue from rodents suggests that preictal discharges at the transition to seizure exhibit a progressive shift in initiation site from CA3a to CA3c, an increase in propagation velocity (Dzhala & Staley, 2003) and a recruitment of increasing numbers of pyramidal cells (Jensen & Yaari, 1997).

Ictal events have been induced in hippocampal and cortical slices by electrical stimulation (Fujiwara-Tsukamoto, Isomura, Imanishi, Fukai, & Takada, 2007), by perfusing the tissue with 0-Mg^{2+} media (Derchansky et al., 2008), by the potassium channel blocker 4-aminopyridine (Ziburkus, 2006), or by partially blocking $GABA_A$ receptors (Gnatkovsky, Librizzi, Trombin, & de Curtis, 2008). At seizure onset, interneurons fire at high frequencies, while pyramidal cell firing is often sparse or delayed although typically occurring before seizure onset (Fujiwara-Tsukamoto et al., 2007; Gnatkovsky et al., 2008; Ziburkus, 2006) and so may contribute to the initiation of ictal events.

3.3. Glutamatergic neurotransmission

Glutamatergic signaling is crucial to ictogenesis, since blockade of AMPA receptors typically suppresses ictal events. In the human epileptic subiculum, NMDA receptor antagonists suppress seizure initiation by preictal discharges, although preictal discharges are not blocked. Equally AMPA receptor antagonists, which do suppress preictal discharges block seizure initiation (Huberfeld et al., 2011). The progressive emergence of preictal discharges is suppressed by NMDA receptor antagonists as in neocortical tissue (D'antuono et al., 2004) and a plasticity of excitatory synapses seems likely to be involved in the protracted preictal transition to seizure. Kainate receptors (Lévesque & Avoli, 2013) and metabotropic receptors (Wong, Bianchi, Taylor, & Merlin, 1999) for glutamate have also been implicated in ictal events.

Bilateral cerebral microdialysis has demonstrated a specific increase in focal glutamate release during the ictogenic period in patients with mesial temporal lobe epilepsy (During & Spencer, 1993). Ambient extracellular glutamate levels were increased at the seizure focus 1.5 min before seizure onset and remained elevated for 16.5 min.

A perturbed glutamate uptake may contribute to ictogenesis. The glutamine–glutamate cycle was shown to be perturbed in the hippocampus of patients with mesial temporal lobe epilepsy compared to autopsy controls and patients without temporal lobe epilepsy (Eid et al., 2004). Glutamine synthetase immunoreactivity was reduced in areas of peak neuronal loss and astroglial proliferation and activity of the enzyme was reduced by about 40%. Defects in glutamine synthetase increase glutamate levels in astrocytes and potentially release of this glutamate pool into extracellular space could trigger seizures (Perez et al., 2012). Reduced activity of this enzyme may also limit local GABA release (Ortinski et al., 2010).

3.4. Interneurons

Interneurons classically provide an inhibitory control of local neuronal ensembles, via efficient, divergent GABAergic synapses. They should therefore exert antiepileptic effects. Reinforcing this idea, antagonists at $GABA_A$ receptors have ictogenic actions in epileptic and nonepileptic patients and animals *in vivo*. Blocking these receptors in slices generates interictal (Miles & Wong, 1983) but usually not ictal discharges. Thus, a reduced efficacy of inhibitory signaling does not favor all forms of epileptiform activity. Recent work suggests "inhibitory" circuits have a more interesting role in ictogenesis.

Some interneurons die during epileptogenesis. This observation was used as evidence for a reduced efficacy of inhibitory circuits. More recent data suggest though that axonal contacts of remaining interneurons are rearranged to innervate new targets. For example, in rats with focal epilepsy after status epilepticus, dendritic inhibition is reduced but somatic inhibition is preserved (Cossart et al., 2001). In human focal cortical dysplasia, the number of interneurons may be reduced but axonal contacts made by remaining interneurons are reorganized (Alonso-Nanclares, 2004). Similarly, in the CA1 region of human sclerotic hippocampus, perisomatic inhibition is preserved (Wittner et al., 2005), dendritic inhibition mediated by calretinin interneurons is reduced (Tóth et al., 2010), while chandelier cells contact new targets (Arellano, Muñoz, Ballesteros-Yáñez, Sola, & DeFelipe, 2004).

Records from surviving interneurons in epileptic tissue suggest they are involved in epileptic synchronies. Thus, interneurons in tissue from patients with focal epilepsies discharge before the onset of interictal events and trigger firing in some pyramidal cells via depolarizing effects of GABA (Cohen et al., 2002; Huberfeld et al., 2007). Interneurons also discharge at high frequency at the transition to seizures. Similar behaviors have been recorded from hippocampal slices (Fujiwara-Tsukamoto et al., 2007; Fujiwara-Tsukamoto, Isomura, Kaneda, & Takada, 2004; Lillis, Kramer, Mertz, Staley, & White, 2012; Ziburkus, 2006), immature hippocampal tissue (Derchansky et al., 2008) and in the whole-brain preparation (Gnatkovsky et al., 2008), during ictal activities induced by 0-Mg^{2+} (Derchansky et al., 2008), by the potassium channel blocker 4-aminopyridine (Lillis et al., 2012; Ziburkus, 2006), by partially blocking $GABA_A$ receptors (Gnatkovsky et al., 2008), and by electrical stimulation (Fujiwara-Tsukamoto et al., 2004, 2007, 2010). In all cases, interneurons discharged at seizure onset at higher frequencies than pyramidal cells. Typically, pyramidal cell firing was reduced, while interneuron firing increased during initial stages of ictal events. Evidence that interneurons contribute to ictogenesis was obtained using the opioid mu agonist DAGO. This agent hyperpolarizes interneurons and blocks seizure-like events (Avoli,

Louvel, Kurcewicz, Pumain, & Barbarosie, 1996). Finally, synchronization of interneuronal networks may be facilitated by gap-junction coupling (Skinner, Zhang, Velazquez, & Carlen, 1999).

What is the effect of intense interneuron firing at the transition to seizure? Does liberated GABA inhibit pyramidal cells to counteract local excitatory actions? Or does it excite pyramidal cells as during interictal discharges? How is GABAergic signaling affected by Cl^- dysregulation and accumulation in pyramidal cells? We will consider these issues in a section of the review that examines GABAergic transmission.

Data on ictal discharges induced in healthy mouse cortex suggest inhibitory actions of GABA are maintained in some circumstances (Trevelyan, Sussillo, Watson, & Yuste, 2006; Trevelyan et al., 2007). Perfusion with 0-Mg^{2+} media induced spreading ictal events preceded by preictal-like discharges. Preictal events decreased with repetition corresponded to an inhibitory restraint overcome eventually by an excitatory drive. Grafts of GABAergic progenitors in epileptic mice reduce seizure frequency (Hunt, Girskis, Rubenstein, Alvarez-Buylla, & Baraban, 2013). Medial ganglionic eminence progenitor cells grafted into amygdala or hippocampus migrate and differentiated into functional interneurons. Grafts reduced seizure frequency by 92% in chronically epileptic animals after pilocarpine treatment.

3.5. GABAergic neurotransmission

The role of $GABA_A$ signaling in ictogenesis may seem paradoxical. Antagonists applied *in vivo* usually trigger seizures in normal and epileptic humans and animals. In slices, the same antagonists typically suppress induced seizures.

Blocking $GABA_A$ receptors *in vivo* is clearly epileptogenic and as such has been used to sustain the concept of an epileptic threshold. Seizures are triggered by lower doses of antagonists in patients with epilepsy. In animal tissues, blocking $GABA_A$ receptors induces interictal-like epileptic activities (Miles & Wong, 1983). However, the pharmacology of ictal-like discharges induced in slices of human epileptic tissue is much more complex than a reduction of GABAergic signaling. In temporal neocortex slices from patients, ictal discharges are elicited by the K^+ channel antagonist 4-aminopyridine (D'Antuono et al., 2004). Blocking $GABA_A$ receptors controls ictal-like events but does not affect transient preictal discharges which precede them (Huberfeld et al., 2011). The classical antiepileptic drug phenobarbital, which reinforces GABAergic signaling, increases both the duration and amplitude of ictal discharges. Strong evidence now indicates that in epileptic tissues depolarizing effects of GABA may contribute both to interictal and ictal activities.

Depolarizing actions of GABA have been shown to be involved in ictal events generated in multiple human epileptic tissues: temporal lobe epilepsy, whether or not associated with hippocampal sclerosis (Cohen et al., 2002; Kohling et al., 1998; Wozny et al., 2003), cortical dysplasia (Cepeda et al., 2007), and glioma-related epilepsy (Conti et al., 2011).

The depolarizing or hyperpolarizing effects of GABA depend on Cl^- flux through $GABA_A$ receptors. At low-intracellular Cl^- concentrations, GABA induces an inward Cl^- flux and produces a hyperpolarization. As intracellular Cl^- concentration increases, the Cl^- reversal potential depolarizes to levels more positive than the neuronal resting potential. Cl^- flows out of the cell inducing a depolarization that may excite the cell to fire if the postsynaptic potential at the axon initial segment reaches spike threshold. Levels of Cl^- inside neurons are controlled in part by cation–chloride–cotransporters which use the chemical gradient established by the Na^+–K^+-ATPase to load or extrude Cl^-. The Na–K–2Cl cotransporter NKCC1 and the K–Cl cotransporter KCC2 are especially important in neuronal Cl homeostasis. NKCC1 loads cells with Cl^-, favoring depolarizing Cl^- fluxes on $GABA_A$ receptor activation, while KCC2 extrudes Cl^- from neurons, reducing Cl^- concentration and contributing to maintain a hyperpolarizing Cl^- flux when $GABA_A$ receptors are activated. KCC2 expression is delayed during development and predominates in mature cells, while NKCC1 is mostly expressed early in development. Cl^- cotransporters regulate basal intracellular Cl^- levels. However, intense inhibitory cell firing, results in sustained opening of $GABA_A$ receptor channels so inducing a significant Cl^- load. If the load cannot be rapidly dissipated, Cl^- accumulates in neurons, the electrochemical gradient collapses, and ionic flux through $GABA_A$ receptors depends largely on HCO_3^- movements which depolarize the neuron (Alger & Nicoll, 1982; Staley, Soldo, & Proctor, 1995). Further, since KCC2 operates close to its thermodynamic equilibrium, even small increases in extracellular K^+ will reverse transport, from Cl^- efflux to influx. At the transition to seizure, however, large increases in internal Cl^- resulting from interneuron firing shift the equilibrium. KCC2 induces large transient increases in external K^+, which can affect local excitability. Cytoplasmic impermeant anions and polyanionic extracellular matrix glycol-proteins have recently been shown to exert major effects on local Cl^- levels (Glykys et al., 2014). We note too that slice preparation may affect internal Cl^- values (Dzhala et al., 2010) increasing the depolarizing effects of GABA.

Cl^- homeostasis is perturbed due to a downregulation of KCC2 and/or an upregulation of NKCC1 in multiple epileptic tissues (Kahle et al., 2008; Noebels et al., 2012).

A consensus seems to be established that depolarizing effects of GABA are crucial at the onset of ictal events in slice work. First, interneurons have been shown to fire at high frequency in multiple models of seizure onset (Fujiwara-Tsukamoto et al., 2007, 2004; Gnatkovsky et al., 2008; Lillis et al., 2012; Zhang et al., 2011). Second, ictal events are initiated by a long-lasting depolarization (Kaila, Lamsa, Smirnov, Taira, & Voipio, 1997; Lamsa & Kaila, 1997; Perreault & Avoli, 1989; Velazquez & Carlen, 1999). This potential reverses at depolarized potentials indicating suggesting it may result from Cl^- and/or HCO_3^- flux through $GABA_A$ receptors (Benini, D'antuono, Pralong, & Avoli, 2003; Gnatkovsky et al., 2008; Lopantsev & Avoli, 1998). Third, interneuron firing during preictal events just before seizure onset causes massive GABA release (Zhang et al., 2012). Fourth, blocking $GABA_A$ receptors may block seizure onset (D'Antuono et al., 2004; Fujiwara-Tsukamoto et al., 2007; Higashima, Kinoshita, Yamaguchi, & Koshino, 1996; Huberfeld et al., 2011; Köhling, Vreugdenhil, Bracci, & Jefferys, 2000; Lillis et al., 2012; Lopantsev & Avoli, 1998), although not in all cases (Derchansky et al., 2008; Zhang et al., 2012). Fifth, high-frequency GABAergic inputs increase intracellular Cl^- in pyramidal cells during preictal spiking (Lillis et al., 2012). This effect is blocked by $GABA_A$ receptors antagonists and associated with a large increase in external K^+ (Avoli, Louvel, et al., 1996).

Some data contradict this scheme. The transition to seizure induced by perfusing 0-Mg^{2+} in the intact hippocampus is characterized by recurrent preictal discharges (Derchansky et al., 2008). These events depend in part on depolarizing synaptic events, independent of GABAergic signaling but rather mediated by potentiating glutamatergic inputs. Similar potentiating glutamatergic inputs act to overwhelm the inhibitory restraint during the preictal phase in neocortex (Trevelyan et al., 2006).

Inhibiting the NKCC1 cotransporter with low doses of the diuretic bumetanide suppresses some seizure-like activities. The efficacy varies according to the stimulus that induces ictal events (Kilb, Sinning, & Luhmann, 2007), possibly due to the involvement of astrocyte cotransporters and especially in the absence of a sclerotic injury (Margineanu & Klitgaard, 2006). Bumetanide is effective in chronically epileptic hippocampal tissue from rats (Bragin, Sanderson, et al., 2009) and from human temporal lobe (G. Huberfeld, personal observations). Bumatenide also blocks seizure-like events in hippocampal mirror foci in the double in vitro hippocampus preparation, but does not abolish the induction of mirror foci (Nardou, Ben-Ari, & Khalilov, 2009). Bumetanide-suppressed seizures induced in the immature brain combined with an activation of $GABA_A$ receptors (Dzhala, Brumback, & Staley, 2008; Dzhala et al., 2005), were only partially effective in an acute model of seizures induced by blocking $GABA_A$ receptors (Mares & Stehlíková, 2010), and had no effect on flurothyl-induced seizures *in vivo* (Minlebaev & Khazipov, 2011). Bumetanide has more consistent effects on seizures induced in mature animals, including kainate-induced status epilepticus (Schwartzkroin, Baraban, & Hochman, 1998), and sound-triggered seizures in postischemic audiogenic seizure-prone rats (Reid, Welin, Wiberg, Terenghi, & Novikov, 2010). Similar results have been obtained in limited trials on epileptic patients (Eftekhari et al., 2013; Kahle, Barnett, Sassower, & Staley, 2009).

Finally, GABA actions at G-protein-coupled $GABA_B$ receptors may also contribute to ictogenesis. Intrathecal injections of $GABA_B$ receptors agonist baclofen in patients treated for spasticity sometimes induce seizures (Kofler, Kronenberg, Rifici, Saltuari, & Bauer, 1994). $GABA_B$ agonists can switch epileptic events induced by 4-AP in rat hippocampal slices from interictal- to ictal-like events (Avoli, Benini, de Guzman, & Omar, 2004; Watts & Jefferys, 1993). Ictogenic effects of $GABA_B$ receptors demonstrated recently in mice with temporal lobe epilepsy (Dugladze et al., 2013) may depend on presynaptic $GABA_B$ effects at synapses made by inhibitory basket cells with pyramidal cells.

4. THE ENVIRONMENT OF NEURONS: IONS, ASTROCYTES, AND THE EXTRACELLULAR SPACE

4.1. Potassium

Extracellular K^+ increases are associated with interictal, preictal, and ictal epileptic events (Avoli, Barbarosie, et al., 1996; Avoli, Louvel, Pumain, & Köhling, 2005; Avoli, Nagao, Kohling, Lücke, & Mattia,, 1996; Benninger, Kadis, & Prince, 1980; D'antuono et al., 2004; Heinemann, Lux, & Gutnick, 1977; Lux, 1974; Lux, Heinemann, & Dietzel, 1986; Morris, Najm, & Kahane, 2008; Timofeev, Bazhenov, Avramescu, & Nita, 2010; Traynelis & Dingledine, 1989).

This K^+ increase may originate in several ways. Firstly, K^+ is released by neuronal firing. Both pyramidal cells and interneurons discharge at high frequency at seizure onset. Even so, this may be a minor source of external K^+ since it persists during focal blockade of voltage-gated Na^+ in a seizure (Barolet & Morris, 1991). Secondly, high-frequency interneuron firing and the resulting repetitive activation of GABAergic synapses may indirectly enhance external K^+. Such K^+ increases are induced by GABA application, abolished by $GABA_A$ receptor antagonists (Barolet & Morris, 1991) and can be reproduced by stimulating inhibitory synapses in the absence of glutamatergic transmission (Kaila et al., 1997). This second source of K^+ extrusion from pyramidal cells occurs via the K–Cl cotransporter KCC2 after the increase in intracellular Cl^- due to inhibitory synaptic activity (Viitanen, Ruusuvuori, Kaila, & Voipio, 2010). It requires functional cellular carbonic anhydrase (Ruusuvuori et al., 2004) and is associated by an outward flux of HCO_3^- (Viitanen et al., 2010). Thirdly, external K^+ increases as a result of multiple processes in astrocytes. K^+ increases after astrocyte activation of GABAergic receptors (Fraser et al., 1995; MacVicar, Tse, Crichton, & Kettenmann, 1989). Glutamate uptake by the astrocyte transporter EAAT2 releases K^+ (Bjørnsen et al., 2007). Kir 4.1 channels as well as the Na/H ATPase in astrocytes normally buffer external K^+, but their effects are downregulated in epileptic tissues (Grisar, Guillaume, & Delgado-Escueta, 1992; Guillaume, Grisar, Delgado-Escueta, Laschet, & Bureau-Heeren, 1991). On the other hand, increased K^+ uptake by the cation chloride transporter NKCC1 in astrocytes may disrupt neurotransmitter metabolism and extracellular distribution. For instance, increased K^+ levels induce astrocyte swelling, increased intracellular Cl^-, and release excitatory amino acids. This effect is blocked by NKCC1 inhibitor bumetanide (Su, Kintner, & Sun, 2001).

Increases in external K^+ favor seizure initiation in several ways. At levels above 8.5 mM, external K^+ induces interictal bursts, and seizure-like events in hippocampal rat slices (Traynelis & Dingledine, 1989), although human epileptic tissue is resistant (Huberfeld et al., 2011). High levels of external K^+ can support a slow spread of epileptic activities when glutamatergic signaling is suppressed (Avoli et al., 2002; Louvel et al., 2001; Uva, Avoli, & de Curtis, 2009). Increasing extracellular K^+ may load neurons with Cl^- via in inverse flux at the KCC2 cotransporter (DeFazio, Keros, Quick, & Habliz, 2000; Kakazu, Uchida, Nakagawa, Akaike, & Nabekura, 2000; Thompson & Gähwiler, 1989), so enhancing the depolarizing actions of GABA. Increase of extracellular K^+ from 3.5 to 10 mM increases intracellular Cl^- by 3 mM, from resting levels close to 10 mM (DeFazio et al., 2000). Finally, increased external K^+ depolarizes neurons, increases firing, and favors burst firing (Jensen, Azouz, & Yaari, 1994), and may generate ectopic axonal action potentials (Pinault, 1992) all of which are likely to contribute to transition to the seizure (Jensen & Yaari, 1997).

4.2. Calcium

Levels of extracellular Ca^{2+} are reduced with the increase in external K^+ at seizure onset. Ca^{2+} may fall to 0.7 mM (Heinemann, Konnerth, Pumain, & Wadman, 1986; Lux & Heinemann, 1978). This reduction may result from Ca^{2+} entry at presynaptic sites, postsynaptic Ca^{2+} entry associated with burst firing, and also Ca^{2+} entry during oscillations in astrocytes. Neuronal excitability is increased at lowered external Ca^{2+} levels (Haas & Jefferys, 1984).

4.3. Extracellular space

The size and properties of extracellular space contribute to the transition to seizure. Manipulating the effective extracellular space by perfusion with hypo- or hyperosmolar solutions *in vitro* or *in vivo* modifies the transition to seizure and seizure incidence (Fox, Bikson, & Jefferys, 2007; Hochman, Baraban, Owens, & Schwartzkroin, 1995). Extracellular space is reduced by the cellular swelling that occurs during epileptiform activities, especially seizures, with additional proictal effects. Cellular swelling in epilepsy has been demonstrated using tetraethylammonium (Buchheim et al., 1999), by measuring tissue resistance (Fox, Bikson, & Jefferys, 2004) and with optical signals. Reducing extracellular space size enhances neuronal excitability by increasing effective extracellular levels of ions particularly K^+ ions with effects already described, maybe as well as affecting neuronal Cl^- and GABAergic signaling (Glykys et al., 2014). Effects on ambient neurotransmitter levels and diffusion tend to increase neuronal excitability and enhanced electric field interactions also tend to synchronize neural firing (Jiruska et al., 2013a, 2013b).

Electric fields generated by firing of single neurons or larger neuronal ensembles are transmitted via the conductive extracellular space to exert ephaptic effects which recruit other neurons to fire (Anastassiou, Perin, Markram, & Koch, 2011; Buzsaki, Anastassiou, & Koch, 2012). These effects occur on a millisecond timescale representing a fast neuronal synchronizing mechanism of speed similar to that for gap-junctional coupling (Jefferys, 1995). Ephaptic field transmission effects are weak in normal low excitability conditions. In contrast, high-amplitude field potentials generated by large neuronal populations can generate electric fields with gradients large enough to modify neuronal excitability and spike timing (Buzsaki et al., 2012). At the transition to seizure, these mechanisms can rapidly synchronize large neuronal populations. Fields generated by large epileptiform discharges facilitate the recruitment of other neurons, enhance neuronal synchrony, and encourage the spread of epileptic activity (Jefferys, 1995; Jiruska et al., 2013a, 2013b; Taylor & Dudek, 1984). Changes in the effective extracellular space and tissue resistance thus modulate electric field interactions. The progressive increase in tissue resistance during seizures seems likely to enhance neuronal coupling via electric field effects (Fox et al., 2004).

Changes in the size and composition of extracellular space resulting from the intense neuronal firing associated with epileptiform events are usually short lived. In contrast, chronic changes, corresponding to a maintained structural reorganization, may result from changes in extracellular matrix proteins and abnormalities of glial cells particularly astrocytes.

4.4. Astrocytes

Astrocytes have been involved in both epileptogenesis and ictogenesis. Their contribution was highlighted by showing interictal epileptic activity was maintained in the absence of neuronal action potentials and synaptic transmission (Tian et al., 2005) However, ictal events were not maintained, suggesting a neuronal–glial interplay is crucial for their generation. Astrocytes may contribute via the Ca^{2+} waves that spread through coupled networks of these cells. Ictal-like events triggered by NMDA in the presence of 4-AP were strictly associated with Ca^{2+} waves, while interictal discharges could occur independently of astrocyte activity. Moreover, ictal events could be abolished by suppressing Ca^{2+} signals in astrocytes (Gómez-Gonzalo et al., 2010).

Astrocytes exert a crucial control on extracellular space notably via the water channel aquaporin-4, and so may contribute to ictogenesis. A reduced extracellular volume favors seizures (Hochman, 2012) and downregulation of aquaporin-4 in mesial temporal lobe epilepsy (Binder et al., 2006) may effectively decrease extracellular space. Potassium channels Kir 4.1, associated with aquaporin-4, are also downregulated in astrocytes from epileptic foci (Hinterkeuser et al., 2000) and so presumably enhance levels of external K^+ with proictal effects on neuronal excitability (Djukic, Casper, Philpot, Chin, & McCarthy, 2007).

Astrocytes liberate glutamate and also regulate excitatory signaling by taking up glutamate, against K^+, via the EAAT1 and EAAT2 transporters. They liberate adenosine, via membrane panexines, which tends to reduce glutamate release via presynaptic A1 receptors (Boison, 2012). The enzyme adenosine kinase is upregulated in astrogliosis and acts to reduce adenosine release from astrocytes thus increasing glutamatergic excitation (Li, Quan Lan, Fredholm, Simon, & Boison, 2007). We considered previously a reduced efficacy of glutamine synthetase, increasing glutamate (Ortinski et al., 2010).

Further contributions of astrocytes to ictogenesis by their structural role in the blood–brain barrier and more generally in inflammatory signaling together with microglia (Vezzani, French, Bartfai, & Baram, 2011) must also be considered.

5. CONCLUSION

A current vision of ictogenesis must include multiple actors, pyramidal cells, interneurons, and astrocytes, interacting via synaptic and nonsynaptic signals, regulated by several ionic species, Cl^-, K^+, Ca^{2+} both inside cells and outside in the extracellular space. It seems probable that these diverse pathways can contribute in distinct ways in diverse epileptic syndromes, with diverse lesions or genetic backgrounds influence by differential effects of endogenous and exogenous factors to produce different seizure onset patterns.

We have attempted to propose a biological basis for ictogenesis in partial epilepsies. Epileptic networks may or may not be associated with an injury, may vary, and presumably involve pyramidal cells and interneurons as well as astrocytes. Interictal events may be triggered by interneurons or pyramidal cells or both. The type of interneuron remains to be defined, as does a possible contribution of interneurons networks coupled by gap junctions, and the contribution of basal or dynamic defects in postsynaptic Cl-homeostasis.

Transition to seizure may occur slowly after a prolonged transition period, associated with a progressive emergence of recruiting discharges or quickly with brutal onset. Recurring preictal discharges, emerging from an increased excitability state, become more complex and longer, recruiting more neurones with successive discharges. High-frequency interneuron firing will progressively increase Cl^- levels in postsynaptic cells leading to a reduction in the amplitude of hyperpolarizing GABAergic signals. Eventually, GABAergic responses may become depolarizing and even excite postsynaptic cells via HCO_3^- currents through $GABA_A$ receptors. Neuronal volume and osmolarity changes also would tend to enhance intracellular Cl^- concentration and favor depolarizing effects of GABA. Cl^-

accumulation should increase Cl^- efflux through the cotransporter KCC2 (although basal levels may be low in some neurons). This efflux also extrudes K^+ into the extracellular space. Extracellular K^+ will also increase as a consequence of interneuron and pyramidal cell firing and glutamate uptake by astrocytes via the transporter EAAT2. The probable downregulation of Kir 4.1 in astrocytes at the epileptic focus should reduce buffering of extracellular K^+. As K^+ levels rise, KCC2 transport should reverse to load rather than extrude Cl^- from neurons favoring depolarizing effects of GABA. The initial increase in intracellular Cl^-, the increase in extracellular K^+, and the secondary Cl^- loading of neurons would tend to overlap in a given brain area. Increased release and poor clearance of glutamate as well as a reduced inhibition of release due to decreased extracellular adenosine should all tend to reinforce the state of heightened neuronal excitability and synchrony.

A disorganized, chaotic neuronal firing will result from this combination of an impaired inhibition and depolarizing actions of GABA together with an enhanced glutamatergic excitation and K-dependent increase in neuronal excitability. Possibly, this combination of cellular events is associated with the high-frequency, low-voltage defections of EEG records. Neurons of the focus will synchronize out of this disorganized excitable state due to excitatory and/or reappearing inhibitory effects of GABA. The emerging synchronous bursts would then correspond to the oscillatory EEG bursts that constitute the core of the seizure. How the seizure stops is another story.

ACKNOWLEDGMENTS

We thank Richard Miles for valuable comments on this manuscript. Supported by grants from the Ministry of Health of the Czech Republic (IGA NT/14489-3), Neuron Fund for Support of Science (Czech Republic, 2012/10) and the Czech Science Foundation (Czech Republic, P303/14-02634S), and UPMC Convergence (France).

REFERENCES

Alger, B. E., & Nicoll, R. A. (1982). Pharmacological evidence for two kinds of GABA receptor on rat hippocampal pyramidal cells studied in vitro. *Journal of Physiology, 328*, 125–141.

Alonso-Nanclares, L. (2004). Microanatomy of the dysplastic neocortex from epileptic patients. *Brain, 128*(1), 158–173.

Anastassiou, C. A., Perin, R., Markram, H., & Koch, C. (2011). Ephaptic coupling of cortical neurons. *Nature Neuroscience, 14*, 217–223.

Arellano, J. I., Muñoz, A., Ballesteros-Yáñez, I., Sola, R. G., & DeFelipe, J. (2004). Histopathology and reorganization of chandelier cells in the human epileptic sclerotic hippocampus. *Brain, 127*(Pt 1), 45–64.

Avoli, M., Barbarosie, M., Lücke, A., Nagao, T., Lopantsev, V., & Kohling, R. (1996). Synchronous GABA-mediated potentials and epileptiform discharges in the rat limbic system in vitro. *Journal of Neuroscience: The Official Journal of the Society for Neuroscience, 16*(12), 3912–3924.

Avoli, M., Benini, R., de Guzman, P., & Omar, A. (2004). GABAb receptor activation and limbic network ictogenesis. *Neuropharmacology, 46*(1), 43–51.

Avoli, M., D'Antuono, M., Louvel, J., Köhling, R., Biagini, G., Pumain, R., et al. (2002). Network and pharmacological mechanisms leading to epileptiform synchronization in the limbic system in vitro. *Progress in Neurobiology, 68*(3), 167–207.

Avoli, M., de Curtis, M., & Köhling, R. (2013). Does interictal synchronization influence ictogenesis? *Neuropharmacology, 69*, 37–44.

Avoli, M., Louvel, J., Kurcewicz, I., Pumain, R., & Barbarosie, M. (1996). Extracellular free potassium and calcium during synchronous activity induced by 4-aminopyridine in the juvenile rat hippocampus. *Journal of Physiology, 493*(Pt 3), 707–717.

Avoli, M., Louvel, J., Pumain, R., & Köhling, R. (2005). Cellular and molecular mechanisms of epilepsy in the human brain. *Progress in Neurobiology, 77*(3), 166–200.

Avoli, M., Nagao, T., Kohling, R., Lücke, A., & Mattia, D. (1996). Synchronization of rat hippocampal neurons in the absence of excitatory amino acid-mediated transmission. *Brain Research, 735*(2), 188–196.

Barbarosie, M., & Avoli, M. (1997). CA3-driven hippocampal-entorhinal loop controls rather than sustains in vitro limbic seizures. *Journal of Neuroscience: The Official Journal of the Society for Neuroscience, 17*(23), 9308–9314.

Barolet, A. W., & Morris, M. E. (1991). Changes in extracellular K^+ evoked by GABA, THIP and baclofen in the guinea-pig hippocampal slice. *Experimental Brain Research, 84*(3), 591–598.

Baumgartner, C., Series, W., Leutmezer, F., Pataraia, E., Aull, S., Czech, T., et al. (1998). Preictal SPECT in temporal lobe epilepsy: Regional cerebral blood flow is increased prior to electroencephalography-seizure onset. *Journal of Nuclear Medicine: Society of Nuclear Medicine, 39*(6), 978–982.

Benini, R., D'antuono, M., Pralong, E., & Avoli, M. (2003). Involvement of amygdala networks in epileptiform synchronization in vitro. *Neuroscience, 120*(1), 75–84.

Benninger, C., Kadis, J., & Prince, D. A. (1980). Extracellular calcium and potassium changes in hippocampal slices. *Brain Research, 187*(1), 165–182.

Bernard, C., Anderson, A., Becker, A., Poolos, N. P., Beck, H., & Johnston, D. (2004). Acquired dendritic channelopathy in temporal lobe epilepsy. *Science, 305*(5683), 532–535, American Association for the Advancement of Science.

Bikson, M., Fox, J. E., & Jefferys, J. G. R. (2003). Neuronal aggregate formation underlies spatiotemporal dynamics of nonsynaptic seizure initiation. *Journal of Neurophysiology, 89*(4), 2330–2333.

Binder, D. K., Yao, X., Zador, Z., Sick, T. J., Verkman, A. S., & Manley, G. T. (2006). Increased seizure duration and slowed potassium kinetics in mice lacking aquaporin-4 water channels. *Glia, 53*(6), 631–636.

Bjørnsen, L. P., Eid, T., Holmseth, S., Danbolt, N. C., Spencer, D. D., & de Lanerolle, N. C. (2007). Changes in glial glutamate transporters in human epileptogenic hippocampus: Inadequate explanation for high extracellular glutamate during seizures. *Neurobiology of Disease, 25*(2), 319–330.

Boison, D. (2012). Adenosine dysfunction in epilepsy. *Glia, 60*(8), 1234–1243.

Bower, M. R., Stead, M., Meyer, F. B., Marsh, W. R., & Worrell, G. A. (2012). Spatiotemporal neuronal correlates of seizure generation in focal epilepsy. *Epilepsia, 53*(5), 807–816.

Bragin, A., Azizyan, A., Almajano, J., & Engel, J. (2009). The cause of the imbalance in the neuronal network leading to seizure activity can be predicted by the electrographic pattern of the seizure onset. *Journal of Neuroscience, 29*(11), 3660–3671.

Bragin, D. E., Sanderson, J. L., Peterson, S., Connor, J. A., & Müller, W. S. (2009). Development of epileptiform excitability in the deep entorhinal cortex after status epilepticus. *European Journal of Neuroscience*, *30*(4), 611–624.

Bragin, A., Wilson, C. L., & Engel, J. (2000). Chronic epileptogenesis requires development of a network of pathologically interconnected neuron clusters: A hypothesis. *Epilepsia*, *41*(Suppl. 6), S144–S152.

Bragin, A., Wilson, C. L., Staba, R. J., Reddick, M., Fried, I., & Engel, J. (2002). Interictal high-frequency oscillations (80–500 Hz) in the human epileptic brain: Entorhinal cortex. *Annals of Neurology*, *52*(4), 407–415.

Buchheim, K., Schuchmann, S., Siegmund, H., Gabriel, H. J., Heinemann, U., & Meierkord, H. (1999). Intrinsic optical signal measurements reveal characteristic features during different forms of spontaneous neuronal hyperactivity associated with ECS shrinkage in vitro. *European Journal of Neuroscience*, *11*, 1877–1882.

Buzsáki, G. (2004). Large-scale recording of neuronal ensembles. *Nature Neuroscience*, 7(5), 446–451.

Buzsaki, G., Anastassiou, C. A., & Koch, C. (2012). The origin of extracellular fields and currents—EEG, ECoG, LFP and spikes. *Nature Reviews Neuroscience*, *13*, 407–420.

Casillas Espinosa, P. M., Powell, K. L., & O'Brien, T. J. (2012). Regulators of synaptic transmission: Roles in the pathogenesis and treatment of epilepsy. *Epilepsia*, *53*(s9), 41–58.

Cepeda, C., André, V. M., Wu, N., Yamazaki, I., Uzgil, B., Vinters, H. V., et al. (2007). Immature neurons and GABA networks may contribute to epileptogenesis in pediatric cortical dysplasia. *Epilepsia*, *48*(Suppl. 5), 79–85.

Cohen, I., Navarro, V., Clemenceau, S., Baulac, M., & Miles, R. (2002). On the origin of interictal activity in human temporal lobe epilepsy in vitro. *Science*, *298*(5597), 1418–1421.

Conti, L., Palma, E., Roseti, C., Lauro, C., Cipriani, R., de Groot, M., et al. (2011). Anomalous levels of Cl- transporters cause a decrease of GABAergic inhibition in human peritumoral epileptic cortex. *Epilepsia*, *52*(9), 1635–1644.

Cook, M. J., O'Brien, T. J., Berkovic, S. F., Murphy, M., Morokoff, A., Fabinyi, G., et al. (2013). Prediction of seizure likelihood with a long-term, implanted seizure advisory system in patients with drug-resistant epilepsy: A first-in-man study. *Lancet Neurology*, *12*(6), 563–571.

Cossart, R., Dinocourt, C., Hirsch, J. C., Merchan-Perez, A., De Felipe, J., Ben-Ari, Y., et al. (2001). Dendritic but not somatic GABAergic inhibition is decreased in experimental epilepsy. *Nature Neuroscience*, *4*(1), 52–62.

Dakos, V., Scheffer, M., van Nes, E. H., Brovkin, V., Petoukhov, V., & Held, H. (2008). Slowing down as an early warning signal for abrupt climate change. *Proceedings of the National Academy of Sciences of the United States of America*, *105*(38), 14308–14312.

D'Antuono, M., Louvel, J., Kohling, R., Mattia, D., Bernasconi, A., Olivier, A., et al. (2004). GABAA receptor-dependent synchronization leads to ictogenesis in the human dysplastic cortex. *Brain*, *127*(7), 1626–1640.

de Curtis, M., & Avanzini, G. (2001). Interictal spikes in focal epileptogenesis. *Progress in Neurobiology*, *63*(5), 541–567.

DeFazio, R. A., Keros, S., Quick, M. W., & Habliz, J. J. (2000). Potassium-coupled chloride cotransport controls intracellular chloride in rat neocortical pyramidal neurons. *Journal of Neuroscience*, *20*(21), 8069–8076.

Derchansky, M., Jahromi, S. S., Mamani, M., Shin, D. S., Sik, A., & Carlen, P. L. (2008). Transition to seizures in the isolated immature mouse hippocampus: A switch from dominant phasic inhibition to dominant phasic excitation. *Journal of Physiology*, *586*(2), 477–494.

Djukic, B., Casper, K. B., Philpot, B. D., Chin, L. S., & McCarthy, K. D. (2007). Conditional knock-out of Kir41 leads to glial membrane depolarization, inhibition of potassium and glutamate uptake, and enhanced short-term synaptic potentiation. *Journal of Neuroscience*, *27*(42), 11354–11365.

Dugladze, T., Maziashvili, N., Bôrgers, C., Gurgenidze, S., Haussler, U., Winkelmann, A., et al. (2013). GABAb autoreceptor-mediated cell type-specific reduction of inhibition in epileptic mice. *Proceedings of the National Academy of Sciences of the United States of America*, *110*(37), 15073–15078.

During, M. J., & Spencer, D. D. (1993). Extracellular hippocampal glutamate and spontaneous seizure in the conscious human brain. *Lancet*, *341*(8861), 1607–1610.

Dzhala, V. I., Brumback, A. C., & Staley, K. J. (2008). Bumetanide enhances phenobarbital efficacy in a neonatal seizure model. *Annals of Neurology*, *63*(2), 222–235.

Dzhala, V. I., Kuchibhotla, K. V., Glykys, J. C., Kahle, K. T., Swiercz, W. B., Feng, G., et al. (2010). Progressive NKCC1-dependent neuronal chloride accumulation during neonatal seizures. *Journal of Neuroscience*, *30*(35), 11745–11761.

Dzhala, V. I., & Staley, K. J. (2003). Transition from interictal to ictal activity in limbic networks in vitro. *Journal of Neuroscience*, *23*(21), 7873–7880.

Dzhala, V. I., Talos, D. M., Sdrulla, D. A., Brumback, A. C., Mathews, G. C., Benke, T. A., et al. (2005). NKCC1 transporter facilitates seizures in the developing brain. *Nature Medicine*, *11*(11), 1205–1213.

Eftekhari, S., Mehvari Habibabadi, J., Najafi Ziarani, M., Hashemi Fesharaki, S. S., Gharakhani, M., Mostafavi, H., et al. (2013). Bumetanide reduces seizure frequency in patients with temporal lobe epilepsy. *Epilepsia*, *54*(1), e9–e12.

Eid, T., Thomas, M. J., Spencer, D. D., Rundén-Pran, E., Lai, J. C. K., Malthankar, G. V., et al. (2004). Loss of glutamine synthetase in the human epileptogenic hippocampus: Possible mechanism for raised extracellular glutamate in mesial temporal lobe epilepsy. *Lancet*, *363*(9402), 28–37.

Engel, J. J., Van Ness Paul, C., Rasmussen, T. B., & Ojemann, L. M. (1996). *Outcome with respect to epileptic seizures* (2nd ed.). (p. 609). Philadelphia: Lippincott-Raven.

Epsztein, J., Represa, A., Jorquera, I., Ben-Ari, Y., & Crepel, V. (2005). Recurrent mossy fibers establish aberrant kainate receptor-operated synapses on granule cells from epileptic rats. *Journal of Neuroscience*, *25*(36), 8229–8239.

Fisher, R. S., van Emde Boas, W., Blume, W., Elger, C., Genton, P., Lee, P., et al. (2005). Epileptic seizures and epilepsy: Definitions proposed by the international league against epilepsy (ILAE) and the international bureau for epilepsy (IBE). *Epilepsia*, *46*(4), 470–472.

Fox, J. E., Bikson, M., & Jefferys, J. G. (2004). Tissue resistance changes and the profile of synchronized neuronal activity during ictal events in the low-calcium model of epilepsy. *Journal of Neurophysiology*, *92*, 181–188.

Fox, J. E., Bikson, M., & Jefferys, J. G. (2007). The effect of neuronal population size on the development of epileptiform discharges in the low calcium model of epilepsy. *Neuroscience Letters*, *411*, 158–161.

Fraser, D. D., Duffy, S., Angelides, K. J., Perez-Velazquez, J. L., Kettenmann, H., & MacVicar, B. A. (1995). GABAA/benzodiazepine receptors in acutely isolated hippocampal astrocytes. *Journal of Neuroscience*, *15*(4), 2720–2732.

Fuerst, D., Shah, J., Shah, A., & Watson, C. (2003). Hippocampal sclerosis is a progressive disorder: A longitudinal volumetric MRI study. *Annals of Neurology*, *53*(3), 413–416.

Fujiwara-Tsukamoto, Y., Isomura, Y., Imanishi, M., Fukai, T., & Takada, M. (2007). Distinct types of ionic modulation of GABA actions in pyramidal cells and interneurons during electrical induction of hippocampal seizure-like network activity. *European Journal of Neuroscience*, *25*(9), 2713–2725.

Fujiwara-Tsukamoto, Y., Isomura, Y., Imanishi, M., Ninomiya, T., Tsukada, M., Yanagawa, Y., et al. (2010). Prototypic seizure activity driven by mature hippocampal fast-spiking interneurons. *Journal of Neuroscience*, *30*(41), 13679–13689.

Fujiwara-Tsukamoto, Y., Isomura, Y., Kaneda, K., & Takada, M. (2004). Synaptic interactions between pyramidal cells and interneuron subtypes during seizure-like activity in the rat hippocampus. *Journal of Physiology*, *557*(Pt 3), 961–979.

Glykys, J., Dzhala, V., Egawa, K., Balena, T., Saponjian, Y., Kuchibhotla, K. V., et al. (2014). Local impermeant anions establish the neuronal chloride concentration. *Science*, *343*(6171), 670–675.

Gnatkovsky, V., Librizzi, L., Trombin, F., & de Curtis, M. (2008). Fast activity at seizure onset is mediated by inhibitory circuits in the entorhinal cortex in vitro. *Annals of Neurology*, *64*(6), 674–686.

Gómez-Gonzalo, M., Losi, G., Chiavegato, A., Zonta, M., Cammarota, M., Brondi, M., et al. (2010). An excitatory loop with astrocytes contributes to drive neurons to seizure threshold. *PLoS Biology*, *8*(4), e1000352.

Grisar, T., Guillaume, D., & Delgado-Escueta, A. V. (1992). Contribution of Na+, K+-ATPase to focal epilepsy: A brief review. *Epilepsy Research*, *12*(2), 141–149.

Guillaume, D., Grisar, T., Delgado-Escueta, A. V., Laschet, J., & Bureau-Heeren, M. (1991). Phosphorylation of brain (Na+, K+)-ATPase alpha catalytic subunits in normal and epileptic cerebral cortex: II. Partial seizures in human epilepsy. *Journal of Neuroscience Research*, *29*(2), 218–224.

Haas, H. L., & Jefferys, J. G. (1984). Low-calcium field burst discharges of CA1 pyramidal neurones in rat hippocampal slices. *Journal of Physiology*, *354*, 185–201.

Heinemann, U., Konnerth, A., Pumain, R., & Wadman, W. J. (1986). Extracellular calcium and potassium concentration changes in chronic epileptic brain tissue. *Advances in Neurology*, *44*, 641–661.

Heinemann, U., Lux, H. D., & Gutnick, M. J. (1977). Extracellular free calcium and potassium during paroxysmal activity in the cerebral cortex of the cat. *Experimental Brain Research*, *27*(3–4), 237–243.

Higashima, M., Kinoshita, H., Yamaguchi, N., & Koshino, Y. (1996). Activation of GABAergic function necessary for after discharge generation in rat hippocampal slices. *Neuroscience Letters*, *207*(2), 101–104.

Hinterkeuser, S., Schröder, W., Hager, G., Seifert, G., Blümcke, I., Elger, C. E., et al. (2000). Astrocytes in the hippocampus of patients with temporal lobe epilepsy display changes in potassium conductances. *European Journal of Neuroscience*, *12*(6), 2087–2096.

Hochman, D. W. (2012). The extracellular space and epileptic activity in the adult brain: Explaining the antiepileptic effects of furosemide and bumetanide. *Epilepsia*, *53*(Suppl. 1), 18–25.

Hochman, D. W., Baraban, S. C., Owens, J. W., & Schwartzkroin, P. A. (1995). Dissociation of synchronization and excitability in furosemide blockade of epileptiform activity. *Science*, *270*, 99–102.

Huberfeld, G., Le Duigou, C., Le Van Quyen, M., Navarro, V., Baulac, M., & Miles, R. (2013). The paradox of the paroxysm: Can seizure precipitants help explain human ictogenesis? *Neuroscientist*, *19*(5), 523–540.

Huberfeld, G., Menendez De La Prida, L., Pallud, J., Cohen, I., Le Van Quyen, M., Adam, C., et al. (2011). Glutamatergic pre-ictal discharges emerge at the transition to seizure in human epilepsy. *Nature Neuroscience*, *14*(5), 627–634.

Huberfeld, G., Wittner, L., Clemenceau, S., Baulac, M., Kaila, K., Miles, R., et al. (2007). Perturbed chloride homeostasis and GABAergic signaling in human temporal lobe epilepsy. *Journal of Neuroscience*, *27*(37), 9866–9873.

Hunt, R. F., Girskis, K. M., Rubenstein, J. L., Alvarez-Buylla, A., & Baraban, S. C. (2013). GABA progenitors grafted into the adult epileptic brain control seizures and abnormal behavior. *Nature Neuroscience*, *16*(6), 692–697.

Jacobs, J., Zelmann, R., Jirsch, J., Chander, R., Dubeau, C. É. C. F., & Gotman, J. (2009). High frequency oscillations (80–500 Hz) in the preictal period in patients with focal seizures. *Epilepsia*, *50*(7), 1780–1792.

Jefferys, J. G. (1995). Nonsynaptic modulation of neuronal activity in the brain: Electric currents and extracellular ions. *Physiological Reviews*, *75*, 689–723.

Jefferys, J. G., Menendez de la Prida, L., Wendling, F., Bragin, A., Avoli, M., Timofeev, I., et al. (2012). Mechanisms of physiological and epileptic HFO generation. *Progress in Neurobiology*, *98*(3), 250–264.

Jensen, M. S., Azouz, R., & Yaari, Y. (1994). Variant firing patterns in rat hippocampal pyramidal cells modulated by extracellular potassium. *Journal of Neurophysiology*, *71*(3), 831–839.

Jensen, M. S., & Yaari, Y. (1988). The relationship between interictal and ictal paroxysms in an in vitro model of focal hippocampal epilepsy. *Annals of Neurology*, *24*(5), 591–598.

Jensen, M. S., & Yaari, Y. (1997). Role of intrinsic burst firing, potassium accumulation, and electrical coupling in the elevated potassium model of hippocampal epilepsy. *Journal of Neurophysiology*, 77(3), 1224–1233.

Jiruska, P., Csicsvari, J., Powell, A. D., Fox, J. E., Chang, W. C., Vreugdenhil, M., et al. (2010). High-frequency network activity, global increase in neuronal activity, and synchrony expansion precede epileptic seizures in vitro. *Journal of Neuroscience*, *30*(16), 5690–5701.

Jiruska, P., de Curtis, M., Jefferys, J. G. R., Schevon, C. A., Schiff, S. J., & Schindler, K. (2013a). Synchronization and desynchronization in epilepsy: Controversies and hypotheses. *Journal of Physiology*, *591*(4), 787–797.

Jiruska, P., de Curtis, M., Jefferys, J. G., Schevon, C. A., Schiff, S. J., & Schindler, K. (2013b). Synchronization and desynchronization in epilepsy: Controversies and hypotheses. *Journal of Physiology*, *591*, 787–797.

Kahle, K. T., Barnett, S. M., Sassower, K. C., & Staley, K. J. (2009). Decreased seizure activity in a human neonate treated with bumetanide, an inhibitor of the Na(+)-K(+)-2Cl(−) cotransporter NKCC1. *Journal of Child Neurology*, *24*(5), 572–576.

Kahle, K. T., Staley, K. J., Nahed, B. V., Gamba, G., Hebert, S. C., Lifton, R. P., et al. (2008). Roles of the cation-chloride cotransporters in neurological disease. *Nature Clinical Practice. Neurology*, *4*(9), 490–503.

Kaila, K., Lamsa, K., Smirnov, S., Taira, T., & Voipio, J. (1997). Long-lasting GABA-mediated depolarization evoked by high-frequency stimulation in pyramidal neurons of rat hippocampal slice is attributable to a network-driven, bicarbonate-dependent K + transient. *Journal of Neuroscience: The Official Journal of the Society for Neuroscience*, *17*(20), 7662–7672.

Kakazu, Y., Uchida, S., Nakagawa, T., Akaike, N., & Nabekura, J. (2000). Reversibility and cation selectivity of the K(+)-Cl(−) cotransport in rat central neurons. *Journal of Neurophysiology*, *84*(1), 281–288.

Kalitzin, S., Velis, D., Suffczynski, P., Parra, J., & da Silva, F. L. (2005). Electrical brain-stimulation paradigm for estimating the seizure onset site and the time to ictal transition in temporal lobe epilepsy. *Clinical Neurophysiology*, *116*(3), 718–728.

Khosravani, H., Pinnegar, C. R., Mitchell, J. R., Bardakjian, B. L., Federico, P., & Carlen, P. L. (2005). Increased high-frequency oscillations precede in vitro low-Mg seizures. *Epilepsia*, *46*(8), 1188–1197.

Kilb, W., Sinning, A., & Luhmann, H. J. (2007). Model-specific effects of bumetanide on epileptiform activity in the in-vitro intact hippocampus of the newborn mouse. *Neuropharmacology*, *53*(4), 524–533.

Kofler, M., Kronenberg, M. F., Rifici, C., Saltuari, L., & Bauer, G. (1994). Epileptic seizures associated with intrathecal baclofen application. *Neurology*, *44*(1), 25–27.
Kohling, R., Lücke, A., Straub, H., Speckmann, E. J., Tuxhorn, I., Wolf, P., et al. (1998). Spontaneous sharp waves in human neocortical slices excised from epileptic patients. *Brain*, *121*(Pt 6), 1073–1087.
Köhling, R., Vreugdenhil, M., Bracci, E., & Jefferys, J. G. (2000). Ictal epileptiform activity is facilitated by hippocampal $GABA_A$ receptor-mediated oscillations. *Journal of Neuroscience*, *20*, 6820–6829.
Kramer, M. A., Truccolo, W., Eden, U. T., Lepage, K. Q., Hochberg, L. R., Eskandar, E. N., et al. (2012). Human seizures self-terminate across spatial scales via a critical transition. *Proceedings of the National Academy of Sciences of the United States of America*, *109*(51), 21116–21121.
Lamsa, K., & Kaila, K. (1997). Ionic mechanisms of spontaneous GABAergic events in rat hippocampal slices exposed to 4-aminopyridine. *Journal of Neurophysiology*, *78*(5), 2582–2591.
Lehnertz, K., Mormann, F., Osterhage, H., Müller, A., Prusseit, J., Chernihovskyi, A., et al. (2007). State-of-the-art of seizure prediction. *Journal of Clinical Neurophysiology*, *24*(2), 147–153.
Lévesque, M., & Avoli, M. (2013). The kainic acid model of temporal lobe epilepsy. *Neuroscience and Biobehavioral Reviews*, *37*(10 Pt 2), 2887–2899.
Lévesque, M., Bortel, A., Gotman, J., & Avoli, M. (2011). Neurobiology of disease. *Neurobiology of Disease*, *42*(3), 231–241.
Li, T., Quan Lan, J., Fredholm, B. B., Simon, R. P., & Boison, D. (2007). Adenosine dysfunction in astrogliosis: Cause for seizure generation? *Neuron Glia Biology*, *3*(4), 353–366.
Lillis, K. P., Kramer, M. A., Mertz, J., Staley, K. J., & White, J. A. (2012). Pyramidal cells accumulate chloride at seizure onset. *Neurobiology of Disease*, *47*(3), 358–366.
Litt, B., & Lehnertz, K. (2002). Seizure prediction and the preseizure period. *Current Opinion in Neurology*, *15*(2), 173–177.
Lopantsev, V., & Avoli, M. (1998). Participation of GABAA-mediated inhibition in ictal like discharges in the rat entorhinal cortex. *Journal of Neurophysiology*, *79*(1), 352–360.
Lopes da Silva, F. H., Blanes, W., Kalitzin, S. N., Parra, J., Suffczynski, P., & Velis, D. N. (2003). Dynamical diseases of brain systems: Different routes to epileptic seizures. *IEEE Transactions on Biomedical Engineering*, *50*(5), 540–548.
Louvel, J., Papatheodoropoulos, C., Siniscalchi, A., Kurcewicz, I., Pumain, R., Devaux, B., et al. (2001). GABA-mediated synchronization in the human neocortex: Elevations in extracellular potassium and presynaptic mechanisms. *Neuroscience*, *105*(4), 803–813.
Lux, H. D. (1974). The kinetics of extracellular potassium: Relation to epileptogenesis. *Epilepsia*, *15*(3), 375–393.
Lux, H. D., & Heinemann, U. (1978). Ionic changes during experimentally induced seizure activity. *Electroencephalography and Clinical Neurophysiology Supplement*, *34*, 289–297.
Lux, H. D., Heinemann, U., & Dietzel, I. (1986). Ionic changes and alterations in the size of the extracellular space during epileptic activity. *Advances in Neurology*, *44*, 619–639.
MacVicar, B. A., Tse, F. W., Crichton, S. A., & Kettenmann, H. (1989). GABA-activated Cl − channels in astrocytes of hippocampal slices. *Journal of Neuroscience*, *9*(10), 3577–3583.
Mares, P., & Stehlíková, M. (2010). Anticonvulsant doses of ganaxolone do not compromise motor performance in immature rats. *Neuroscience Letters*, *469*(3), 396–399.
Margineanu, D. G., & Klitgaard, H. (2006). Differential effects of cation-chloride co-transport-blocking diuretics in a rat hippocampal slice model of epilepsy. *Epilepsy Research*, *69*(2), 93–99.
Miles, R., & Wong, R. K. (1983). Single neurones can initiate synchronized population discharge in the hippocampus. *Nature*, *306*(5941), 371–373.

Minlebaev, M., & Khazipov, R. (2011). Antiepileptic effects of endogenous beta-hydroxybutyrate in suckling infant rats. *Epilepsy Research*, *95*(1–2), 100–109.

Morman, F., & Jefferys, J. (2013). Neuronal firing in human epileptic cortex: The ins and outs of synchrony during seizures. *Epilepsy Current*, *13*(2), 100–102.

Mormann, F., Andrzejak, R. G., Elger, C. E., & Lehnertz, K. (2007). Seizure prediction: The long and winding road. *Brain*, *130*(Pt 2), 314–333.

Mormann, F., Kreuz, T., Andrzejak, R. G., David, P., Lehnertz, K., & Elger, C. E. (2003). Epileptic seizures are preceded by a decrease in synchronization. *Epilepsy Research*, *53*(3), 173–185.

Morris, H., Najm, I., & Kahane, P. (2008). Epilepsy surgery: Patient selection. In H. O. Lüders (Ed.), *Textbook of epilepsy surgery* (pp. 230–238). London: Informa Healthcare.

Nardou, R., Ben-Ari, Y., & Khalilov, I. (2009). Bumetanide, an NKCC1 antagonist, does not prevent formation of epileptogenic focus but blocks epileptic focus seizures in immature rat hippocampus. *Journal of Neurophysiology*, *101*(6), 2878–2888.

Noebels, J. L., Avoli, M., Rogawski, M. A., Olsen, R. W., Delgado-Escueta, A. V., Miles, R., et al. (2012). *Chloride homeostasis and GABA signaling in temporal lobe epilepsy* (4th ed.). Bethesda, MD: National Center for Biotechnology Information (US).

Ortinski, P. I., Dong, J., Mungenast, A., Yue, C., Takano, H., Watson, D. J., et al. (2010). Selective induction of astrocytic gliosis generates deficits in neuronal inhibition. *Nature Neuroscience*, *13*(5), 584–591.

Pacia, S. V., & Ebersole, J. S. (1999). Intracranial EEG in temporal lobe epilepsy. *Journal of Clinical Neurophysiology*, *16*(5), 399–407.

Perez, E. L., Lauritzen, F., Wang, Y., Lee, T.-S. W., Kang, D., Zaveri, H. P., et al. (2012). Evidence for astrocytes as a potential source of the glutamate excess in temporal lobe epilepsy. *Neurobiology of Disease*, *47*(3), 331–337.

Perreault, P., & Avoli, M. (1989). Effects of low concentrations of 4-aminopyridine on CA1 pyramidal cells of the hippocampus. *Journal of Neurophysiology*, *61*(5), 953–970.

Pinault, D. (1992). Ectopic axonal firing in an epileptic cortical focus is not triggered by thalamocortical volleys during the interictal stage. *Brain Research*, *576*(1), 175–180.

Pitkänen, A., Kharatishvili, I., Karhunen, H., Lukasiuk, K., Immonen, R., Nairismägi, J., et al. (2007). Epileptogenesis in experimental models. *Epilepsia*, *48*(s2), 13–20.

Reddy, D. S. (2013). Role of hormones and neurosteroids in epileptogenesis. *Frontiers in Cellular Neuroscience*, *31*(7), 115.

Reid, A. J., Welin, D., Wiberg, M., Terenghi, G., & Novikov, L. N. (2010). Peripherin and ATF3 genes are differentially regulated in regenerating and non-regenerating primary sensory neurons. *Brain Research*, *1310*, 1–7.

Richardson, M. P., & Jefferys, J. G. R. (2011). Introduction—Epilepsy research UK workshop 2010 on "preictal phenomena" *Epilepsy Research*, *97*(3), 229–230.

Rogawski, M. A., & Löscher, W. (2004). The neurobiology of antiepileptic drugs. *Nature Reviews. Neurosciences*, *5*(7), 553–564.

Rosenow, F., & Lüders, H. (2001). Presurgical evaluation of epilepsy. *Brain*, *124*(Pt 9), 1683–1700.

Rudrauf, D., Douiri, A., Kovach, C., Lachaux, J.-P., Cosmelli, D., Chavez, M., et al. (2006). Frequency flows and the time-frequency dynamics of multivariate phase synchronization in brain signals. *NeuroImage*, *31*(1), 209–227.

Ruusuvuori, E., Li, H., Huttu, K., Palva, J. M., Smirnov, S., Rivera, C., et al. (2004). Carbonic anhydrase isoform VII acts as a molecular switch in the development of synchronous gamma-frequency firing of hippocampal CA1 pyramidal cells. *Journal of Neuroscience*, *24*(11), 2699–2707.

Scheffer, M., Bascompte, J., Brock, W. A., Brovkin, V., Carpenter, S. R., Dakos, V., et al. (2009). Early-warning signals for critical transitions. *Nature*, *461*(7260), 53–59.

Scheffer, M., Carpenter, S. R., Lenton, T. M., Bascompte, J., Brock, W., Dakos, V., et al. (2012). Anticipating critical transitions. *Science*, *338*(6105), 344–348.

Schevon, C. A., Goodman, R. R., McKhann, G., Jr., & Emerson, R. G. (2010). Propagation of epileptiform activity on a submillimeter scale. *Journal of Clinical Neurophysiology*, *27*(6), 406–411.

Schevon, C. A., Ng, S. K., Cappell, J., Goodman, R. R., McKhann, G., Jr., Waziri, A., et al. (2008). Microphysiology of epileptiform activity in human neocortex. *Journal of Clinical Neurophysiology*, *25*(6), 321–330.

Schevon, C. A., Weiss, S. A., McKhann, G., Goodman, R. R., Yuste, R., Emerson, R. G., et al. (2012). Evidence of an inhibitory restraint of seizure activity in humans. *Nature Communications*, *3*, 1060.

Schwartzkroin, P. A., Baraban, S. C., & Hochman, D. W. (1998). Osmolarity, ionic flux, and changes in brain excitability. *Epilepsy Research*, *32*(1–2), 275–285.

Skinner, F. K., Zhang, L., Velazquez, J. L., & Carlen, P. L. (1999). Bursting in inhibitory interneuronal networks: A role for gap-junctional coupling. *Journal of Neurophysiology*, *81*(3), 1274–1283.

Sloviter, R. S. (1991). Feedforward and feedback inhibition of hippocampal principal cell activity evoked by perforant path stimulation: GABA-mediated mechanisms that regulate excitability in vivo. *Hippocampus*, *1*(1), 31–40.

Staley, K. J., Soldo, B. L., & Proctor, W. R. (1995). Ionic mechanisms of neuronal excitation by inhibitory GABAA receptors. *Science*, *269*(5226), 977–981.

Stead, M., Bower, M., Brinkmann, B. H., Lee, K., Marsh, W. R., Meyer, F. B., et al. (2010). Microseizures and the spatiotemporal scales of human partial epilepsy. *Brain*, *133*(9), 2789–2797.

Su, G., Kintner, D. B., & Sun, D. (2001). Contribution of Na + − K + − Cl − cotransporter to high-[K +]o-induced swelling and EAA release in astrocytes. *American Journal of Physiology: Cell Physiology*, *282*(5), C1136–C1146.

Su, H., Sochivko, D., Becker, A., Chen, J., Jiang, Y., Yaari, Y., et al. (2002). Upregulation of a T-type Ca^{2+} channel causes a long-lasting modification of neuronal firing mode after status epilepticus. *Journal of Neuroscience*, *22*(9), 3645–3655.

Suffczynski, P., Kalitzin, S., da Silva, F. L., Parra, J., Velis, D., & Wendling, F. (2008). Active paradigms of seizure anticipation: Computer model evidence for necessity of stimulation. *Physical Review E Statistical, Nonlinear and Soft Matter Physics*, *78*(5 Pt. 1), 051917.

Tao, J. X., Baldwin, M., Hawes-Ebersole, S., & Ebersole, J. S. (2007). Cortical substrates of scalp EEG epileptiform discharges. *Journal of Clinical Neurophysiology*, *24*(2), 96–100.

Taylor, C. P., & Dudek, F. E. (1984). Synchronization without active chemical synapses during hippocampal after discharges. *Journal of Neurophysiology*, *52*, 143–155.

Thompson, S. M., & Gähwiler, B. H. (1989). Activity-dependent disinhibition. III. Desensitization and GABAB receptor-mediated presynaptic inhibition in the hippocampus in vitro. *Journal of Neurophysiology*, *61*(3), 524–533.

Thrush, S. F., Hewitt, J. E., Dayton, P. K., Coco, G., Lohrer, A. M., Norkko, A., et al. (2009). Forecasting the limits of resilience: Integrating empirical research with theory. *Proceedings of Biological Sciences/The Royal Society*, *276*(1671), 3209–3217.

Tian, G.-F., Azmi, H., Takano, T., Xu, Q., Peng, W., Lin, J., et al. (2005). An astrocytic basis of epilepsy. *Nature Medicine*, *11*(9), 973–981.

Timofeev, I., Bazhenov, M., Avramescu, S., & Nita, D. A. (2010). Posttraumatic epilepsy: The roles of synaptic plasticity. *Neuroscientist: A Review Journal Bringing Neurobiology, Neurology and Psychiatry*, *16*(1), 19–27.

Tóth, K., Eross, L., Vajda, J., Halász, P., Freund, T. F., & Maglóczky, Z. (2010). Loss and reorganization of calretinin-containing interneurons in the epileptic human hippocampus. *Brain*, *133*(9), 2763–2777.

Traynelis, S. F., & Dingledine, R. (1989). Modification of potassium-induced interictal bursts and electrographic seizures by divalent cations. *Neuroscience Letters*, *98*(2), 194–199.

Trevelyan, A. J., Sussillo, D., Watson, B. O., & Yuste, R. (2006). Modular propagation of epileptiform activity: Evidence for an inhibitory veto in neocortex. *Journal of Neuroscience*, *26*(48), 12447–12455.

Trevelyan, A. J., Sussillo, D., & Yuste, R. (2007). Feedforward inhibition contributes to the control of epileptiform propagation speed. *Journal of Neuroscience*, *27*(13), 3383–3387.

Truccolo, W., Donoghue, J. A., Hochberg, L. R., Eskandar, E. N., Madsen, J. R., Anderson, W. S., et al. (2011). Single-neuron dynamics in human focal epilepsy. *Nature Neuroscience*, *14*(5), 635–641.

Uva, L., Avoli, M., & de Curtis, M. (2009). Synchronous GABA-receptor-dependent potentials in limbic areas of the in-vitro isolated adult guinea pig brain. *European Journal of Neuroscience*, *29*(5), 911–920.

van Nes, E. H., & Scheffer, M. (2007). Slow recovery from perturbations as a generic indicator of a nearby catastrophic shift. *American Naturalist*, *169*(6), 738–747.

Velazquez, J. L., & Carlen, P. L. (1999). Synchronization of GABAergic interneuronal networks during seizure-like activity in the rat horizontal hippocampal slice. *European Journal of Neuroscience*, *11*(11), 4110–4118.

Vezzani, A., French, J., Bartfai, T., & Baram, T. Z. (2011). The role of inflammation in epilepsy. *Nature Reviews. Neurology*, 7(1), 31–40.

Vezzani, A., Friedman, A., & Dingledine, R. J. (2013). Neuropharmacology. *Neuropharmacology*, *69*(C), 16–24.

Viitanen, T., Ruusuvuori, E., Kaila, K., & Voipio, J. (2010). The K+ − Cl cotransporter KCC2 promotes GABAergic excitation in the mature rat hippocampus. *Journal of Physiology*, *588*(Pt 9), 1527–1540.

Watts, A. E., & Jefferys, J. G. (1993). Effects of carbamazepine and baclofen on 4-aminopyridine-induced epileptic activity in rat hippocampal slices. *British Journal of Pharmacology*, *108*, 819–823.

Weinand, M. E., Carter, L. P., el-Saadany, W. F., Sioutos, P. J., Labiner, D. M., & Oommen, K. J. (1997). Cerebral blood flow and temporal lobe epileptogenicity. *Journal of Neurosurgery*, *86*(2), 226–232.

Weiss, S. A., Banks, G. P., McKhann, G. M., Goodman, R. R., Emerson, R. G., Trevelyan, A. J., et al. (2013). Ictal high frequency oscillations distinguish two types of seizure territories in humans. *Brain*, *36*(Pt 12), 3796–3808.

Williams, P. A., Hellier, J. L., White, A. M., Staley, K. J., & Dudek, F. E. (2007). Development of spontaneous seizures after experimental status epilepticus: Implications for understanding epileptogenesis. *Epilepsia*, *48*(s5), 157–163.

Williams, P. A., White, A. M., Clark, S., Ferraro, D. J., Swiercz, W., Staley, K., et al. (2009). Development of spontaneous recurrent seizures after kainate-induced status epilepticus. *Journal of Neuroscience*, *29*(7), 2103–2112.

Wittner, L., Eross, L., Czirják, S., Halász, P., Freund, T. F., & Maglóczky, Z. (2005). Surviving CA1 pyramidal cells receive intact perisomatic inhibitory input in the human epileptic hippocampus. *Brain*, *128*(Pt 1), 138–152.

Wong, R. K., Bianchi, R., Taylor, G. W., & Merlin, L. R. (1999). Role of metabotropic glutamate receptors in epilepsy. *Advances in Neurology*, *79*, 685–698.

Worrell, G. A., Jerbi, K., Kobayashi, K., Lina, J. M., Zelmann, R., & Le Van Quyen, M. (2012). Recording and analysis techniques for high-frequency oscillations. *Progress in Neurobiology*, *98*(3), 265–278.

Worrell, G. A., Parish, L., Cranstoun, S. D., Jonas, R., Baltuch, G., & Litt, B. (2004). High-frequency oscillations and seizure generation in neocortical epilepsy. *Brain*, *127*(7), 1496–1506.

Wozny, C., Kivi, A., Lehmann, T.-N., Dehnicke, C., Heinemann, U., & Behr, J. (2003). Comment on "on the origin of interictal activity in human temporal lobe epilepsy in vitro" *Science*, *301*(5632), 463, author reply 463.

Wright, M.-A. S. Y., Orth, M., Patsalos, P. N., Smith, S. J. M., & Richardson, M. P. (2006). Cortical excitability predicts seizures in acutely drug-reduced temporal lobe epilepsy patients. *Neurology*, *67*(9), 1646–1651.

Zhang, Z. J., Koifman, J., Shin, D. S., Ye, H., Florez, C. M., Zhang, L., et al. (2012). Transition to seizure: Ictal discharge is preceded by exhausted presynaptic GABA release in the hippocampal CA3 region. *Journal of Neuroscience*, *32*(7), 2499–2512.

Zhang, X., Tokoglu, F., Negishi, M., Arora, J., Winstanley, S., Spencer, D. D., et al. (2011). Social network theory applied to resting-state fMRI connectivity data in the identification of epilepsy networks with iterative feature selection. *Journal of Neuroscience Methods*, *199*(1), 129–139.

Ziburkus, J. (2006). Interneuron and pyramidal cell interplay during in vitro seizure-like events. *Journal of Neurophysiology*, *95*(6), 3948–3954.

Zijlmans, M., Jacobs, J., Zelmann, R., Dubeau, F., & Gotman, J. (2009). High-frequency oscillations mirror disease activity in patients with epilepsy. *Neurology*, *72*(11), 979–986.

CHAPTER EIGHT

Seizure Termination

Frédéric Zubler, Andreas Steimer, Heidemarie Gast, Kaspar A. Schindler[1]
Department of Neurology, Inselspital, University Hospital Bern, University of Bern, Bern, Switzerland
[1]Corresponding author: e-mail address: kaspar.schindler@insel.ch

Contents

1. Introduction 188
2. Metabolic Mechanisms of Seizure Termination 189
   2.1 Oxygen, glucose, and neurotransmitter depletion 189
   2.2 Acidosis 190
   2.3 Extracellular potassium concentration 191
   2.4 Neuromodulators 192
3. Network Aspects of Seizure Termination 193
   3.1 Synchronization 193
   3.2 Graphs and functional networks 197
   3.3 Seizure termination as a critical transition 199
4. Conclusions 202
References 203

Abstract

A better understanding of the mechanisms by which most focal epileptic seizures stop spontaneously within a few minutes would be of highest importance, because they could potentially help to improve existing and develop novel therapeutic measures for seizure control. Studies devoted to unraveling mechanisms of seizure termination often take one of the two following approaches. The first approach focuses on metabolic mechanisms such as ionic concentrations, acidity, or neuromodulator release, studying how they are dependent on, and in turn affect changes of neuronal activity.

The second approach uses quantitative tools to derive functional networks from electrophysiological recordings and analyzes these networks with mathematical methods, without focusing on actual details of cell biology. In this chapter, we summarize key results obtained by both of these approaches and attempt to show that they are complementary and equally necessary in our aim to gain a better understanding of seizure termination.

International Review of Neurobiology, Volume 114
ISSN 0074-7742
http://dx.doi.org/10.1016/B978-0-12-418693-4.00008-X

1. INTRODUCTION

Epileptic seizures are transient and intrinsically dynamic phenomena with an onset, propagation, and ending phase (Cash, 2013; Jiruska et al., 2013; Timofeev & Steriade, 2004). While the time of seizure occurrence is notoriously difficult to predict, the evolution of seizures is usually highly stereotypical, in that they follow the same clinical and electroencephalographic sequences of symptoms, signs, and signals within individuals (Schindler et al., 2011). At the end of this sequence, seizure termination is more than the mere cessation of epileptic activity. It results from numerous mechanisms, reliably unfolding on many different spatiotemporal scales, and in part even culminating during the postictal time period. On a smaller scale, several of the factors leading to seizure termination directly result from metabolic changes following sustained neuronal activity. The sensitivity of neurons to self-induced modifications of their environment—modifications that they may cause themselves by increased activity—provides them with very effective feedback mechanisms.

The brain, however, cannot be reduced to the individual cells it contains, and seizures are due to more complex phenomena than a monotonous increase of firing rates or synchronization. Epilepsy is therefore increasingly considered as a disorder of brain networks (Engel et al., 2013; Kramer & Cash, 2012; Stam & Van Straaten, 2012) whereby both, the localized dynamics of spatially restricted regions and the collective behavior of the "system as a whole," are put into context when studying seizure initiation, propagation, and termination.

We have structured this review as follows: Section 2 is devoted to "smaller scale" or local, principally metabolic mechanisms of seizure termination. These mechanisms have been recently discussed in two comprehensive reviews (Lado & Moshe, 2008; Löscher & Köhling, 2010); therefore, we concentrate here on a few aspects for which there is overwhelming evidence for relevance in seizure termination and refer the interested reader to the two reviews mentioned above. Then in Section 3, we discuss epileptic seizures from a "larger scale" network perspective and introduce the mathematical methods typically invoked in such an approach.

Finally, we conclude by discussing the important links existing between phenomena occurring on smaller and larger scales.

2. METABOLIC MECHANISMS OF SEIZURE TERMINATION

2.1. Oxygen, glucose, and neurotransmitter depletion

Epileptic seizures are associated with an increase in cerebral metabolic activity (Chapman, Meldrum, & Siesjö, 1977; Freund, Buzsáki, Prohaska, Leon, & Somogyi, 1989; Ingvar, 1986). Because restoration of the resting membrane potential after firing of action potentials or postsynaptic currents is energetically expensive (Attwell & Laughlin, 2001), a decrease in metabolic reserve could hypothetically be a good candidate for promoting seizure termination. However, experimental work could not confirm a significant decrease in glucose or oxygen during epileptic seizures. The main reason is that an increase in neuronal activity leads to an increase in cerebral blood flow. This effect has also been observed under ictal and interictal conditions (Glynn & Detre, 2013; Schwartz, 2007).

Even when occurring, changes in metabolite concentration do not induce rapid termination of sustained epileptiform activity. For example, in a bicuculline administered rat-model of epilepsy, cortical glycogen concentration fell to 23% of control levels in the first minutes after seizure onset, but returned to control concentration after 120 min of seizure activity; the glucose concentration remained close to 50% of control value for more than 1 h; the ATP concentration was only minimally reduced during the first seconds and was found to be even normal thereafter (Chapman et al., 1977). Similar observations were made in other animal models, as well as in humans during status epilepticus (Shorvon, 1994). In patients, ictal hypoxemia might have deleterious effects and could contribute to sudden unexpected deaths in epilepsy (Bateman, Li, & Seyal, 2009; Blum et al., 2000) before limiting epileptiform activity. In fact, there is evidence that hypoglycemia and hypoxia are proconvulsive (Delanty, Vaughan, & French, 1998).

If seizures do not halt because neurons run out of oxygen or glucose, could it then be that increased neuronal activity causes depletion of neurotransmitters and hence seizure termination? Exhaustion of presynaptic glutamate has been discussed as a possible mechanism limiting the duration and rate of bursts in a rat hippocampal slice model under high extracellular potassium concentration (Staley, Longacher, Bains, & Yee, 1998). However, in these experiments epileptiform activity continued as long as the slice was maintained under ictogenic conditions. Therefore, vesicle depletion would at most restrict the firing of

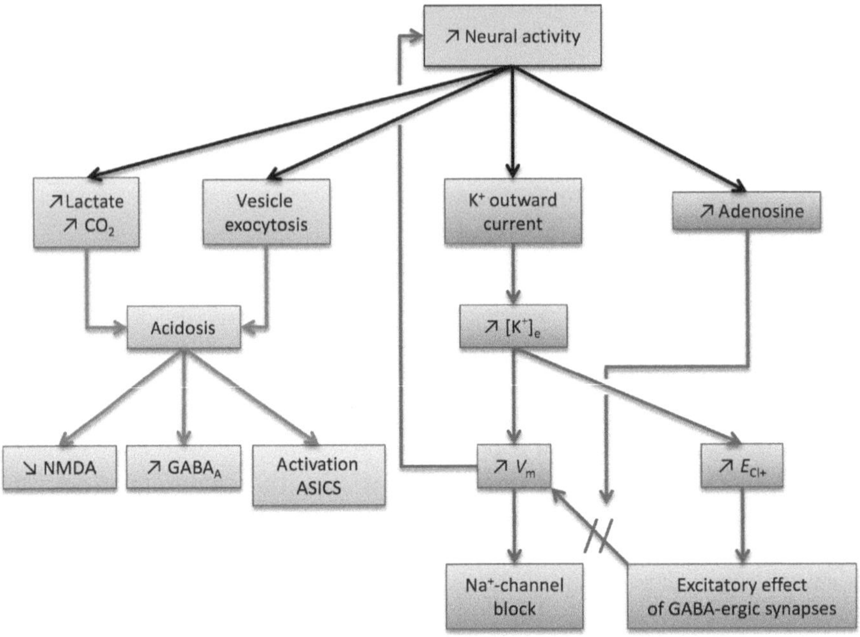

Figure 8.1 Summary of postulated metabolic mechanisms of seizure termination. Sustained neuronal activity during seizures leads to the following: (1) Acidosis resulting from an increase in lactate and CO_2 release, as well as from the exocytosis of the acidic contents of synaptic vesicles; acidosis changes the dynamics of several ion channels, which reduce neuronal activity. (2) An outward potassium current, increasing the extracellular potassium concentration, which depolarizes the cell membrane; this depolarization first increases and then inhibits (through inactivation of sodium channels) neuronal firing. The increase of extracellular potassium leads to an intracellular increase of chloride, switching GABA-ergic transmission from inhibitory to excitatory. (3) Adenosine is released by firing neurons; it promotes seizure termination by reducing the depolarizing effect of GABA-ergic activity (ASICS, acid-sensing ion channels).

individual cells, but would not be sufficient to halt the seizure as a whole (Löscher & Köhling, 2010).

In summary, it is unlikely that seizures typically stop following depletion of factors essential for neuronal functions. We discuss in the next section more plausible hypothesis whereby epileptiform activity itself produces factors, which promote seizure termination (Fig. 8.1).

2.2. Acidosis

Several lines of evidence imply an important role for acidosis in seizure termination. Raising extracellular PCO_2 or lowering extracellular pH in a low

Mg^{++} slice model of epilepsy increased the interval between and the amplitude of seizure-like events (Velísek, Dreier, Stanton, Heinemann, & Moshé, 1994). The seizure suppressive action of CO_2 inhalation has been demonstrated in numerous human and animal studies as early as 1928 (Tolner et al., 2011—and references therein). In addition, several seizure suppressive drugs such as acetazolamide and topimarate have a carbonic anhydrase blocking effect, which reduces the pH value in the brain.

Acidity influences neuronal activity mainly by changing the dynamics of several types of ion channels. A low pH has been shown to inhibit NMDA receptors (Traynelis & Cull-Candy, 1990), while enhancing $GABA_A$ receptors (Dietrich & Morad, 2010). Acidosis also opens acid-sensing ion channels 1a (ASIC1a), triggering Na^+ and Ca^+ inward currents inducing neuronal firing. Because ASIC1a are mainly found on inhibitory interneurons, it has been postulated that they may contribute to seizure termination. Supporting this hypothesis, disruption of ASIC1a in mice increased the severity of seizures in a mouse model of epilepsy, whereas overexpression of this receptor had the opposite effect (Ziemann et al., 2008).

There are several mechanisms by which neuronal activity can reduce brain pH. Aerobic and anaerobic metabolism, through production of CO_2 and lactate contribute to brain acidity. Another mechanism is the fusion of synaptic vesicles, whose membrane contains H^+-ATPase, which acidifies the vesicle content. Vesicle exocytosis frees the acidic vesicle content and integrates the H^+-ATPase into the cell membrane leading to a transient pH drop within the synaptic cleft after synaptic transmission (Forgac, 2007). As further mechanisms, activity-evoked HCO_3^- transport or electrolyte modifications (Magnott, Heo, Dlouhy, & Dahdaleh, 2012) affect pH.

In summary, sustained synaptic transmission as occurs during an epileptic seizure induces transient acidosis, which in turn reduces neuronal excitability (Sinning & Hübner, 2013; Wemmie, Taugher, & Collin, 2013).

2.3. Extracellular potassium concentration

Under physiological conditions, the potassium concentration in the cytoplasm ($[K^+]_i$) is larger than the concentration in the extracellular space ($[K^+]_e$) and the reversal potential for potassium (E_K) is negative. When K^+-gates open, potassium currents re- or hyperpolarize neurons (for instance at the end of an action potential, or after a paroxysmal depolarization shift). According to the potassium accumulation hypothesis (for review, see Fröhlich, Bazhenov, Iragui-Madoz, & Sejnowski, 2008), during sustained neuronal firing the sum of individual potassium outward currents leads to an increase of $[K^+]_e$, which in turn affects neuronal excitability through at least two different mechanisms. First, higher $[K^+]_e$ values shift E_K and consequently the neurons' membrane potential moves toward more positive values, changing the cells' firing properties. Second, the potassium-chloride cotransporter KCC2, which under normal conditions uses the difference in potassium concentrations to extrude Cl^- out of the cell, changes directionality and imports Cl^-, while transporting K^+ back inside the cell. As a result, E_{Cl} increases, thereby rendering the inhibitory effect of GABA-ergic synapses less effective, which may even become excitatory (Blaesse, Airaksinen, Rivera, & Kaila, 2009; Kaila, Lamsa, Smirnov, Taira, & Voipio, 1997; Lillis, Kramer, Mertz, Staley, & White, 2012; Payne, Rivera, Voipio, & Kaila, 2003). As a consequence, the neurons become more excitable, which possibly further increases their firing rates. If $[K^+]_e$ continues to increase, the repolarized potential necessary for the deinactivation of voltage-gated Na^+ channels after action potentials (Armstrong, 2006; Armstrong & Gilly, 1977) can no longer be attained, which prevents further firing of the neuron (depolarization block).

In summary, it seems plausible that $[K^+]_e$ increase might initially participate in seizure continuation and/or propagation through a positive feedback mechanism, but also eventually promote seizure termination by inactivating Na^+ channels.

2.4. Neuromodulators

Adenosine is a degradation product of ATP predominantly released by action potential firing neurons into the synaptic cleft (Lovatt et al., 2012), which has a long-known seizure-suppression effect (Dunwiddie, 1980; Young & Dragunow, 1994), probably by reducing the excitatory effect of $GABA_A$ receptors as often detected during seizures (leading to depolarizing block of Na-channels; Ilie, Raimondo, & Akerman, 2012). Sustained neuronal activity thus leads to a release of adenosine, which then may promote seizure termination (Boison, 2013; Von Gompel et al., 2014). Apart from adenosine, several substances acting as nonclassic neurotransmitters such as neuropeptide Y have been proposed to play a role in seizure termination, because they are released during sustained neuronal stimulation and have been shown to exert various seizure suppressive effects (for review, see Löscher & Köhling, 2010).

3. NETWORK ASPECTS OF SEIZURE TERMINATION

3.1. Synchronization

Feedback mechanisms, which tend to reduce neuronal firing rates in response to an increased activity, as described in the previous section, fail to explain two important phenomena. The first one is sustained epileptic activity (i.e., status epilepticus). If the brain can support ongoing epileptic activity for an hour and more, why do most seizures stop within a few minutes? The second phenomenon is the observation that EEG correlates of epileptiform activity often simultaneously stop in many channels (Fig. 8.2A). It is highly unlikely that metabolic feedback mechanisms (or any other "local" process) by chance become effective in different areas at the same time. The often abrupt termination of seizures suggests that counter mechanisms are coordinated across larger scales and are rapidly and reliably activated in the course of a seizure. One important way to coordinate local processes occurring within a larger network is through synchronization.

The concept of synchronization can be traced back to the observation of Huygens in the seventeenth century, that two pendulum clocks hung on a wooden beam tend to swing in opposite directions (technically referred to as "antiphase synchronization"). Huygens found that the two clocks "communicated" by transmitting kinetic energy of their oscillating pendulum through discrete movements of the beam (Bennett, Schatz, Rockwood, & Wiesenfeld, 2002). For systems with a dominant oscillatory mode, such as Huygens pendulum watches, synchronization has been defined as an adjustment of rhythms, that is, of both phase and frequency, between subsystems due to weak interactions (Pikovsky, Rosenblum, & Kurths, 2002). This relatively narrow definition of synchronization has been extended to general systems without regular oscillatory behavior: generalized synchronization between two subsystems occurs when the state of the first one is determined by (but is not necessarily equal to) the state of the second one (Rulkov, Sushchik, & Tsimring, 1995). According to this latter definition, synchronization is close to the notion of interdependence. Synchronization usually implies one of the three following relationships between the subsystems: (1) either one influences the other, or (2) they both influence themselves reciprocally, or (3) both are driven by a third process.

How can we quantify synchronization in practice? In the context of neuroscience—and in particular, when assessing the dynamics of epileptic

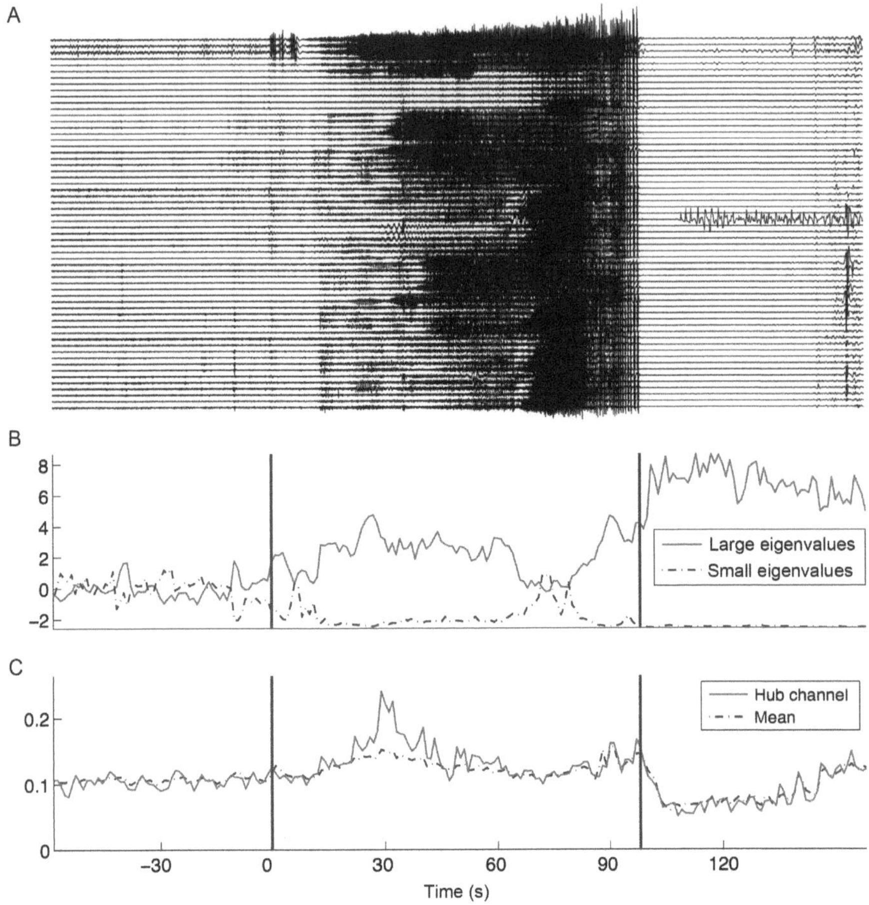

Figure 8.2 Epileptic seizure in a 27-year-old male patient with temporal lobe epilepsy. Recorded during evaluation for epilepsy surgery with intracranial EEG (iEEG), the seizure begins with a spike-wave visible in the top-most three channels (depth electrode in the left hippocampus), followed by high-frequency low-amplitude oscillations. In the course of the seizure, the amplitude of oscillation increases, the frequency decreases, and epileptiform signals propagate to other groups of electrodes, ipsi- and contralateral. Before seizure termination, the EEG waveform appears synchronized; the seizure ends at all channels simultaneously (A). To quantify global synchronization between the intracranial EEG signals, we follow (Schindler, Elger, & Lehnertz, 2007; Schindler, Leung, Elger, & Lehnertz, 2007) and compute the eigenvalues of the synchronization matrix (see text). High correlation is typically associated with high values of the largest eigenvalues and low values of the smallest eigenvalues. We see that synchronization increases already in the preictal phase, decreases during the second third of the seizure, and finally increases dramatically before and shortly after seizure termination. The global synchronization is maximal in the postictal time period (B). The global information flow between iEEG signals, which we assessed here by symbolic transfer entropy, Zubler, Gast, Abela et al., 2014 is maximal both during propagation and at seizure termination. In the hub channel (the channel with highest connection degree), the maximum is reached during seizure propagation, with a small peak before seizure termination (C).

seizures—synchronization depends on the spatial scale we investigate (Jiruska et al., 2013). Synchrony between two neurons is usually assessed by quantifying the similarity between their action potential trains, which are considered as discrete signals (Chicharro, Kreuz, & Andrzejak, 2011; Kreuz, Haas, Morelli, Abarbanel, & Politi, 2007; Victor, 2005). Neuronal synchronization at larger spatial scale, between two brain areas, for example, is estimated by comparing continuous signals derived from the electrical (EEG, MEG) or metabolic (fMRI) activity of large cell populations. Several synchronization measures exist to compare these averaged signals (Kreuz, 2013; Quian Quiroga, Kraskov, Kreuz, & Grassenberg, 2002; Wendling, Ansari-Asl, Bartolomei, & Senhadji, 2009), giving account of only a few here. A commonly used and computationally highly efficient measure is cross-correlation, which quantifies linear aspects of the temporal similarities between signals. Other methods quantify the similarity of signals in the frequency domain, such as cross spectrum coherence, which compares the amplitude at different frequencies, or phase coherence methods, which apply the original definition of synchronization to general signals by comparing their so-called instantaneous phases. Along with the definition of generalized synchronization, information theoretical methods detect general interdependencies between signals treated as random variables. Mutual information, for example, measures how much our uncertainty about a random variable is reduced by knowing a second one, or put more bluntly, how much one signal informs us about another. Transfer entropy measures not only the magnitude but also the direction of information flow between signals, and has been used in neuroscience to infer directed connectivities.

Synchronization of neuronal oscillations is essential to the normal functioning of the brain (Uhlhaas & Singer, 2006) and has been postulated to play an important role in the integration of information from distributed subnetworks (for review, see Varela, Lachaux, Rodriguez, & Martinerie, 2001). Impairment of synchrony is observed in several psychiatric and neurocognitive diseases such as schizophrenia (Uhlhaas et al., 2006) and Alzheimer disease (Jeong, Gore, & Peterson, 2001; Pijnenburg et al., 2004), or following brain lesions (Engel, König, Kreiter, & Singer, 1991). Also in epilepsy, reductions of large-scale synchronization seem to be detrimental.

One of the first evidence that impaired synchronization on larger spatial scales may be relevant for the pathophysiology of epilepsy came from the work of Mormann, Lehnertz, David, and Elger (2000), showing that compared to interictal values, phase coherence computed between EEG signals recorded in the mesial temporal lobe with depth-electrodes was reduced in the preictal time period and increased during the seizure; in the postictal period phase coherence reached "normal" interictal values again. The authors postulate that a certain degree of desynchronization might be a necessary condition for seizures to be generated and that seizures could even act as a reset mechanism to restore normal synchronization. Despite a few methodological caveats (the instantaneous phase was computed for a broad-band signal, which does not allow for a straightforward interpretation as in the case of a narrow-band signal; the interictal phase control periods did not take vigilance into account), desynchronization before or at seizure onset was corroborated by other studies. For example, Wendling, Bartolomei, Bellanger, Bourien, and Chauvel (2003) demonstrated that cross-correlation between intracranial electrodes in the seizure-onset zone (SOZ) (defined according to the presence of low amplitude high-frequency oscillations) dropped at seizure onset, and increased during the seizure (the postictal period was not investigated). Schindler, Leung, et al. (2007) studied the dynamics of global (and hence not only pairwise) synchronization during epileptic seizures, using results from linear algebra based on principal component analysis, i.e., the eigenvalues of the correlation matrix, which is an array containing the correlations between all pairs of EEG signals (Fig. 8.2). In essence, if a few eigenvalues are dominant, global synchronization is considered to be high; conversely, if all eigenvalues have similar magnitudes, global synchronization is low (signals are decorrelated). Repeatedly evaluating the correlation matrix for a moving time window, the authors found a gradual increase of global synchronization starting before and continuing after seizure termination into the postictal time period. A similar method was used in Fig. 8.1B: unlike in the original work (Schindler, Leung, et al., 2007), we used a pseudo-monopolar montage to avoid correlation between next-neighbor derivations. In this example, synchronization increases slightly around seizure onset, transiently decreases during the second third of the seizure, and then dramatically increases before seizure termination. This dynamics can be confirmed using pairwise methods to compute synchrony (Fig. 8.1C; Zubler, Gast, Abela et al., 2014). In a follow-up study (Schindler, Elger, et al., 2007), it was demonstrated that also in status epilepticus, an increase in global synchronization occurred at or right before the end of epileptiform activity. Moreover, it was shown that in a few cases seizure suppressive drugs were associated with an increase of synchronization. The authors postulated that this increase of synchronization could be a mechanism to terminate seizure activity by simultaneously driving extended neuronal networks into a refractory state.

Interestingly, similar findings were made at the single-neuron level by Truccolo et al. (2011), who found a high level of heterogeneity of spiking activity during seizure initiation and propagation, while neuronal activity became more homogenous before, during, and after seizure termination.

In summary, focal seizure termination seems to be associated with an overall increase in synchronization. Next, we will examine if this is a generic phenomenon, or if it is confined to some particular parts of the ictogenic network.

3.2. Graphs and functional networks

Mathematically, networks can be represented by graphs. A graph is formally defined as a set of nodes (also called vertices) and a set of links (or edges) connecting some of the nodes in a pairwise manner. A graph is said to be weighted if a numerical value (other than zero or one, in which case the graph is referred to a "binary") is associated with each edge and it is directed if the edges have a direction, that is, if the edge connecting node A to node B is different from the edge connecting B to A. An illustrative example of a graph is an airline route map, where nodes represent cities, and edges are direct flights between cities. This graph is directed (it might be possible to fly directly from New York City to Boston, but not from Boston to New York City); it can be weighted, by assigning to each edge the distance between the linked cities, or the price of the flight ticket. In neuroscience, two major types of graph have been considered: structural/anatomical or functional (Kramer & Cash, 2012; Rubinov & Sporns, 2010; Stam & Van Straaten, 2012; Van Diessen, Diederen, Braun, Jansen & Stam, 2013; Van Diessen, Hanemaaijer, et al., 2013). In structural networks, each node typically represents a brain area, and edges represent the existence of an anatomical connection that links two areas—for instance, an axonal bundle assessed by diffusion tensor imaging. On the other hand, in functional networks, nodes represent recording sites of fMRI, EEG, or MEG signals, and edges indicate the interdependence between the two signals recorded at these sites. The structure of such a graph depends on the location and number of recording sites, as well as on the measure used to define synchronization between the signals. Another important parameter is the threshold of synchronization to define an edge: when applied to signals of finite duration, most of the synchronization measures presented above will typically not yield a value of exactly zero, even if the signals are completely unrelated. It is thus necessary to define a significance level above which we accept the existence of "true" interdependence/synchronization between two signals. The higher the threshold, the fewer edges will be in the graph (Kramer, Eden, Cash, & Kolaczyk, 2009). Once a graph is obtained, several characteristics can be computed to quantitatively characterize specific nodes, specific edges, or the general structure of the network (for review, see Rubinov & Sporns, 2010).

Kramer et al. (2010) studied the evolution of such a graph derived from electrocorticographic signals in patients with pharmacoresistant epilepsy, with edges representing statistically significant cross-correlations. At seizure onset, the graph structure mainly consisted of one large component (meaning that most nodes were connected through a sequence of edges). In the course of the seizure, this large component first fractured into smaller subnetworks, which then merged to form a single dominant component again before seizure termination. Other studies confirmed that the ictal decrease in synchrony was not a homogenous weakening of all links, but rather a fragmentation resulting in clusters, loosely connected with each other (Bialonski & Lehnertz, 2013). This effect is also present in the seizure of Fig. 8.2A (unlike Kramer et al., we used zero-lag correlation for consistency with the method used in Fig. 8.2B; level for drawing a link between two nodes was set to the 95th percentile of pairwise correlations between 60 surrogate EEG signals). One minute before seizure onset, the structure is dominated by a component of 50 electrodes, the remaining 10 electrodes are isolated or for two of them connected into a very small component of size 2 (Fig. 8.3A). One minute after seizure onset, the structure of the graph has dramatically changed, with smaller components of maximal size of 12 electrodes. Interestingly, 8 of the 11 electrodes recording from brain areas that were resected later (belonging to the visually identified SOZ) are grouped in an isolated cluster. Finally, 30 s after seizure termination, most electrodes have coalesced into one dominant component (43 nodes), while the isolated subnetwork of eight electrodes of the SOZ has disappeared. This fragmentation of functional brain networks and epileptic seizures is consistent with studies showing a functional disconnection of the SOZ from the rest of the brain in ictal (Van Diessen, Diederen, et al., 2013; Van Diessen, Hanemaaijer, et al., 2013) and interictal period (Warren et al., 2010). Together with experimental work showing that cortical deafferentation promotes epileptiform activity (Topolnik, Steriade, & Timofeev, 2003), these results are consistent with the hypothesis that synchronization may promote seizure termination by restoring connectivity between functionally disconnected brain networks.

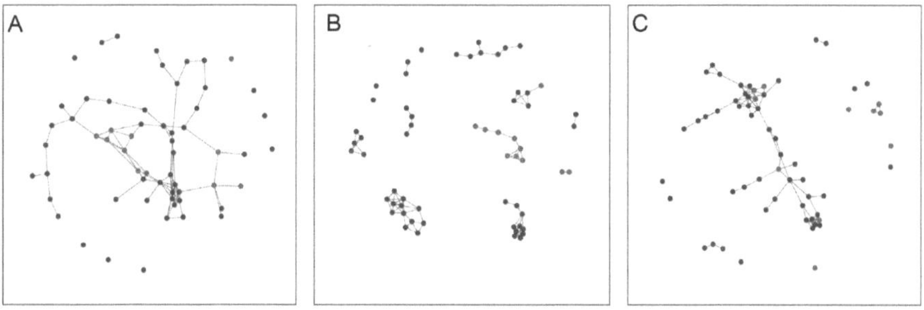

Figure 8.3 Fragmentation and coalescence in an ictal functional network, initially described in Kramer et al. (2010). Three epochs of 2.5 s of the seizure displayed in Fig. 8.2A are represented as a graph, with nodes representing individual iEEG electrodes and links denoting the existence of functional coupling between pairs of channels as measured by pair-wise zero-lag cross-correlation. 60 s before seizure onset, the network structure is dominated by a single connected component of 50 electrodes (A). 60 s after seizure onset, the network has fractured into several disjoint components (B). 30 s after seizure termination, most of the components have merged again (C). Nodes displayed in grey correspond to channels located in a brain area that was resected during epilepsy surgery; the length of the link is inversely proportional to its strength; distance between nonconnected nodes is arbitrary.

However, synchronization does not only have positive effects, at least at seizure onset. Constructing a functional graph based on band-filtered phase coherence during ictal periods, Ibrahim et al. (2013) confirmed that at fast frequencies (beta and above), electrodes in the SOZ were functionally disconnected from other electrodes until about the middle of the seizure, but reconnected at seizure termination (disconnection was quantified with a graph theoretical property, called clustering coefficient). However, at lower frequencies, the SOZ electrodes were central nodes within the network at seizure onset (as measured by the mean of another graph property, the so-called eigenvector centrality). According to the authors' interpretation, this central position of SOZ could promote seizure spreading and even play a causal role in consciousness impairment.

In summary, restoration of connectivity between brain subnetworks that have become functionally independent could promote seizure termination, but could also play a deleterious role at seizure onset by allowing epileptic activity to spread from the SOZ.

3.3. Seizure termination as a critical transition

Graphs, in which each element of a system is represented individually as a node, and each interaction as an edge, are highly informative representations of network states. The downside of graphs is that to represent the temporal evolution of a system, one has to repeatedly compute and then compare new graphs. In contrast, the formalism of dynamic systems theory focuses on the qualitative behavior of systems over time. In this theory, the state of a whole system at any given time is represented as a single point (the so-called phase point) in a multidimensional space (the phase space). As the system evolves, its corresponding phase point moves, and the long-term behavior of the system is represented by a trajectory in phase space. For instance, we can conceptualize (though not visualize) the 60 EEG signals of Fig. 8.2A as a single trace in a 60-dimensional space.

In dynamic systems theory, the internal interactions in a system (often related to the edges in the graph representation) correspond to the forces acting on the phase point, moving it along its trajectory. One way to imagine the systems internal dynamics is as a landscape, the potential, on which the phase point rolls like a small ball (Strogatz, 2001). Only in very simple cases, such as, for instance, the vibration of a point mass attached to a spring, is it possible to actually derive the exact potential. But the strength of dynamic systems theory is that even for very complex systems we can often understand enough of the internal dynamics to qualitatively predict the behavior of the system. By qualitative behavior, we refer to properties such as: under which conditions the system is in equilibrium, will display periodic oscillations, or will adopt a chaotic behavior. In some cases, not only the system's state (the position of the phase point) but also the system's dynamics (the landscape on which the point phase rolls) changes with time. A modification of the internal dynamics leading to a qualitative change in the system's behavior is called a bifurcation. Fundamental changes in systems when they pass a bifurcation are called critical transitions (Scheffer et al., 2009) and have several generic properties. These properties have been recognized during dramatic changes in systems as diverse as climate transition, species extinction, or financial markets.

It has recently been shown by Kramer et al. (2012) how seizure termination at the macroscopic level, that is, at the spatial scale assessed with EEG, exhibits several classical features of critical transitions. These features are: slowing down of the recorded signals, as quantified by spectrum analysis; an increase in the variance of the signals; an increase in spatiotemporal correlations between the signals; and finally flickering or bistability, that is, repetitive switching between two modes.

These concepts are illustrated in Fig. 8.4 for four EEG sections; each section represents 2.5 s from the first signal of Fig. 8.1A (first contact of a depth electrode in the left hippocampus); the potentials displayed here are

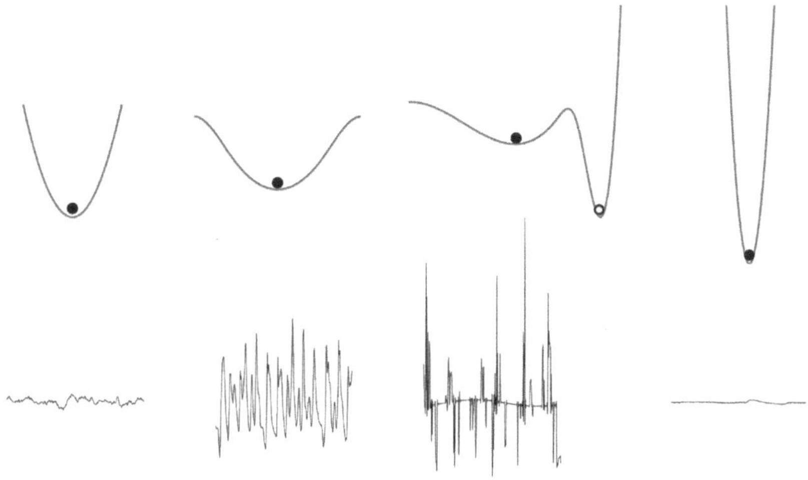

Figure 8.4 Seizure termination as a critical transition. According to dynamic systems theory, the state of a system can be visualized as a small ball (the state point) rolling over a landscape; dramatic changes in the system behavior are often associated with a modification of the landscape. Applied to the analysis of epileptic seizures, the oscillations recorded by an EEG electrode can be associated with the movement of the state point rolling back and forth in a trough. During seizure development, as the trough flattens, the oscillations increase in amplitude and decrease in frequency. Shortly before seizure termination, a second trough appears, which is much steeper. Now the system jumps between two states: either very large oscillations or an almost flat line. Seizure stops with further modification of the landscape, when now only the deeper well persists, and only minor oscillations are possible (top: schematic illustration; bottom: four epochs of 2.5 s taken from the first channel in Fig. 8.2A, from left to right: 140 s before seizure onset, 45 s after seizure onset, 10 s before seizure termination, 14 s after seizure termination).

schematic representations. In the preictal time period, the systems internal dynamics is represented by a relatively deep trough at the bottom of which the state point rolls back and forth, along with oscillations recorded by the electrode. Throughout seizure onset and continuation, the landscape flattens, allowing for oscillations of higher amplitude and lower frequency. Before seizure termination, a second trough appears. This changes qualitatively the behavior of the system, which now alternates between two qualitatively different behaviors: either the state point is in the now flatter trough, resulting in larger oscillation, or at the bottom of the second, profounder dwell in which it can almost not oscillate. Finally, in the postictal time period, only the deeper trough subsists, from which the system cannot escape.

Critical transitions are often associated with positive feedback, implying that under certain circumstances the state of the system modifies the dynamics of the system, while the dynamics influences the state, moving unidirectionally toward the critical transition (Scheffer et al., 2009). In case of status epilepticus, the system was found to approach a critical transition, without reaching it; instead, it retracted back to the ictal configuration. In a computational model, the authors could reproduce the finding of seizure termination by increasing the strength of excitatory synapses, which constitutes a plausible mechanism for positive feedback (Kramer et al., 2012).

On a smaller scale (single-cell level), the only characteristic of a critical transition found was slowing down (without correlation increases or flickering), suggesting that termination through critical transition is a property of the larger scale network, depending on the coordinated behaviors of individual cells (Kramer et al., 2012).

In summary, the interplay between network state and network dynamics could rapidly promote seizure termination, once the system reaches an appropriate threshold. When the system fails to reach the threshold on its own, epileptiform activity persists, unless an external factor (medication, electrical stimulation) provides the additional "push" to end the seizure.

4. CONCLUSIONS

According to the International League Against Epilepsy, an epileptic seizure is "a transient occurrence of signs and/or symptoms due to abnormal excessive or synchronous neuronal activity in the brain" (Fisher et al., 2005). Formally and implicitly, this definition requires the abnormal excessive neural activity and/or the abnormal synchrony to end, for a seizure to stop. However, as discussed in this review, seizure termination is not due to a reemergence of the preictal state, but to mechanisms building up during the seizure and lasting into the postictal time period, such as large scale synchronization. On a smaller spatial scale, seizure-related increase in neural metabolic activity has several consequences, such as changes in pH or electrolytes, which already lead to a decrease in neuronal excitability prior to seizure termination. Synchronization also evolves during the course of a seizure. However, the situation here is far more complex. First, there are several types of synchronization amongst other things depending on the spatial scale of observation. Second, an initial change of synchronization—probably even in the preictal phase—has a permissive effect on seizure generation and propagation. And third, because the exact mechanisms whereby synchronization might contribute to seizure termination (for example, by coordination of inhibition over extended cortical areas, by functional reconnection of disconnected regions, etc.) are not yet fully elucidated.

Although we have so far treated local metabolic effects of neural activity and large-scale functional networks separately, these processes are deeply closely interrelated. The electric signals recorded by EEG electrodes, and which we assess by different synchronization measures, are caused by the summed transmembrane currents (mostly but not exclusively due to active synapses) in numerous individual cells (Buzsáki et al., 2012). Conversely, the larger scale patterns of activity constrain and modulate local neuronal firing. For example, the global electrical field resulting from the collective activity of a large number of neurons and glial cells may influence transmembrane currents in individual neurons, affecting their excitabilities (Fröhlich et al., 2010; Jefferys, 1995; Schevon et al., 2012).

In summary, seizure termination is a highly complex phenomenon, influenced and dependent on smaller scale, larger scale, top-down, and bottom-up effects, which is far from being understood. Our motivation to study these intricate mechanisms is based on the hope that a deeper understanding of the multiscale dynamics of seizure termination will ultimately lead to advances in diagnostics and treatment of epilepsy.

REFERENCES

Armstrong, C. M. (2006). Na channel inactivation from open and closed states. *Proceedings of the National Academy of Sciences of the United States of America, 103*(47), 17991–17996.

Armstrong, C. M., & Gilly, W. F. (1977). Fast and slow steps in the activation of sodium channels. *The Journal of General Physiology, 74*(6), 691–711.

Attwell, A., & Laughlin, S. B. (2001). An energy budget for signaling in the grey matter of the brain. *Journal of Cerebral Blood Flow & Metabolism, 21*, 1133–1145.

Bateman, L. M., Li, C.-S., & Seyal, M. (2009). Ictal hypoxemia in localization-related epilepsy: Analysis of incidence, severity and risk factors. *Brain, 131*(12), 3239–3245.

Bennett, M., Schatz, M., Rockwood, H., & Wiesenfeld, K. (2002). Huygens's clocks. *Proceedings of the Royal Society A, 458*, 123–321.

Bialonski, S., & Lehnertz, K. (2013). Assortative mixing in functional brain networks during epileptic seizures. *Chaos, 23*(3), 033139.

Blaesse, P., Airaksinen, M. S., Rivera, C., & Kaila, K. (2009). Cation-chloride cotransporters and neuronal function. *Neuron, 61*, 820–838.

Blum, A. S., Ives, J. R., Goldberger, A. L., Al-Aweel, I. C., Krishnamurthy, K. B., Drislane, F. W., et al. (2000). Oxygen desaturations triggered by partial seizures: Implications for cardiopulmonary instability in epilepsy. *Epilepsia, 41*(5), 536–541.

Boison, D. (2013). Adenosine and seizure termination: Endogenous mechanisms. *Epilepsy Currents, 13*(1), 35–37.

Buzsáki, G., Anastassiou, C. A., & Koch, C. (2012). The origin of extracellular fields and currents–EEG, ECoG, LFP and spikes. *Nature Review Neuroscience, 13*(6), 407–420.

Cash, S. S. (2013). Status epilepticus as a system disturbance: is status epilepticus due to synchronization or desynchronization? *Epilepsia, 54*(Suppl. 6), 37–39.

Chapman, A. G., Meldrum, B. S., & Siesjö, B. K. (1977). Cerebral metabolic changes during prolonged epileptic seizures in rats. *Journal of Neurochemistry*, *28*(5), 1025–1035.

Chicharro, D., Kreuz, T., & Andrzejak, R. G. (2011). What can spike train distances tell us about the neural code? *Journal of Neuroscience Methods*, *199*, 146–165.

Delanty, N., Vaughan, C. J., & French, J. A. (1998). Medical causes of seizures. *The Lancet*, *352*(9125), 383–390.

Dietrich, C. J., & Morad, M. (2010). Synaptic acidification enhances GABAA signaling. *The Journal of Neuroscience*, *30*(47), 16044–16052.

Dunwiddie, T. V. (1980). Endogenously released adenosine regulates excitability in the in vitro hippocampus. *Epilepsia*, *21*, 541–548.

Engel, A. K., König, P., Kreiter, A. K., & Singer, W. (1991). Interhemispheric synchronization of oscillatory neuronal responses in cat visual cortex. *Science*, *252*, 1177–1179.

Engel, J., Jr., Thompson, P. M., Stern, J. M., Staba, R. J., Bragin, A., & Mody, I. (2013). Connectomics and epilepsy. *Current Opinion in Neurology*, *26*(2), 186–194.

Fisher, R. S., van Emde, B. W., Blume, W., Elger, C., Genton, P., Lee, P., et al. (2005). Epileptic seizures and epilepsy: Definitions proposed by the International League Against Epilepsy (ILAE) and the International Bureau for Epilepsy (IBE). *Epilepsia*, *46*(4), 470–472.

Forgac, M. (2007). Vacuolar ATPases: Rotary proton pumps in physiology and pathophysiology. *Nature Reviews Molecular Cell Biology*, *8*(11), 917–929.

Freund, T. F., Buzsáki, G., Prohaska, O. J., Leon, A., & Somogyi, P. (1989). Simultaneous recording of local electrical activity, partial oxygen tension and temperature in the rat hippocampus with a chamber-type microelectrode. Effects of anaesthesia, ischemia and epilepsy. *Neuroscience*, *28*(3), 539–549.

Fröhlich, F., Bazhenov, M., Iragui-Madoz, V., & Sejnowski, T. J. (2008). Potassium dynamics in the epileptic cortex: New insights on an old topic. *The Neuroscientist*, *14*(5), 422–433.

Fröhlich, F., & McCormick, D. A. (2010). Endogenous electric fields may guide neocortical network activity. *Neuron*, *67*(1), 129.

Glynn, S. M., & Detre, J. A. (2013). Imaging epilepsy and epileptic seizures using fMRI. In S. Ulmer, & O. Jansen (Eds.), *fMRI—Basic and clinical applications* (pp. 177–189). Heidelberg: Springer-Verlag.

Ibrahim, G. H., Anderson, R., Akiyama, T., Ochi, A., Otsubo, H., Singh-Cadieux, G., et al. (2013). Neocortical pathological high-frequency oscillations are associated with frequency-dependent alterations in functional network topology. *Journal of Neurophysiology*, *110*, 2475–2483.

Ilie, A., Raimondo, J. V., & Akerman, C. J. (2012). Adenosine release during seizures attenuates GABAA receptor-mediated depolarization. *Journal of Neuroscience*, *32*(15), 5321–5332.

Ingvar, I. (1986). Cerebral blood flow and metabolic rate during seizures. Relationship to epileptic brain damage. *Annals of the New York Academy of Science*, *462*, 194–206.

Jefferys, J. G. (1995). Non synaptic modulation of neuronal activity in the brain: Electric currents and extracellular ions. *Physiology Review*, *75*, 689–723.

Jeong, J., Gore, J. C., & Peterson, B. S. (2001). Mutual information analysis of the EEG in patients with Alzheimer's disease. *Clinical Neurophysiology*, *112*, 827–835.

Jiruska, P., de Curtis, M., Jefferys, J. G., Schevon, C. A., Schiff, S. J., & Schindler, K. (2013). Synchronization and desynchronization in epilepsy: controversies and hypotheses. *The Journal of Physiology*, *591*(Pt 4), 787–797.

Kaila, K., Lamsa, K., Smirnov, S., Taira, T., & Voipio, J. (1997). Long-lasting GABA-mediated depolarization evoked by high-frequency stimulation in pyramidal neurons of rat hippocampal slice is attributable to a network-driven, bicarbonate-dependent K+ transient. *The Journal of Neuroscience*, *17*(20), 7662–7672.

Kramer, M. A., & Cash, S. S. (2012). Epilepsy as a disorder of cortical network organization. *The Neuroscientist, 18*(4), 360–372.
Kramer, M. A., Eden, U. T., Cash, S. S., & Kolaczyk, E. D. (2009). Network inference with confidence from multivariate time series. *Physical Review E, 79*, 061916.
Kramer, M. A., Eden, U. T., Kolaczyk, E. D., Zepeda, R., Eskandar, E. N., & Cash, S. S. (2010). Coalescence and fragmentation of cortical networks during focal seizures. *The Journal of Neuroscience, 30*, 10076–10085.
Kramer, M. A., Truccolo, W., Eden, U. T., Lepage, K. Q., Hochberg, L. R., Eskandar, E. N., et al. (2012). Human seizures self-terminate across spatial scales via a critical transition. *Proceedings of the National Academy of Sciences of the United States of America, 109*(51), 21116–21121.
Kreuz, T. (2013). Synchronization measures. In R. Quian Quiroga, & S. Panzeri (Eds.), *Principles of neural codings* (pp. 97–120). Boca Raton: CRC Press.
Kreuz, T., Haas, J. S., Morelli, A., Abarbanel, H. D. I., & Politi, A. (2007). Measuring spike train synchrony. *Journal of Neuroscience Methods, 165*, 151–161.
Lado, F. A., & Moshé, S. L. (2008). How do seizures stop? *Epilepsia, 49*(10), 1651–1664.
Lillis, K. P., Kramer, M. A., Mertz, J., Staley, K. J., & White, J. A. (2012). Pyramidal cells accumulate chloride at seizure onset. *Neurobiology of Disease, 47*(3), 358–366.
Löscher, W., & Köhling, R. (2010). Functional, metabolic, and synaptic changes after seizures as potential targets for antiepileptic therapy. *Epilepsy & Behavior, 19*, 105–113.
Lovatt, D., Xu, Q., Liu, W., Takano, T., Smith, N. A., Schnermann, J., et al. (2012). Neuronal adenosine release, and not astrocytic ATP release, mediates feedback inhibition of excitatory activity. *Proceedings of the National Academy of Sciences of the United States of America, 109*(16), 6265–6270.
Magnott, V. A., Heo, H.-Y., Dlouhy, B. J., & Dahdaleh, N. S. (2012). Detecting activity-evoked pH changes in human brain. *Proceedings of the National Academy of Sciences of the United States of America, 109*(21), 8270–8273.
Mormann, F., Lehnertz, K., David, P., & Elger, C. E. (2000). Mean phase coherence as a measure for phase synchronization and its application to the EEG of epilepsy patients. *Physica D: Nonlinear Phenomena, 144*(3–4), 358–369.
Payne, J. A., Rivera, C., Voipio, J., & Kaila, K. (2003). Cation-chloride co-transporters in neuronal communication, development and trauma. *Trends in Neuroscience, 26*, 199–206.
Pijnenburg, Y. A., vd Madeb, Y., van Cappellen van Walsum, A. M., Knol, D. L., Scheltens, Ph., & Stam, C. J. (2004). EEG synchronization likelihood in mild cognitive impairment and Alzheimer's disease during a working memory task. *Clinical Neurophysiology, 115*, 1332–1339.
Pikovsky, A., Rosenblum, M., & Kurths, M. (2002). *Synchronization: A universal concept in nonlinear science.* New York: Cambridge University Press.
Quian Quiroga, R., Kraskov, A., Kreuz, T., & Grassenberg, P. (2002). Performance of different synchronization measures in real data: A case study on electroencephalographic signals. *Physical Review E, 65*, 0419031–0419039.
Rubinov, M., & Sporns, O. (2010). Complex network measures of brain connectivity: uses and interpretations. *Neuroimage, 52*(3), 1059–1069.
Rulkov, N. F., Sushchik, M. M., & Tsimring, L. S. (1995). Generalized synchronization of chaos in directionally coupled chaotic systems. *Physical Review E, 51*, 980–994.
Scheffer, M., Bascompte, J., Brock, W. A., Brovkin, V., Carpenter, S. R., et al. (2009). Early-warning signals for critical transitions. *Nature, 461*, 53–59.
Schevon, C. A., Weiss, S. A., McKhann, G. L., Goodman, R. R., Yuste, R., Emerson, R. G., et al. (2012). Evidence of an inhibitory restraint of seizure activity in humans. *Nature Communications, 3*, 1600.
Schindler, K., Elger, C. E., & Lehnertz, K. (2007). Increasing synchronization may promote seizure termination: Evidence from status epilepticus. *Clinical Neurophysiology, 118*, 1955–1968.

Schindler, K., Leung, H., Elger, C. E., & Lehnertz, K. (2007). Assessing seizure dynamics by analysing the correlation structure of multichannel intracranial EEG. *Brain, 130*(1), 65–77.

Schwartz, T. H. (2007). Neurovascular coupling and epilepsy: Hemodynamic markers for localizing and predicting seizure onset. *Epilepsy Currents*, 7(4), 91–94.

Schindler, K., Gast, H., Stieglitz, L., Stibal, A., Hauf, M., et al. (2011). Forbidden ordinal patterns of periictal intracranial EEG indicate deterministic dynamics in human epileptic seizures. *Epilepsia*, 1528–1167.

Shorvon, S. D. (1994). *Status epilepticus: its clinical form and treatment in children and adults.* Cambridge: Cambridge University Press.

Sinning, A., & Hübner, C. A. (2013). Minireview: pH and synaptic transmission. *FEBS Letters, 587*(13), 1923–1928.

Staley, K. J., Longacher, M., Bains, J. S., & Yee, A. (1998). Presynaptic modulation of CA3 network activity. *Nature Neuroscience, 1*(3), 201–209.

Stam, C. J., & Van Straaten, E. C. W. (2012). The organization of physiological brain networks. *Clinical Neurophysiology, 123*(6), 1067–1087.

Strogatz, S. H. (2001). *Nonlinear dynamics and chaos with applications to physics, biology, chemistry, and engineering.* (pp. 241–297) Cambridge MA: Westview Press.

Timofeev, I., & Steriade, M. (2004). Neocortical seizures: Initiation, development and cessation. *Neuroscience, 123*(2), 299–336.

Tolner, E. A., Hochman, D. W., Hassinen, P., Otáhal, J., Gaily, E., Haglund, M. M., et al. (2011). Five percent CO_2 is a potent, fast-acting inhalation anticonvulsant. *Epilepsia, 52*, 104–114.

Topolnik, L., Steriade, M., & Timofeev, I. (2003). Partial cortical deafferentation promotes development of paroxysmal activity. *Cerebral Cortex, 13*(8), 883–893.

Traynelis, S. F., & Cull-Candy, S. G. (1990). Proton inhibition of N-methyl-D-aspartate receptors in cerebellar neurons. *Nature, 345*(6273), 347–350.

Uhlhaas, P. J., Linden, D. E. J., Singer, W., Haenschel, C., Lindner, M., Maurer, K., et al. (2006). Dysfunctional long-range coordination of neural activity during Gestalt perception in schizophrenia. *The Journal of Neuroscience, 26*(31), 8168–8175.

Uhlhaas, P. J., & Singer, W. (2006). Neural synchrony in brain disorders: Relevance for cognitive dysfunctions and pathophysiology. *Neuron, 52*, 155–168.

Van Diessen, E., Diederen, S. J. H., Braun, K. P. J., Jansen, F. E., & Stam, C. J. (2013). Functional and structural brain networks in epilepsy: What have we learned? *Epilepsia, 54*(11), 1855–1865.

Van Diessen, E., Hanemaaijer, J. I., Otte, W. M., Zelmann, R., Jacobs, J., Jansen, F. E., et al. (2013). Are high frequency oscillations associated with altered network topology in partial epilepsy? *NeuroImage, 82C*, 564–573.

Varela, F., Lachaux, J.-P., Rodriguez, E., & Martinerie, J. (2001). The brainweb: Phase synchronization and large-scale integration. *Nature Review Neuroscience, 2*, 229–239.

Velísek, L., Dreier, J. P., Stanton, P. K., Heinemann, U., & Moshé, S. L. (1994). Lowering of extracellular pH suppresses low-Mg(2+)-induces seizures in combined entorhinal cortex-hippocampal slices. *Experimental Brain Research, 101*(1), 44–52.

Victor, J. D. (2005). Spike train metrics. *Current Opinion in Neurobiology, 15*, 585–592.

Von Gompel, J. J., Bower, M. R., Worrell, G. A., Stead, M., Chang, S. Y., Goerss, S. J., et al. (2014). Increased cortical extracellular adenosine correlates with seizure termination. *Epilepsia, 55*(2), 233–244.

Warren, C. P., Hu, S., Stead, M., Brinkmann, B. H., Bower, M. R., & Worrell, G. A. (2010). Synchrony in normal and focal epileptic brain: the seizure onset zone is functionally disconnected. *Journal of Neurophysiology, 104*(6), 3530–3539.

Wemmie, J. A., Taugher, J., & Collin, J. K. (2013). Acid-sensing ion channels in pain and disease. *Nature Reviews Neuroscience, 14*, 461–471.

Wendling, F., Ansari-Asl, K., Bartolomei, F., & Senhadji, L. (2009). From EEG signals to brain connectivity: A model-based evaluation of interdependence measures. *Journal of Neuroscience Methods*, *183*, 9–18.

Wendling, F., Bartolomei, F., Bellanger, J. J., Bourien, J., & Chauvel, P. (2003). Epileptic fast intracerebral EEG activity: Evidence for spatial decorrelation at seizure onset. *Brain*, *126*, 1449–1459.

Young, D., & Dragunow, M. (1994). Status epilepticus may be caused by loss of adenosine anticonvulsant mechanisms. *Neuroscience*, *58*, 245–261.

Ziemann, A. E., Schnizler, M. K., Albert, G. W., Severson, M. A., Howard, M. A., Welsh, M. J., et al. (2008). Seizure termination by acidosis depends on ASIC1a. *Nature Neuroscience*, *11*(7), 816–822.

Zubler, F., Gast, H., Abela E., Rummel, C., Hauf, M, Wiest, R., Pollo, C., Schindler. (2014). Detecting Functional Hubs of Ictogenic Networks. Brain Topography, 1–13

CHAPTER NINE

Epileptic Focus and Alteration of Metabolism

Jakub Otáhal[*,1], **Jaroslava Folbergrová**[*], **Richard Kovacs**[†], **Wolfram S. Kunz**[‡], **Nicola Maggio**[§,¶]

[*]Institute of Physiology, v.v.i., Academy of Sciences of the Czech Republic, Prague, Czech Republic
[†]Institute for Neurophysiology, Charité–Medical University Berlin, Berlin, Germany
[‡]Department of Epileptology, University of Bonn, Bonn, Germany
[§]Department of Neurology, The Joseph Sagol Neuroscience Center, The Chaim Sheba Medical Center, Tel HaShomer, Israel
[¶]Talpiot Medical Leadership Program, The Chaim Sheba Medical Center, Tel HaShomer, Israel
[1]Corresponding author: e-mail address: jotahal@epilepsy.biomed.cas.cz

Contents

1. Introduction 210
2. Acute Changes in Energy Metabolism During Seizures 212
3. Metabolic Dysfunction of Chronic Epileptic Tissue 216
   3.1 Oxidative stress 217
   3.2 Mitochondrial dysfunction 217
4. Alteration of Mitochondrial Genome in Chronic Epileptic Tissue 220
5. Changes in Blood Flow Regulation in Epileptic Tissue 224
6. Alteration of Blood Brain Barrier 228
   6.1 Blood brain barrier breakdown: Cause or consequence of epilepsy? 228
   6.2 Blood brain barrier: A site of pharmacoresistance in epilepsy? 231
7. Conclusions 232

Acknowledgments 233

References 233

Abstract

Epilepsy is one of the most common neurologic disorders affecting a substantial part of the population worldwide. Epileptic seizures represent the situation of increased neuronal activity associated with the enhanced demands for sufficient energy supply. For that purpose, very efficient regulatory mechanisms have to operate to ensure that cerebral blood flow, delivery of oxygen, and nutrients are continuously adapted to the local metabolic needs. The sophisticated regulation has to function in concert at several levels (systemic, tissue, cellular, and subcellular). Particularly, mitochondria play a key role not only in the energy production, but they are also central to many other processes including those leading to neuronal death. Impairment of any of the involved pathways can result in serious functional alterations, neurodegeneration, and potentially in epileptogenesis. The present review will address some of the important issues concerning vascular and metabolic changes in pathophysiology of epilepsy.

International Review of Neurobiology, Volume 114
ISSN 0074-7742
http://dx.doi.org/10.1016/B978-0-12-418693-4.00009-1

1. INTRODUCTION

Epilepsy is a disorder of the brain characterized by enduring predisposition to generate unprovoked epileptic seizures. Due to the general agreement that epilepsy and/or seizures are result of an unbalance between excitatory and inhibitory neurotransmitter systems, epilepsy research focused historically on alterations of synaptic transmission. However, due to recent dramatic progress in diagnostic and imaging techniques and their wide use in both clinical and experimental epileptology, we have now increasing body of evidence that epilepsy should be considered more complex than faulty synaptic function. Indeed, many nonsynaptic and non-neuronal changes of the brain tissue have been shown to accompany epileptogenesis and epilepsy. Alteration of cellular metabolism, vascular reorganization, impairment of blood brain barrier (BBB), inflammatory response including astrocytic transformation, and remodeling of neuronal circuits are some of these nonsynaptic alterations of the epileptic brain. These changes can be found in all—from systemic to molecular—levels. Moreover, some of these alterations have been shown to be not only the consequence of seizures, but they can render the tissue seizure prone. Therefore, better understanding of these processes is crucial not only for elucidation of principles that underlie epileptogenesis and ictogenesis but also for developing better diagnostic and therapeutic approaches in epileptology.

In contrast to the other peripheral capillaries cerebral vasculature has complex structure to fulfill specific functions. On one hand, the regional cerebral blood flow (CBF) has to follow metabolic demands of the surrounding nervous tissue; on the other hand, brain capillary wall represents wide area for controlled bidirectional exchange of specific substrates between brain and blood. Both functions have been shown to be altered during both acute epileptic seizures and chronic epilepsy. Moreover, modified substrate exchange on BBB interferes with antiepileptic drug treatment. The precise mechanisms of these changes are not known; however, astrocytic inflammatory transformation and severe oxidative stress are suspected causes of epileptic vascular transformation. The major site of production of superoxide, the primary reactive oxygen species (ROS), is considered to be the respiratory chain in the mitochondria. Mitochondria are dynamic intracellular organelles with multiple functions. The central role of mitochondria is the synthesis of ATP by oxidative phosphorylation, carried out by a series of five large multi-subunit complexes (I–V) embedded in the inner mitochondrial membrane

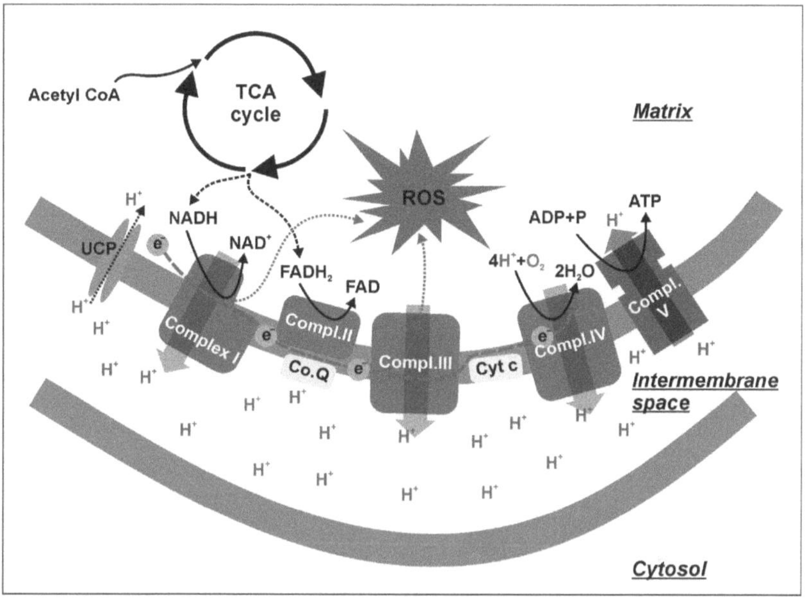

Figure 9.1 A schematic drawing of the mitochondrial respiratory chain and important sites of the ROS production.

(see Fig. 9.1). In addition to this energy-providing function, mitochondria are also central to many other processes, like neurotransmitter synthesis, maintenance of calcium homeostasis, redox signaling, the production and modulation of ROS, and they are also intimately involved in pathways leading to neuronal death (Davis & Williams, 2012; Duchen, 2000; Ott, Gogvadze, Orrenius, & Zhivotovsky, 2007; Wallace, Fan, & Procaccio, 2010). Besides short-term regulation of oxidative metabolism, mitochondrial function also depends on their assembly, maintenance, and mitochondrial network dynamics. Mitochondrial proteins have dual origin, being encoded by both the mitochondrial DNA (mtDNA) and by nuclear genes, translated in the cytoplasm, and then imported into mitochondria. Mitochondria continuously undergo fission and fusion cycles, which is intimately linked to refurbishment and removal of dysfunctional mitochondria. Mitochondrial biogenesis and turnover is regulated and coordinated by multiple transcription factors and coactivators that enable mitochondria to respond to alterations in metabolic demand. The regulation of anterograde/retrograde mitochondrial trafficking is important to ensure energy supply to the distant parts of the cells (Safiulina & Kaasik, 2013). Disruption to mitochondrial assembly, turnover, and function contributes to many human pathologies, raising the need for therapies and making mitochondria a potentially important therapeutic target (Smith, Hartley, Cocheme, & Murphy, 2012). Accumulating evidence indicates that oxidative stress and mitochondrial dysfunction play an important role in the pathogenesis of epilepsy (Folbergrová & Kunz, 2012; Patel, 2004; Rowley & Patel, 2013; Waldbaum & Patel, 2010).

In the following, we review current knowledge concerning metabolic and vascular changes not only during the acute phase of seizures but also during epileptogenesis and chronic epilepsy. Other nonsynaptic and/or nonneuronal changes including astrocytic transformation and inflammation in epilepsy are reviewed in other recent articles (Kovacs, Heinemann, & Steinhauser, 2012; Marchi et al., 2014; Vezzani, Friedman, & Dingledine, 2013).

2. ACUTE CHANGES IN ENERGY METABOLISM DURING SEIZURES

Increases in neuronal activity result in enhancement of brain energy metabolism in order to keep pace with the ATP-demand for synaptic signaling and maintenance of transmembrane ion gradients. Neurometabolic coupling operates via "pull" and "push"-mechanisms, i.e., either as a consequence of alterations in ATP or other energy metabolism intermediates, or in a feed-forward manner by activity-dependent changes in intracellular and intramitochondrial [Ca^{2+}], [K^{+}], and nitric oxide (NO) concentrations. A new level of feed-forward regulation was added by the discovery of mitochondrial G-protein-coupled neurotransmitter receptors (Benard et al., 2012). Changes in [Ca^{2+}], [K^{+}], and NO can immediately modify mitochondrial membrane potential ($\Delta\Psi_m$) and regulate the activity of electron transport chain complexes and mitochondrial dehydrogenases. On the long run, neuronal activity might alter mitochondrial network by influencing mitochondrial mobility, distribution, as well as fusion and fission of individual mitochondria. Metabolic segregation occurs in close proximity of markedly different subcellular metabolic compartments (Hertz et al., 2013; Liotta et al., 2012, for a detailed description of the mechanism of neurometabolic coupling, the reader is referred to current reviews: Andreux, Houtkooper, & Auwerx, 2013; Gellerich et al., 2013; Kann & Kovacs, 2007; Nicholls & Budd, 2000).

As seizures represent an extraordinary metabolic burden, energy metabolism has to be pushed to its limits. The alterations in the mechanisms of neurometabolic coupling might contribute to seizure-induced mitochondrial and cell damage (Kann, Huchzermeyer, Kovacs, Wirtz, & Schuelke, 2011; Kann & Kovacs, 2007; Kovacs et al., 2002). Animal models of epilepsy allow measurement of metabolism-related parameters, such as the ratio of reduced/oxidized flavin and adenine dinucleotides ($FAD/FADH_2$, $NAD^+/NADH$, $NADP^+/NADPH$), the tissue partial oxygen tension (pO_2), or the level of lactate and high-energy phosphates. However, the spatial (and often the temporal) resolution of such recordings is limited. Recently, the transgenic animals harboring different mutations offer the possibility to investigate the mitochondrial contribution to the development of epilepsy (Kudin, Zsurka, Elger, & Kunz, 2009; Liang & Patel, 2004). Isolated mitochondria and cell cultures are widely used to study the regulation of neuronal energy metabolism. Unfortunately, only a subset of conditions occurring during seizures can be modeled in these systems (Malinska et al., 2010). With increasing complexity of the epilepsy models (i.e., primary neuronal cultures, brain slices, and slice cultures), epileptiform activity more and more resembles seizures at the expense of losing information about the heterogeneity of mitochondrial responses. In the following, we will summarize recent experimental data on epileptiform activity-induced changes in energy metabolism in *in vitro* and animal models of epilepsy.

Seizures are associated with massive ion movements and ion gradients have to be restored by active pump transports (consuming the majority of energy) in order to protect against cell swelling and necrotic cell death (Erecinska & Dagani, 1990; Hertz et al., 2013; Lux, Heinemann, & Dietzel, 1986). Potassium is an important determinant of $\Delta\Psi_m$ and consequently of the proton motive force that drives ATP synthesis (Nicholls & Budd, 2000). Several types of potassium channels are expressed on the mitochondrial membrane (Bednarczyk, 2009; Szewczyk, Jarmuszkiewicz, & Kunz, 2009). Potential neuroprotective role has been attributed to these channels in ischemic preconditioning but their involvement in epilepsy is rather hypothetic (Busija, Gaspar, Domoki, Katakam, & Bari, 2008). Although the exact mechanism of neuroprotection is still elusive, potassium-dependent depolarization of mitochondria might prevent deleterious mitochondrial Ca^{2+}-overload and ROS formation (see below and Szewczyk et al., 2009).

Seizures and seizure-like events (SLEs) *in vitro* are associated with massive intracellular Ca^{2+} ($[Ca^{2+}]_i$) transients (Kovacs, Szilagyi, Barabas, Heinemann, & Kardos, 2000). Intracellular Ca^{2+}-load is discussed as the initiator of excitotoxic neuron loss in epilepsy and mitochondrial Ca^{2+}-buffering is a key component in the cascade leading to cell death.

Mitochondrial Ca^{2+}-accumulation readily occurs during physiological stimulation of brain slices and slice cultures (Kann, Kovacs, & Heinemann, 2003; Kann et al., 2012; Williams, Boyman, Chikando, Khairallah, & Lederer, 2013) and also during spontaneous seizure-like activity (Kovacs, Kardos, Heinemann, & Kann, 2005; Kovacs, Schuchmann, Gabriel, Kardos, & Heinemann, 2001). Mitochondrial Ca^{2+}-uptake is driven by $\Delta\Psi_m$ and mediated by the electrogenic mitochondrial calcium uniporter (MCU), whose molecular identity was revealed recently (Pan et al., 2013). Ca^{2+} represents a feed-forward regulator of metabolic load, as Ca^{2+} increases the activity of mitochondrial dehydrogenases, the electron transport chain, and the malate/aspartate exchanger (Denton, 2009; Gellerich et al., 2013; Territo, Mootha, French, & Balaban, 2000). The extrusion pathways for Ca^{2+} involve a Na–Ca (NCX) and Ca–H exchangers that also use $\Delta\Psi_m$ for Ca^{2+}-clearance (Nicholls & Budd, 2000). Thus, mitochondrial Ca^{2+}-signaling occurs in expense of a decrease in $\Delta\Psi_m$ during physiological stimuli (Gerencser et al., 2012) and even more so during SLEs (Kovacs et al., 2001). A futile Ca^{2+}-cycling seems to be responsible for $\Delta\Psi_m$ loss during SLEs, as depolarization of the mitochondrial network could be prevented by blocking the MCU and the NCX (Kovacs et al., 2005). However, a contribution of mitoBK$_{Ca}$ cannot be excluded completely. In contrast to the short-lasting, mitochondrial depolarization observed during physiological stimuli, the loss of $\Delta\Psi_m$ outlasted SLEs by minutes indicating lasting restriction of the proton motive force (Gerencser et al., 2012; Kovacs et al., 2005; Schuchmann, Lückermann, Kulik, Heinemann, & Ballanyi, 2000). In excitotoxicity, mitochondrial Ca^{2+}-load is the main initiating event for opening the mitochondrial permeability transition pore (mPTP). mPTP opening leads to release of proapoptotic factors, restriction of energy metabolism, and finally to cell death. Therefore, it is tempting to speculate, that mPTP is a potential factor causing cell damage following sustained seizure activity such as status epilepticus (SE) (Kudin et al., 2009; Meldrum, 1993).

The regulatory effect of Ca^{2+} on mitochondrial dehydrogenases can be studied by monitoring redox changes of the electron carrier flavin and adenine nucleotides. Reduced NADH and NADPH (referred as NAD(P)H) as well as oxidized FAD are more fluorescent then their respective counterparts (Kovacs et al., 2001; Mayevsky & Chance, 1982; Schuchmann, Kovacs, Kann, Heinemann, & Buchheim, 2001; Shuttleworth, 2010). Pharmacological or electrical stimulation of brain slices results in biphasic changes in the NAD(P)H and FAD fluorescence. An initial short-lasting oxidative shift of the nucleotides is followed by a lasting reduction ("dip" and "overshoot" or "peak" and "undershoot") for NAD(P)H and FAD, respectively (Brennan, Connor, & Shuttleworth, 2006; Huchzermeyer et al., 2008; Kasischke, Vishwasrao, Fisher, Zipfel, & Webb, 2004; Liotta et al., 2012). In analogy to physiological stimuli, SLEs were associated with biphasic redox changes with a reducing shift outlasting the activity by several minutes (Kovacs et al., 2001; Malinska et al., 2010). Although we could not unambiguously determine whether the components of the biphasic fluorescent transients originate from the same cell type (i.e., neurons or glia), the data strongly suggest that a surplus of reducing equivalents are available during and following seizures. This is of particular importance, as over-reduction of the electron transport chain complexes would favor ROS formation (Malinska et al., 2010). Indeed, the amplitude of the biphasic NAD(P)H fluorescence responses decayed during the course of recurrent SLEs (Kovacs et al., 2002; Schuchmann et al., 2001) and this decay could be prevented by ROS scavengers, suggesting that ROS-mediated metabolic dysfunction and cell damage occur during intense seizure activity (Kovacs et al., 1999, 2002). Similar deterioration of NAD(P)H fluorescence transients could be observed in chronic epileptic brain tissue, obtained either from the animal models of epilepsy or from surgery of human epilepsy (Kann et al., 2005; Kudin et al., 2009).

In the *in vitro* brain slices with constant carbogen (95% O_2 5% CO_2) aeration, tissue pO_2 is determined solely by diffusion distance from the surface and the rate of respiration. Thus, oxygen consumption during physiological or epileptiform activity can be compared by simply monitoring changes in pO_2 (Hajos et al., 2009; Huchzermeyer, Berndt, Holzhutter, & Kann, 2013). SLEs were associated with massive decrease in pO_2 of $\sim$120 mm Hg in an approximate depth of 100–150 μm from the slice surface (Kann et al., 2011). Nevertheless, the tissue was still hyperoxic ($\sim$150 mm Hg) during SLEs as compared to pO_2 *in vivo*, suggesting that oxygen availability is not limiting respiration. Interestingly, the amplitude of pO_2 transients during carbachol-induced gamma oscillations were comparable to the changes induced by seizure-like activity, indicating that there is a "ceiling" level for enhancement of oxidative energy metabolism in brain slices (Kann et al., 2011). *In vivo* CBF and oxygen delivery are continuously adapted to the local metabolic needs by neuronal activity-dependent release of vasoactive agents (Zhao et al., 2011) (see below), which complicates the measurement of the actual oxygen consumption.

There is a discrepancy between the kinetics of pO_2 changes and the NAD(P)H fluorescence signals. While in the latter the oxidative "dip" was rather short, pO_2 decreased during the whole SLE and recovered only after cessation of the activity. The lasting overshoot of reduced NAD(P)H during and following SLEs indicates that the delivery of reducing equivalents overwhelms oxidation by the electron transport chain. High $[Ca^{2+}]_m$ and NAD(P)H levels might explain enhanced ROS formation in the *in vitro* models as well as in animal models of epilepsy (Folbergrová, Otáhal, & Druga, 2012; Frantseva, Velazquez, Hwang, & Carlen, 2000; Kovacs et al., 2002, 2001; Liang, Ho, & Patel, 2000; Malinska et al., 2010). The main ROS-forming sites within mitochondria are the electron chain complexes I and III, although other flavoprotein enzymes such as α-ketoglutarate dehydrogenase and α-glycerphosphate dehydrogenase might also contribute to superoxide formation (Kudin, Bimpong-Buta, Vielhaber, Elger, & Kunz, 2004; Malinska et al., 2010) (see Fig. 9.1). The increased superoxide formation during SLEs in brain slices and slice cultures has been demonstrated using fluorescence probes such as dihydroethidine or MitoSox (Kovacs et al., 2009, 2001; Robinson, Janes, & Beckman, 2008). By contrast to biphasic NAD(P)H transients, ROS formation increased during the course of recurrent SLEs, suggesting that ROS-induced metabolic dysfunction fosters ROS formation in a vicious cycle.

Besides ROS, also nitrogen-centered radical species (RNS) might contribute to the deleterious effect of seizures. Enhanced NO formation was shown in both *in vivo* and *in vitro* models of epilepsy (Kaneko, Itoh, Berliner, Miyasaka, & Fujii, 2002; Kovacs et al., 2009; Schuchmann, Albrecht, Heinemann, & von Bohlen und Halbach, 2002). Additionally to its regulatory effect on vascular tone, NO immediately alters mitochondrial metabolism, as it inhibits electron transport chain complexes and activates mitochondrial potassium channels (Brown, 2010; Szewczyk et al., 2009). In the presence of superoxide, NO forms the highly toxic peroxynitrite that might contribute to neurotoxicity, impairment of neurovascular coupling, and the seizure-induced opening of the BBB (Danjo, Ishihara, Watanabe, Nakamura, & Itoh, 2013; Winkler et al., 2012).

3. METABOLIC DYSFUNCTION OF CHRONIC EPILEPTIC TISSUE

Many metabolic changes have been reported to occur following acute seizures and during various phases of epileptogenesis and chronic epilepsy. As already mentioned above, there is increasing evidence that oxidative stress and mitochondrial dysfunction are implicated in the pathogenesis of epilepsy. Therefore, interest in the present section will be focused on some aspects of these issues.

3.1. Oxidative stress

Prolonged seizure activity increases superoxide anion production in several brain regions, as demonstrated by its detection *in situ* using dihydroethidine (Folbergrová et al., 2012; Peterson, Morrow, Liu, & Liu, 2002). The enhanced ROS formation was also demonstrated during SLEs *in vitro* (see Section 2). Furthermore, the increase in mitochondrial hydrogen peroxide production has been reported following kainate (KA)- and lithium–pilocarpine-induced epileptogenesis (Jarrett, Liang, Hellier, Staley, & Patel, 2008; Waldbaum, Liang, & Patel, 2010). Existence of oxidative stress is also evident by the increased lipid peroxidation, and by the modification (inactivation) of sensitive targets in brain, such as the TCA cycle enzymes aconitase (Folbergrová et al., 2007; Liang et al., 2000; Sleven, Gibbs, Heales, Thom, & Cock, 2006; Vielhaber et al., 2008) and α-KDH (Cock et al., 2002), mitochondrial electron transport chain complex I (Folbergrová et al., 2007; Folbergrová, Ješina, Haugvicová, Lisý, & Houštěk, 2010; Kudin et al., 2002; Kunz et al., 2000; Ryan, Backos, Reigan, & Patel, 2012), or by the increased mitochondrial markers of oxidative damage, like protein carbonyl groups (Bruce & Baudry, 1995; Folbergrová et al., 2010; Ryan et al., 2012), 3-nitrotyrosine (3-NT), and 4-hydroxynonenal (4-HNE) (Folbergrová et al., 2010).

3.2. Mitochondrial dysfunction

The role of mitochondrial function/dysfunction in relation to epilepsy has been studied in humans and in several animal models of experimentally induced seizures. Severe impairment of respiratory chain complex I activity was detected in hippocampal CA3 neurons from patients with chronic temporal lobe epilepsy (TLE) (Kunz et al., 2000), as well as in CA1 and CA3 hippocampal fields of rats after pilocarpine-induced (Kudin et al., 2002) or kainic acid-induced (Chuang, Chang, Lin, Hsu, & Chan, 2004; Ryan et al., 2012) SE. The marked decrease of complex I activity was also demonstrated in cerebral cortex mitochondria of immature rats after SE induced by homocysteic acid (Folbergrová et al., 2007, 2010). This decrease persisted during long periods of survival (up to 5 weeks) periods, corresponding in this model to the development of spontaneous seizures (epileptogenesis). The decrease was selective for complex I and it was not associated either with changes in the size of the assembled complex I or with changes in complex I content of mitochondria (Folbergrová et al., 2007, 2010). The decrease was substantially reduced by treatment with selected free radical scavengers, namely, by the SOD mimetics MnTMPYP and Tempol, and by a selective peroxynitrite scavenger and decomposition catalyst FeTPPS. Concurrently with the decreased activities of complex I, significant increases in three markers of mitochondrial oxidative damage (3-NT, 4-HNE, and protein carbonyls) were detected (Folbergrová et al., 2010). All these findings strongly suggest that oxidative modification of complex I, localized potentially on some critical subunit of complex I, is very likely responsible for the sustained deficiency of complex I activity, in accordance with extreme sensitivity of this enzyme to ROS and RNS (Folbergrová et al., 2010 and references therein). A decreased activity of complex I during both the acute and the chronic phases of epileptogenesis after kainate-induced SE in adult rats was accompanied by increased protein carbonyls (Ryan et al., 2012). Mass spectrometry analysis identified specific metal catalyzed carbonylation to arginine 76 within the 75 kDa subunit of complex I. Computational-based molecular modeling studies predicted that carbonylation at this site can induce substantial structural alterations of the protein complex leading to the impaired function.

Most likely also the other posttranslational oxidative modifications of complex I can occur, e.g., by nitration of tyrosine (and/or tryptophan) residues within the complex, apparently due to the peroxynitrite- or peroxynitrite-derived radicals (Murray, Taylor, Zhang, Ghosh, & Capaldi, 2003; Radi, Cassina, Hodara, Quijano, & Castro, 2002; Riobo et al., 2001; Singh, Sullivan, & Hall, 2007; Yamamoto et al., 2002). Pearce, Kanai, Epperly, and Peterson (2005) demonstrated that irreversible inhibition of complex I activity was accompanied by formation of 3-NT in 3 out of 46 subunits. Interestingly, tyrosine nitration has also been detected on glutamine synthetase, accompanied by the reduction of enzyme activity in the pentylenetetrazol model of epilepsy (Bidmon et al., 2008). However, other potential mechanisms can be considered, such as S-nitrosation of some of complex I protein thiols, oxidation of iron–sulfur clusters and others. It can be expected that such a marked inhibition of complex I activity will have significant implications for mitochondrial function. In immature rats, it has been shown that pronounced inhibition of complex I was not accompanied by impaired ATP production, apparently due to the excess capacity of complex I demonstrated by energy thresholds, i.e., the extent to which mitochondrial complex activities can be inhibited before significant changes in oxidative phosphorylation occur (Folbergrová et al., 2007, 2010 and references therein). There are, however, other potential consequences of the complex I inhibition. It is well established that complex I is not only a target for ROS and RNS, but it is also the important source of ROS and/or RNS production, especially when partially inhibited (Fato, Bergamini, Leoni,

Strocchi, & Lenaz, 2008; Kudin et al., 2004; Kussmaul & Hirst, 2006; Parihar, Parihar, Villamena, Vaccaro, & Ghafourifar, 2008; Sipos, Tretter, & Adam-Vizi, 2003). Therefore, it can be assumed that increased ROS and/or RNS formation may contribute to neuronal injury demonstrated in several models of epilepsy (Folbergrová & Kunz, 2012; Waldbaum & Patel, 2010).

It has been shown that high-affinity astroglial and neuronal glutamate transporters, which are important for maintaining low levels of synaptic glutamate, are extremely sensitive to oxidative damage, resulting in reduced uptake function (Trotti, Danbolt, & Volterra, 1998). Their inhibition (similarly as inhibition of glutamine synthetase, Bidmon et al., 2008) can potentially contribute to the increased excitability observed during the development of spontaneous seizures, thus contributing to epileptogenesis. Simantov et al. (1999) reported downregulation of the neuronal transporter EAAT3 in rat hippocampus within hours following kainate-induced seizures. Genetically altered mice lacking the glial transporter GLT-1 (EAAT2) have been shown to develop spontaneous seizures (Tanaka et al., 1997) and a decrease in EAAT2 expression was observed in patients with TLE and hippocampal sclerosis (Mathern et al., 1999). Decreased expression both of glial and neuronal glutamate transporters was detected in the cortex of rats with genetic absence epilepsy (Dutuit et al., 2002). Interestingly, increased EAAT2 glial glutamate transporter expression was shown recently to reduce epileptogenic process following pilocarpine-induced SE (Kong et al., 2012).

Polyunsaturated fatty acids present in phospholipids of biological membranes and cardiolipin localized within mitochondrial inner membrane are highly susceptible to oxidation by ROS. Oxidative damage to lipids, evident as an increase of several markers of lipid peroxidation (increase of F(2) isoprostanes, isofuranes, thiobarbituric acid reactive substances, 4-HNE) has been detected following SE induced both in adult (Dal-Pizzol et al., 2000; Patel et al., 2008; Patel, Liang, & Roberts, 2001) and in immature animals (Folbergrová et al., 2007, 2010).

Another target sensitive to oxidative damage is DNA, particularly mitochondrial (mtDNA), as will be addressed in Section 4.

Finally, important findings concern redox status. Changes in cellular and mitochondrial redox status, evaluated by measuring GSH/GSSG and/or CoASH/CoASSG ratios, were detected following acute SE. These ratios decreased in hippocampus of adult rats 24 h after lithium–pilocarpine-induced SE and the alteration persisted throughout the epileptogenesis and chronic epilepsy (Waldbaum et al., 2010). Interestingly, a decrease of both GSH levels and glutathione reductase activity has been observed in brain regions of patients with epilepsy (Mueller, Trabesinger, Boesiger, & Wieser, 2001). These findings suggest that redox status shifts to a more oxidized state and the persisting oxidized environment may thus favor oxidative posttranslational modification of sensitive targets as mentioned above.

4. ALTERATION OF MITOCHONDRIAL GENOME IN CHRONIC EPILEPTIC TISSUE

In recent literature, there are accumulating hints for the contribution of oxygen radicals to the process of epileptogenesis and possibly also to later changes, responsible for development of chronic epilepsy (Folbergrová & Kunz, 2012). There is a direct evidence for the increased generation of oxygen radicals following SE, as mentioned in Section 3.1. Within the scenario of epileptogenesis, the initial insult-related oxygen radical-induced damage of proteins, lipids, and nucleic acids may result in serious alterations of normal central nervous system (CNS) function. Particularly mtDNA damage by ROS escaping repair is critical, since resulting mutations can be propagated and can expand to levels affecting the function of enzymes containing mtDNA-encoded subunits of the respiratory chain complexes I and IV, and the F0F1-ATP synthase (Wiesner, Zsurka, & Kunz, 2006).

After reaction with mtDNA, oxygen radicals cause the formation of 8-hydroxyguanosine (which leads to base exchange; Giulivi, Boveris, & Cadenas, 1995), of thymine glycol and formamidopyrimidine (which leads to replicative block) and of the so-called "AP-sites"—deletions of purine bases (Demple & Harrison, 1994). These mtDNA alterations can be repaired by base excision (Wiesner et al., 2006). However, the repair capacity of mtDNA is considered to be rather limited, since the nucleotide excision repair pathway, additionally present in the nucleus, is apparently missing within mitochondria. Extremely problematic are radical-induced DNA-strand breaks due to destruction of deoxyglucose by attack of hydroxyl radicals, which are formed from hydrogen peroxide by Fenton reaction (Imlay & Linn, 1988). mtDNA molecules containing double-strand breaks can be "misrepaired" by ligation with wrong ends. This mechanism explains the occurrence of large-scale rearrangements of mtDNA (deletions and duplications) as a consequence of double-strand breaks. Linear mtDNA molecules not undergoing double-strand break repair are prone to rapid destruction, which results in depletion of mtDNA (Bacman, Williams, &

Moraes, 2009). The spontaneously occurring deletions can thereafter clonally expand and after exceeding a certain threshold can lead to a segmental impairment of oxidative phosphorylation. In tissues consisting of postmitotic cells, like neurons, the cells harboring high levels of mutant mtDNA are not replaced, but accumulate in the tissue. This mechanism can explain high-mutation loads in aged single cells (fibroblasts, cardiomyocytes; Michikawa, Mazzucchelli, Bresolin, Scarlato, & Attardi, 1999; Nekhaeva et al., 2002). There have been reports about the age-dependent accumulation of mtDNA mutations, mainly deletions, in human brain (and particularly in substantia nigra), but in the whole brain tissue specimen only very low degrees of mtDNA heteroplasmy were detectable (Corral-Debrinski et al., 1992; Cortopassi, Shibata, Soong, & Arnheim, 1992). More recent single-cell studies revealed a substantial accumulation of mtDNA deletions in single substantia nigra neurons of old individuals, which levels were found to be sufficient to impair mitochondrial oxidative phosphorylation (Bender et al., 2006; Kraytsberg et al., 2006). The preference of deletion accumulation in dopaminergic neurons appears to be related to oxygen radicals formed in catecholamine metabolism (Neuhaus et al., 2014). However, the detailed cellular distribution and the influence of these mutations on the function of the human brain and their possible relationship to the pathogenesis of certain CNS disorder like Parkinsonism or epilepsy remains to be elucidated yet.

An impairment of mitochondrial oxidative phosphorylation has been demonstrated for hippocampal neurons, as described in Section 3.2. In the chronic epileptic tissue samples, decreased specific activities (in respect to the mitochondrial marker enzyme citrate synthase) of partially mitochondrial-encoded enzyme complexes I and IV of respiratory chain (NADH:CoQ oxidoreductase and cytochrome *c* oxidase) were detected. These activity changes lead to physiological relevant alterations of mitochondrial energy metabolism of pyramidal neurons (Kudin et al., 2002; Vielhaber et al., 2003). As possible cause of the local alterations of hippocampal mitochondrial function in the pilocarpine model of chronic TLE in rats (Kudin et al., 2002) and in human hippocampal sclerosis (Baron, Kudin, & Kunz, 2007) an about twofold decreased copy number of mtDNA was delineated. A plausible mechanism of mtDNA damage in TLE explaining the local impairment of oxidative phosphorylation is, therefore, very likely an oxygen radical-mediated increased frequency of mtDNA-strand breaks leading to mtDNA fragmentation and subsequently to mtDNA depletion (Baron et al., 2007; Folbergrová & Kunz, 2012). This hypothesis is supported by findings of lowered activity of mitochondrial aconitase, containing an extremely oxygen radical-sensitive iron–sulfur cluster, in hippocampal subfields showing decreased mtDNA copy numbers and decreased activities of respiratory chain complexes I and IV (Baron et al., 2007). More recent data obtained by single-molecule PCR and breakpoint sequencing showed that additionally to the observed mtDNA depletion also abundant mtDNA deletions can be detected in tissue samples from chronic epileptic hippocampus of TLE patients with hippocampal sclerosis (Guo et al., 2010). The distribution of deletional breakpoints in chronic epileptic tissue is apparently different to the deletional spectrum of the aging brain: deletions with 3′ breakpoints around np 16070 are in the chronic epileptic hippocampus extremely abundant, while they are much rare in the normal aging brain. Moreover, hippocampi from patients with HS contain over 10-fold more mtDNA deletions than normal hippocampi of similar age. It is very likely that these high levels of mtDNA deletions, similar to the abundant deletions in dopaminergic neurons (Neuhaus et al., 2014), are related to increased oxidative stress, relevant for development of chronic focal epilepsy (Kudin et al., 2009). These results based on single-molecule PCR and conventional-breakpoint sequencing were confirmed by unbiased deletion detection applying deep sequencing of mtDNA: Fig. 9.2, left panel, shows the deletional spectrum of a control brain sample (obtained from a patient with a parahippocampal lesion) and Fig. 9.2, right panel, shows the deletional

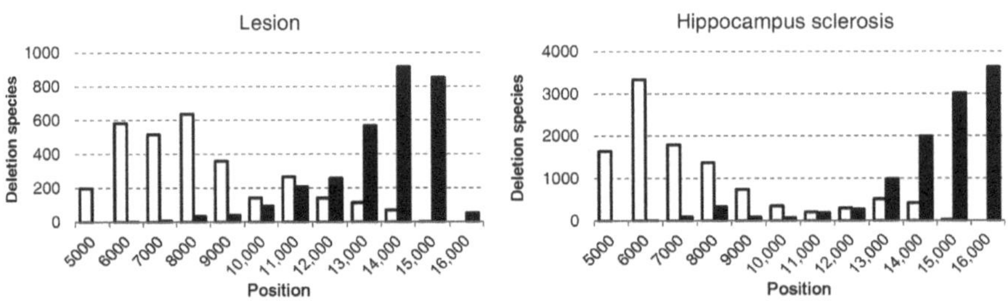

Figure 9.2 Deletional spectra in the CA3 region of the human hippocampus obtained by deep sequencing of the mitochondrial genome. Left panel: Breakpoints of deletions in a patient with a parahippocampal lesion show a distribution that corresponds to previously described deletional spectra from control human brain tissue obtained by single-molecule PCR (Guo et al., 2010). Right panel: Higher abundance of deletions is detected in a patient with hippocampus sclerosis. The highest bar on the right corresponds to deletions with endpoint around nucleotide position 16070. Empty bars, 5′ deletion start points; filled bars, 3′ deletion end points. Note the similarity of the next-generation deletion spectra here to previously published data obtained by single-molecule PCR (Guo et al., 2010; cf. Fig. 9.1A–D, respectively).

spectrum from a patient with hippocampal sclerosis of similar age. Importantly, not only the total number of deletions was much higher in the HS sample but also much more deletions with 3′ breakpoints around np 16070 have been observed. This breakpoint hot spot is preferentially generated by mtDNA double-strand breaks (Srivastava & Moraes, 2005) that can be formed by attacks of ROS (Imlay & Linn, 1988).

Although the direct relevance of observed alterations of mitochondrial genome (depletion and deletions) for seizure generation in chronic epilepsy remains to be elucidated yet, they provide a plausible explanation for the observed decline of activities of mitochondrial respiratory chain enzymes in the epileptic focus from patients with hippocampal sclerosis and in the pilocarpine rat model of TLE (Kudin et al., 2002; Kunz et al., 2000). In this context, it is noteworthy to mention that the impairment of mitochondrial oxidative phosphorylation is restricted not only to the pyramidal cells, but it is also detectable in the CA4 hilar interneurons (Blümcke et al., 1999; Kudin et al., 2009), which are important to maintain the sufficient inhibition of the entire network of hippocampal neurons. As shown in Fig. 9.3, surviving hilar hippocampal neurons from patients with TLE and hippocampal sclerosis are marked in blue in the cytochrome *c* oxidase–succinate dehydrogenase double staining, which is indicative for cytochrome *c* oxidase

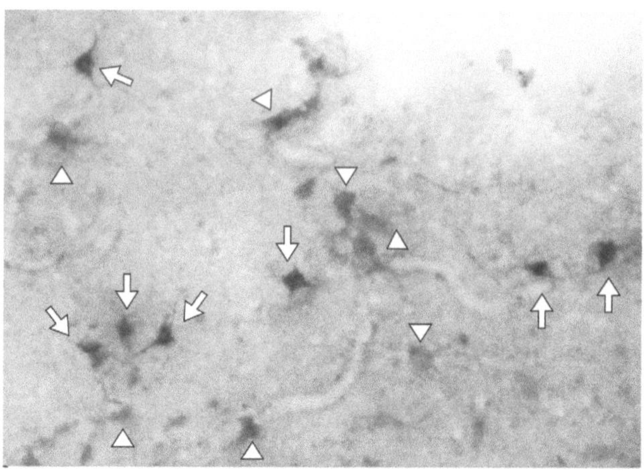

Figure 9.3 Cytochrome *c* oxidase-negative hilar neurons in the CA4 hippocampal region of a patient with temporal lobe epilepsy and hippocampal sclerosis. A 60 μm vibratome slice from the hippocampus has been stained simultaneously for the activities of cytochrome *c* oxidase (brown staining) and succinate dehydrogenase (blue staining). Cytochrome *c* oxidase negative neurons are colored in blue (arrows) while cytochrome *c* oxidase positive neurons are colored in brown (arrow heads).

deficiency. Since, cytochrome *c* oxidase contains three mitochondrially encoded subunits, which are essential for catalysis, its deficiency is usually related to mtDNA mutations. It is therefore reasonable to speculate that the mitochondrial dysfunction of particular mitochondria-rich interneurons (parvalbumin-positive basket cells) is related to high-mtDNA mutation loads, which might be an important contributor to the altered synchronization of network activity in chronic epileptic tissue. Thus, accumulating pathogenic alterations of the mitochondrial genome in hippocampal interneurons could promote the development of spontaneous seizures in the process of epileptogenesis, leading to the development of chronic epilepsy, which is proposed to be highly relevant for the development of hippocampal sclerosis (Folbergrová & Kunz, 2012; Kudin et al., 2009).

5. CHANGES IN BLOOD FLOW REGULATION IN EPILEPTIC TISSUE

Brain tissue is highly dependent on continuous and appropriate supply of nutrients (such as glucose and oxygen) and effective removal of final products of energy metabolism. Indeed, CBF is precisely regulated on both systemic and organ level to fulfill immediate needs of the brain (Iadecola & Nedergaard, 2007; Otáhal et al., 2007). To maintain CBF to certain limits independent on systemic blood pressure fluctuations (~60–150 mm Hg) cerebral vasculature has potent autoregulatory mechanisms (Iadecola & Nedergaard, 2007). However, decline in the CBF leads to functional alterations or even to disintegration of the nervous tissue (Hossmann, 2006). A chronic change in systemic blood pressure, especially hypertension, causes deleterious changes of morphology and function of the cerebral microvasculature. Hypertension is known to be one of the main risk factors of the cognitive decline and dementia (Iadecola & Davisson, 2008) and other neurological diseases including epilepsy (Chung et al., 2013; Li et al., 1997).

These cerebral arteries and arterioles have ordinary vessel structure consisting of endothelial, smooth muscle, and connective cells, including innervation arising from cranial autonomic and sensory ganglia (Iadecola & Davisson, 2008; Iadecola & Nedergaard, 2007). Different structure of the vessel wall is found on smaller cerebral arterioles (<100 μm) and capillaries. These vessels are fully covered by astrocytic processes forming together with remaining structures of the mature vessel semipermeable BBB. The schematic drawing of cross-section of typical cerebral capillary (neurovascular unit) is illustrated in Fig. 9.4. Neurons, astrocytes, pericytes, and vascular

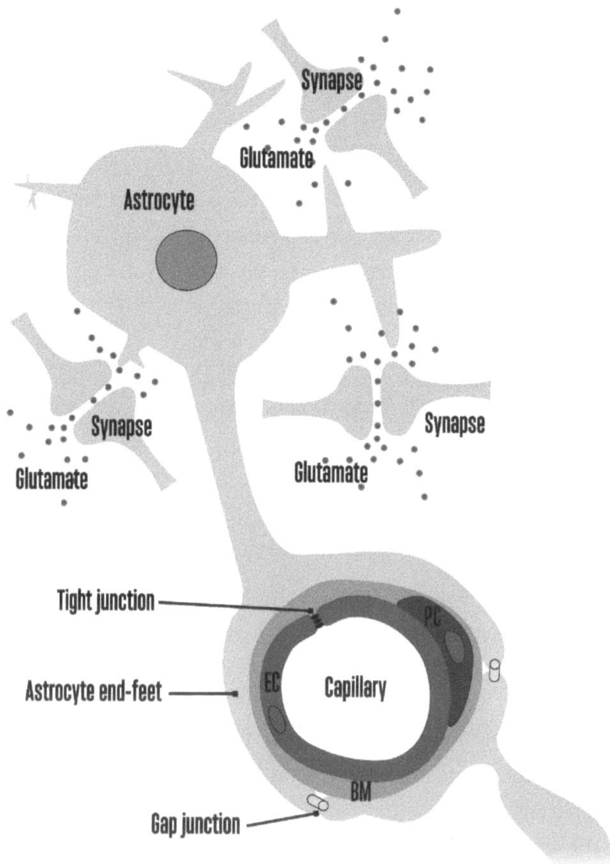

Figure 9.4 The neurovascular unit is morphological basis of blood brain barrier and neurovascular coupling. The endothelial cells (ECs), which line lumen of the brain capillaries, are interconnected by impermeable tight junctions. Basal lamina (BM) covers endothelial cells from abluminal side where contractile cells pericytes (PC) are attached to the endothelial cells. The whole capillary is enclosed by astrocytic end-feet which are connected with gap junctions. Besides the direct neuronal effect on vasculature, astrocytes are critically involved in regulation of hemodynamic responses. Astrocytic processes contact both the synapses and the abluminal surface of vessels. Astrocytes equipped with neurotransmitter receptors and transporters may release vasoactive substances regulating tonus of the vascular wall by influencing contractile pericytes.

cells are all components of the neurovascular unit and work in concert to maintain the homeostasis of the brain microenvironment (Iadecola & Nedergaard, 2007). Endothelial cells (ECs) of the BBB lack fenestrations and they are interconnected together with tight junctions forming impermeable monolayer (Kniesel & Wolburg, 2000). The molecular biology of the tight junctions of BBB is quite complex and integral proteins (occludins and claudins) are often affected during acute and chronic diseases of the brain (Zlokovic, 2008). The hydrophilic molecules must cross the endothelial wall to reach their neuronal targets or leave the brain using active transcellular transport systems, therefore, ECs of the BBB contain increased mitochondrial content. In addition, ECs and astrocytes provide a metabolic barrier (enzymatic BBB) by expressing various enzymes that modify endogenous and exogenous molecules which pass to the brain (Nico & Ribatti, 2012; Zlokovic, 2008). Activity of these enzymes is of interest in pharmacoresistant epilepsies where it can negatively influence delivery of antiepileptic drug and thus their levels in the brain (see Section 6). Astrocytes were shown to modulate permeability of BBB through influencing tight junctions (Abbott, 2002) and play crucial role in neurovascular coupling, i.e., regulation of the regional perfusion according to local neuronal activity by sensing synaptic activity and releasing vasoactive molecules (Gordon, Mulligan, & MacVicar, 2007). Pericytes, contractile cells which are attached in irregular intervals to the abluminal membrane of the ECs have important roles in a range of physiological functions, including local blood flow regulation, angiogenesis, vessel stabilization, EC regulation, and maintenance of the BBB integrity (Hamilton, Attwell, & Hall, 2010).

It should be emphasized that BBB is developed during cerebral vessel maturation. Angiogenesis, the process of formation of new blood vessels from preexisting vessels, is induced in many neurological disorders including epilepsy by releasing vascular endothelial growth factor (VEGF), prominent vascular endothelial growth inducer from astrocytes. The production and release of VEGF is generally started in oxygen-deficient cells by production of hypoxia inducible factor. The BBB is formed in late maturation phase of the angiogenic process, and thus in brain regions where angiogenesis occurs increased permeability of BBB is detected (Liebner, Czupalla, & Wolburg, 2011). Significant increase of vascular density, loss of tight junction proteins, and extravasation of blood proteins into the hippocampal tissue was reported in the pharmacoresistant TLE (Marchi & Lerner-Natoli, 2013). The angiogenic process was confirmed in these patients by detecting high levels of VEGF and overexpression of the receptor tyrosine kinase VEGFR-2 in resected hippocampal tissue (Rigau et al., 2007). Similar results, i.e., increased microvascular density, disruption of BBB, and increased VEGF expression in neurons and astrocytes were found during experimentally induced epileptogenesis in pilocarpine model in rat (Rigau et al., 2007).

The increased energy demand of active neuronal tissue is followed by local increases in blood flow in the area of increased neuronal activity. This functional local hyperemia is result of fast dilation of both cerebral arterioles and capillaries as consequence of increased neuronal activity. Velocity of response of regional blood flow after neuronal activation suggests prompt regulatory mechanism triggered directly by neuronal activity. Indeed, results obtained over the last years provide conclusive support for the view that blood flow is directly coupled to neuronal activity rather than to local energy needs (Brožíčková & Otáhal, 2013; Haydon & Carmignoto, 2006; Hoffmeyer, Enager, Thomsen, & Lauritzen, 2007). The present knowledge on the signaling pathways that leads to the hemodynamic response suggest highly complex molecular mechanisms in which astrocytes and neuron–astrocyte signaling play a central role. NO, carbon monoxide, prostanoids, cytochrome P450 metabolites, adenosine, and K^+ ions are major vasoactive agents released by astrocytes or neurons (Brožíčková & Otáhal, 2013; Iadecola & Davisson, 2008). Some of these signaling molecules were found to significantly influence excitability of the nervous tissue and their levels are altered during epileptic activity. Endogenous NO was recently reported to play important role in initiation of SLEs in hippocampal and entorhinal cortex slices and experimentally induced SE in mice. Moreover, NO levels were elevated in both models mainly due to the increased activity of neuronal NO synthase. NO deprivation by NOS inhibitors and NO scavengers prevented initiation of SLEs in slice cultures and hippocampal–entorhinal cortex slices. Additional evidence was obtained from knockout nNOS (−/−) mice (Kovacs et al., 2009). Moreover, we have shown anticonvulsive effect of a highly selective neuronal NO synthase inhibitor, N(w)-propyl-L-arginine (L-NPA), on kainic acid-induced SE, and early epileptogenesis in mice *in vivo* (Beamer, Otáhal, Sills, & Thippeswamy, 2012).

Epileptic activity dramatically increases brain metabolism and CBF in the epileptic focus during seizure initiation and propagation. This was shown consistently in various animal models and humans using different detection methods (Folbergrová, Ingvar, & Siesjö, 1981; Ma, Zhao, & Schwartz, 2013; Patel, Zhao, Ma, & Schwartz, 2013; Zhao, Ma, Suh, & Schwartz, 2009; Zhao et al., 2011, 2007). Recent progress in knowledge about neurovascular and neurometabolic coupling in seizure focus was accelerated by utilization of advanced optical imaging techniques with high spatio-temporal resolution. Schwartz and his colleagues have shown important properties of neurovascular coupling during epileptic activity induced by 4-AP in rat cerebral cortex by combining electrophysiology with direct measurements of tissue oxygenation and optical methods including laser Doppler flowmetry, intrinsic optical imaging, flavoprotein fluorescence imaging, and two-photon confocal microscopy of surface cortical vasculature (Ma et al., 2013; Patel et al., 2013; Zhao et al., 2009, 2011, 2007). They demonstrated that the decrease in hemoglobin signal in the epileptic focus correlates with a profound but transient decrease in tissue oxygenation despite a large increase in CBF (Zhao et al., 2009). Furthermore, ictal events are preceded by preictal vasoconstriction of blood vessels in the surround of epileptic focus, occurring 1–5 s before seizure onset, which may serve to actively shunt oxygenated blood to the imminently hypermetabolic focus or may be due to small local decreases in metabolism in the surrounding tissue (Zhao et al., 2011). These events are followed by lesser focal increase in CBF, cerebral blood volume, and hyperoxygenation that correlate with the whole duration of the seizure (Zhao et al., 2009). Early ictal hypometabolism, transient decreases in cell swelling and cerebral blood volume in the surround are consistent with early ictal inhibition at seizure onset that shapes the evolving propagating ictal wavefront (Schevon et al., 2012). The exact mechanism of these cerebrovascular and metabolic changes is currently unknown (Zhao et al., 2011). Voltage-sensitive dyes imaging of 4-AP-induced cortical epileptic focus has shown that corresponding functional hyperemia is a parallel phenomenon, uncoupled in both space and time, which exhibits varying degrees of spatial and temporal overlaps with the underlying electrical events and is likely governed by the independent mechanisms (Ma et al., 2013).

During epileptogenesis some of the main components of the neurovascular unit undergo serious changes, which affect regulation of the regional CBF. Namely, epileptic transformation of astrocytes is one of the most important processes modifying regulatory mechanisms of regional CBF at different levels (Kovacs et al., 2012). However, alterations of neurovascular coupling in chronic epileptic patients have to be described in more details to fully understand underlying processes.

6. ALTERATION OF BLOOD BRAIN BARRIER

6.1. Blood brain barrier breakdown: Cause or consequence of epilepsy?

As illustrated in Section 5, the BBB is a highly complex structure, which separates the extracellular fluid of the CNS from the blood of CNS vessels (Ruhrberg & Bautch, 2013). The common consequence of a BBB dysfunction is increased permeability, leading to extravasation of plasma constituents and vasogenic brain edema (Stanimirovic & Friedman, 2012). The BBB impairment can persist for long periods, being involved in secondary inflammation and neuronal dysfunction, thus contributing to disease pathogenesis (Abbott & Friedman, 2012; Vezzani et al., 2013).

Currently, a debate exists on whether BBB breakdown may be a cause or a consequence of seizures and epilepsy. Indeed, a wide range of neurologic conditions, including stroke (Khatri, McKinney, Swenson, & Janardhan, 2012; Schoknecht & Shalev, 2012), epilepsy (Maggio, Blatt, et al., 2013), Alzheimer's disease (Kelleher & Soiza, 2013), migraine (Dreier et al., 2005), and brain tumors (Fortin, 2012; Rosenberg, 2012; Zhan & Lu, 2012), are associated with perturbations of the BBB which contribute to their pathology. Several studies have so far showed that a localized and transient BBB breakdown may trigger both acute epileptiform activity and chronic seizures (Friedman, 2011; Maggio, Blatt, et al., 2013), mainly through changes in brain extracellular ionic environment, albumin, and TGF-beta-mediated astrocytic transformation and inflammatory processes (Ivens et al., 2007; Marchi et al., 2007). On the other hand, the loss of BBB integrity may directly cause epilepsy. In this respect, serine proteases normally expressed in the brain at very low level (Luo, Wang, & Reiser, 2007) may abnormally increase in their "brain" concentrations following BBB breakdown. In particular, although a paucity of information still exists concerning the amount of thrombin crossing the BBB in determined neurological conditions, it has been demonstrated that thrombin levels increase more than 200-fold (from 100 pM to 25 nM) in the cerebrospinal fluid of patients with subarachnoid cerebral hemorrhage (Suzuki et al., 1992). Definitely, if bleeding occurs directly within the brain tissue, active thrombin, and the other proteases can freely diffuse into the brain parenchyma. Seizures are a direct consequence of the contact of thrombin with the brain. Indeed, intracerebral injections of thrombin result in focal motor seizures (Lee, Drury, Vitarbo, & Hoff, 1997), whereas application of thrombin together with its inhibitor alpha-(2-naphthylsulfonyl-glycyl)-4-amidinophenylalanine piperidide (alpha-NAPAP) does not cause any sign of either clinical or electrographic seizures (Lee et al., 1997). Likely, knockout mice for protein Nexin-1, an endogenous thrombin inhibitor, have an increased susceptibility to kainic acid-induced seizures (Luthi et al., 1997). We (Maggio, Blatt, et al., 2013) and others (Isaeva, Hernan, Isaev, & Holmes, 2012) have provided evidence that thrombin modulates synaptic transmission (Maggio, Itsekson, et al., 2013) and induces seizures

(Maggio, Cavaliere, et al., 2013, Maggio, Shavit, Chapman, & Segal, 2008) through activation of PAR-1. Specifically, in hippocampal slices, thrombin at a concentration of 5 nM (1 U/ml) has been shown to increase spontaneous firing of CA3 pyramidal cells (Maggio et al., 2008) and induce epileptic discharges in conditions mimicking BBB breakdown in the slice such as exposure to elevated $[K^+]o$ or low levels of glutamate (Beart & O'Shea, 2007; Chen & Swanson, 2003). The facilitatory action of thrombin on the production of seizure-like activity is independent on NMDARs, and it is most likely due to the effects on synaptic transmission in hippocampal CA3 neurons. Here, thrombin has been shown to enhance both frequency and amplitude of mEPSCs while reducing frequency and amplitude of mIPSCs (Maggio, Cavaliere, et al., 2013). An additional mechanism by which BBB breakdown could promote epileptogenesis, apart from extravasation/brain accumulation of serum proteins and edema formation, is the triggering specific signaling pathways in glial and neuronal cells. In this respect, serum albumin may potently induce calcium signaling and DNA synthesis in astrocytes both in the culture (Nadal, Fuentes, Pastor, & McNaughton, 1995) and in the brain slices (Nadal, Sul, Valdeolmillos, & McNaughton, 1998). Furthermore, upon local BBB disruption, albumin is selectively transported into astrocytes via TGF-beta receptors, where it triggers transcriptional and functional changes, resulting in neuronal hyperexcitability (Ivens et al., 2007). Finally, in the presence of BBB dysfunction, astrocytic signaling involves transcriptional changes which include down-regulation of potassium inward rectifying channels (Kir4.1), water channels (AQP4), gap junction proteins and glutamate transporters (David et al., 2009), proinflammatory molecules such as cyclooxygenase-2 (COX-2) and interleukin-1β (Marchi et al., 2014).

Seizures drastically affect the neurovascular unit, resulting in (1) dramatic increases in brain metabolism and CBF during the ictal period, (2) disruption of the BBB, (3) an acute loss of cerebral pressure autoregulation, and (4) a delayed impairment of cerebrovascular reactivity to various stimuli (Zimmermann, Domoki, & Bari, 2008). Regional patterns of BBB breakdown have been reported during epileptiform seizures induced in animal models by various convulsive agents (Danjo et al., 2013; Nitsch & Klatzo, 1983). This may be due to the direct consequence of the increase in CBF occurring during a seizure (Kovacs et al., 2012). In this setting, besides the direct neuronal effect on vasculature exerted by catecholamines, acetylcholine, neuropeptides or NO, and astrocytes are critically involved in regulation of hemodynamic responses (Filosa, 2010; Gordon, Choi, Rungta,

Ellis-Davies, & MacVicar, 2008). Indeed, astrocytic processes contact both the synapses and the abluminal surface of vessels (see Section 5 and Fig. 9.4) (Haydon & Carmignoto, 2006). Astrocytes equipped with neurotransmitter receptors and transporters may release vasoactive substances such as NO, arachidonic acid derivatives, adenosine, ATP, and influence vascular permeability (Haydon & Carmignoto, 2006). This seems to be a peculiar property of activated astrocytes at the epileptic focus whose altered intracellular calcium release may ultimately lead to BBB breakdown (Carmignoto & Haydon, 2012). Indeed, astrocyte–pericyte interactions have been identified as determinants of functional hyperemia at the microcirculation level (Peppiatt, Howarth, Mobbs, & Attwell, 2006); they may induce vascular remodeling as well, by alteration of thrombospondin signaling and regulation of the activity of matrix metalloproteases (Risher & Eroglu, 2012). It is therefore interesting to speculate what the contribution of BBB breakdown in chronic epilepsy might be, in fact, mediated by an enduring BBB breakdown due to uncontrolled seizures that leads to a continuous leak of thrombin into the brain, which in turn may sustain the epileptic process. If thrombin is then the major reason for seizures in this condition, it is tempting to speculate that PAR-1 antagonists and/or thrombin inhibitors could possibly act as potential antiepileptic drugs (Maggio, Blatt, et al., 2013; Marchi et al., 2014).

6.2. Blood brain barrier: A site of pharmacoresistance in epilepsy?

Among the hypotheses explaining why approximately one-third of individuals with epilepsy will continue to have seizures despite the drug treatment, one concerns the existence of metabolic mechanisms such as biotransformation of xenobiotics by cytochrome P450. An additional hypothesis takes into consideration that a possible upregulation of multidrug transporters in the BBB limits access of antiepileptic drugs to their targets in the brain (Kovacs & Heinemann, 2014; Löscher & Potschka, 2005). This hypothesis follows the observations that multidrug transporters are overexpressed in capillary endothelial tissue resected from patients with intractable epilepsy (Ak et al., 2007) and modulation of P-glycoprotein (P-gp) has been shown to restore sensitivity to the antiepileptic effects of phenobarbital in a rat model of TLE (Brandt, Bethmann, Gastens, & Loscher, 2006). P-gp is a 170-kD transmembrane glycoprotein which undergoes to several posttranslational modifications, such as N-glycosylation, phosphorylation, and ubiquitination (Stepien, Tomaszewski, Tomaszewska, & Czuczwar,

2012). P-gp pumps xenobiotics from intracellular space back to the capillary lumen, thereby maintaining the integrity of BBB and reducing the cerebral accumulation of substrate drugs (Kwan, Schachter, & Brodie, 2011). In drug-resistant epilepsy, the mechanisms leading to P-gp overexpression are probably related to the excessive glutamate release associated with seizure-related activity and activation of NMDA-receptors, leading to the increased intracellular calcium. As a result, the level of phospholipase A2 increases thus mobilizing arachidonic acid from the cell membrane. In the presence of COX-2, arachidonic acid is converted to prostaglandin E2, which then acts on its nuclear receptors, EP1 through EP4, leading to overexpression of P-gp (Potschka, 2010). Due to its fundamental role in pharmacoresistance, P-gp may be a major target both for therapy as well as for the developing novel neuroimaging tools in order to assess the potential of pharmacoresistance in determined epileptic patients.

7. CONCLUSIONS

Epileptic seizures represent the situation of increased neuronal activity associated with the enhanced demand for sufficient energy supply. Efficient regulatory mechanisms have to operate in concert at several levels. They have to ensure that CBF, delivery of oxygen, and nutrients are continuously adapted to the local metabolic needs. Regulation of energy metabolism has to be coupled to neuronal activity. Several factors involved in neurometabolic coupling, such as intracellular and intramitochondrial ion changes, NO, redox changes, pH, etc., are described. The evidence reviewed indicates that regulation of regional CBF, BBB, and metabolism are affected by acute epileptic seizures and chronic epilepsy. The key factor is the increased production of reactive oxygen and nitrogen species and their damaging effects. Concerning mitochondrial dysfunction, inhibition of mitochondrial respiratory chain complexes I and IV was demonstrated. Importantly, inhibited mitochondrial complexes can lead to additional increase of ROS and/or RNS production resulting in further deterioration. Mechanisms of mtDNA damage by ROS, genome modifications, and limited repair capacity of mtDNA are discussed. An increase of several markers of lipid peroxidation has been shown. ROS are also involved in the BBB impairment, astrocytic transformation, inflammation, and dysregulation of neurovascular unit. On the other hand, all the listed disturbances can lead to the enhanced ROS production forming thus a vicious circle. The reported changes are supposed to be a consequence of epileptic seizures but they can also render the brain seizure prone. One of the important problems is pharmacoresistant epilepsy and the potential metabolic mechanisms involved are discussed.

All the above-mentioned alterations can lead to serious impairments of normal CNS function and to neuronal injury. Therefore, there is a need for novel efficient therapeutic strategies, targeting these disturbed pathways. It can be speculated that the substances with antioxidant properties combined with conventional therapies might provide a beneficial effect in the treatment of epilepsy.

ACKNOWLEDGMENTS

J. O. was supported by grants from Ministry of Health of the Czech Republic (IGA NT/14489-3), Neuron Fund for Support of Science (Czech Republic, 2012/10), and the Czech Science Foundation (Czech Republic, P303/10/0999, and P303/14/02634S).

R. K. was supported by the DFG grant KO 3814/1-1.

J. F. was supported by grants 309/05/2015, 309/08/0292, and P302/10/0971 from the Czech Science Foundation, ERC CZ LL 1204 from MEYES CR, by research project AV0 750110509, and by the long-term strategic development financing of the Institute of Physiology ASCR (RV0:67985823).

W. S. K. was supported by grants from DFG (KU 911/21-1), the European Commission (FP7 project EpiPGX, grant 279062), and the Stiftung für Medizinische Wissenschaft (Frankfurt am Main).

REFERENCES

Abbott, N. J. (2002). Astrocyte-endothelial interactions and blood-brain barrier permeability. *Journal of Anatomy*, *200*, 629–638.

Abbott, N. J., & Friedman, A. (2012). Overview and introduction: The blood-brain barrier in health and disease. *Epilepsia*, *53*(Suppl. 6), 1–6.

Ak, H., Ay, B., Tanriverdi, T., Sanus, G. Z., Is, M., Sar, M., et al. (2007). Expression and cellular distribution of multidrug resistance-related proteins in patients with focal cortical dysplasia. *Seizure*, *16*, 493–503.

Andreux, P. A., Houtkooper, R. H., & Auwerx, J. (2013). Pharmacological approaches to restore mitochondrial function. *Nature Reviews Drug Discovery*, *12*, 465–483.

Bacman, S. R., Williams, S. L., & Moraes, C. T. (2009). Intra- and inter-molecular recombination of mitochondrial DNA after in vivo induction of multiple double-strand breaks. *Nucleic Acids Research*, *37*, 4218–4226.

Baron, M., Kudin, A. P., & Kunz, W. S. (2007). Mitochondrial dysfunction in neurodegenerative disorders. *Biochemical Society Transactions*, *35*, 1228–1231.

Beamer, E., Otáhal, J., Sills, G. J., & Thippeswamy, T. (2012). N(w)-propyl-L-arginine (L-NPA) reduces status epilepticus and early epileptogenic events in a mouse model of epilepsy: Behavioural, EEG and immunohistochemical analyses. *The European Journal of Neuroscience*, *36*, 3194–3203.

Beart, P. M., & O'Shea, R. D. (2007). Transporters for L-glutamate: An update on their molecular pharmacology and pathological involvement. *British Journal of Pharmacology*, *150*, 5–17.

Bednarczyk, P. (2009). Potassium channels in brain mitochondria. *Acta Biochimica Polonica, 56*, 385–392.

Benard, G., Massa, F., Puente, N., Lourenco, J., Bellocchio, L., Soria-Gomez, E., et al. (2012). Mitochondrial CB(1) receptors regulate neuronal energy metabolism. *Nature Neuroscience, 15*, 558–564.

Bender, A., Krishnan, K. J., Morris, C. M., Taylor, G. A., Reeve, A. K., Perry, R. H., et al. (2006). High levels of mitochondrial DNA deletions in substantia nigra neurons in aging and Parkinson disease. *Nature Genetics, 38*, 515–517.

Bidmon, H. J., Gorg, B., Palomero-Gallagher, N., Schleicher, A., Haussinger, D., Speckmann, E. J., et al. (2008). Glutamine synthetase becomes nitrated and its activity is reduced during repetitive seizure activity in the pentylenetetrazol model of epilepsy. *Epilepsia, 49*, 1733–1748.

Blümcke, I., Zuschratter, W., Schewe, J. C., Suter, B., Lie, A. A., Riederer, B. M., et al. (1999). Cellular pathology of hilar neurons in Ammon's horn sclerosis. *The Journal of Comparative Neurology, 414*, 437–453.

Brandt, C., Bethmann, K., Gastens, A. M., & Loscher, W. (2006). The multidrug transporter hypothesis of drug resistance in epilepsy: Proof-of-principle in a rat model of temporal lobe epilepsy. *Neurobiology of Disease, 24*, 202–211.

Brennan, A. M., Connor, J. A., & Shuttleworth, C. W. (2006). NAD(P)H fluorescence transients after synaptic activity in brain slices: Predominant role of mitochondrial function. *Journal of Cerebral Blood Flow and Metabolism, 26*, 1389–1406.

Brown, G. C. (2010). Nitric oxide and neuronal death. *Nitric Oxide, 23*, 153–165.

Brožíčková, C., & Otáhal, J. (2013). Effect of an inhibitor of neuronal nitric oxide synthase 7-nitroindazole on cerebral hemodynamic response and brain excitability in urethane-anesthetized rats. *Physiological Research, 62*(Suppl. 1), S57–S66.

Bruce, A. J., & Baudry, M. (1995). Oxygen free radicals in rat limbic structures after kainate-induced seizures. *Free Radical Biology & Medicine, 18*, 993–1002.

Busija, D. W., Gaspar, T., Domoki, F., Katakam, P. V., & Bari, F. (2008). Mitochondrial-mediated suppression of ROS production upon exposure of neurons to lethal stress: Mitochondrial targeted preconditioning. *Advanced Drug Delivery Reviews, 60*, 1471–1477.

Carmignoto, G., & Haydon, P. G. (2012). Astrocyte calcium signaling and epilepsy. *Glia, 60*, 1227–1233.

Chen, Y., & Swanson, R. A. (2003). Astrocytes and brain injury. *Journal of Cerebral Blood Flow and Metabolism, 23*, 137–149.

Chuang, Y. C., Chang, A. Y., Lin, J. W., Hsu, S. P., & Chan, S. H. (2004). Mitochondrial dysfunction and ultrastructural damage in the hippocampus during kainic acid-induced status epilepticus in the rat. *Epilepsia, 45*, 1202–1209.

Chung, T. T., Lin, C. Y., Huang, W. Y., Lin, C. L., Sung, F. C., & Kao, C. H. (2013). Risks of subsequent epilepsy among patients with hypertensive encephalopathy: A nationwide population-based study. *Epilepsy & Behavior, 29*, 374–378.

Cock, H. R., Tong, X., Hargreaves, I. P., Heales, S. J., Clark, J. B., Patsalos, P. N., et al. (2002). Mitochondrial dysfunction associated with neuronal death following status epilepticus in rat. *Epilepsy Research, 48*, 157–168.

Corral-Debrinski, M., Horton, T., Lott, M. T., Shoffner, J. M., Beal, M. F., & Wallace, D. C. (1992). Mitochondrial DNA deletions in human brain: Regional variability and increase with advanced age. *Nature Genetics, 2*, 324–329.

Cortopassi, G. A., Shibata, D., Soong, N. W., & Arnheim, N. (1992). A pattern of accumulation of a somatic deletion of mitochondrial DNA in aging human tissues. *Proceedings of the National Academy of Sciences of the United States of America, 89*, 7370–7374.

Dal-Pizzol, F., Klamt, F., Vianna, M. M., Schroder, N., Quevedo, J., Benfato, M. S., et al. (2000). Lipid peroxidation in hippocampus early and late after status epilepticus induced by pilocarpine or kainic acid in Wistar rats. *Neuroscience Letters*, *291*, 179–182.

Danjo, S., Ishihara, Y., Watanabe, M., Nakamura, Y., & Itoh, K. (2013). Pentylentetrazole-induced loss of blood-brain barrier integrity involves excess nitric oxide generation by neuronal nitric oxide synthase. *Brain Research*, *1530*, 44–53.

David, Y., Cacheaux, L. P., Ivens, S., Lapilover, E., Heinemann, U., Kaufer, D., et al. (2009). Astrocytic dysfunction in epileptogenesis: Consequence of altered potassium and glutamate homeostasis? *The Journal of Neuroscience*, *29*, 10588–10599.

Davis, R. E., & Williams, M. (2012). Mitochondrial function and dysfunction: An update. *The Journal of Pharmacology and Experimental Therapeutics*, *342*, 598–607.

Demple, B., & Harrison, L. (1994). Repair of oxidative damage to DNA: Enzymology and biology. *Annual Review of Biochemistry*, *63*, 915–948.

Denton, R. M. (2009). Regulation of mitochondrial dehydrogenases by calcium ions. *Biochimica et Biophysica Acta*, *1787*, 1309–1316.

Dreier, J. P., Jurkat-Rott, K., Petzold, G. C., Tomkins, O., Klingebiel, R., Kopp, U. A., et al. (2005). Opening of the blood-brain barrier preceding cortical edema in a severe attack of FHM type II. *Neurology*, *64*, 2145–2147.

Duchen, M. R. (2000). Mitochondria and calcium: From cell signalling to cell death. *The Journal of Physiology*, *529*(Pt. 1), 57–68.

Dutuit, M., Touret, M., Szymocha, R., Nehlig, A., Belin, M. F., & Didier-Bazes, M. (2002). Decreased expression of glutamate transporters in genetic absence epilepsy rats before seizure occurrence. *Journal of Neurochemistry*, *80*, 1029–1038.

Erecinska, M., & Dagani, F. (1990). Relationships between the neuronal sodium/potassium pump and energy metabolism. Effects of K+, Na+, and adenosine triphosphate in isolated brain synaptosomes. *The Journal of General Physiology*, *95*, 591–616.

Fato, R., Bergamini, C., Leoni, S., Strocchi, P., & Lenaz, G. (2008). Generation of reactive oxygen species by mitochondrial complex I: Implications in neurodegeneration. *Neurochemical Research*, *33*, 2487–2501.

Filosa, J. A. (2010 Aug 16). Vascular tone and neurovascular coupling: Considerations toward an improved in vitro model. *Front Neuroenergetics*, *2*. pii: 16. http://dx.doi.org/10.3389/fnene.2010.00016. eCollection 2010.

Folbergrová, J., Ingvar, M., & Siesjö, B. K. (1981). Metabolic changes in cerebral cortex, hippocampus, and cerebellum during sustained bicuculline-induced seizures. *Journal of Neurochemistry*, *37*, 1228–1238.

Folbergrová, J., Ješina, P., Drahota, Z., Lisý, V., Haugvicová, R., Vojtíšková, A., et al. (2007). Mitochondrial complex I inhibition in cerebral cortex of immature rats following homocysteic acid-induced seizures. *Experimental Neurology*, *204*, 597–609.

Folbergrová, J., Ješina, P., Haugvicová, R., Lisý, V., & Houštěk, J. (2010). Sustained deficiency of mitochondrial complex I activity during long periods of survival after seizures induced in immature rats by homocysteic acid. *Neurochemistry International*, *56*, 394–403.

Folbergrová, J., & Kunz, W. S. (2012). Mitochondrial dysfunction in epilepsy. *Mitochondrion*, *12*, 35–40.

Folbergrová, J., Otáhal, J., & Druga, R. (2012). Brain superoxide anion formation in immature rats during seizures: Protection by selected compounds. *Experimental Neurology*, *233*, 421–429.

Fortin, D. (2012). The blood-brain barrier: Its influence in the treatment of brain tumors metastases. *Current Cancer Drug Targets*, *12*, 247–259.

Frantseva, M. V., Velazquez, J. L., Hwang, P. A., & Carlen, P. L. (2000). Free radical production correlates with cell death in an in vitro model of epilepsy. *The European Journal of Neuroscience*, *12*, 1431–1439.

Friedman, A. (2011). Blood-brain barrier dysfunction, status epilepticus, seizures, and epilepsy: A puzzle of a chicken and egg? *Epilepsia*, *52*(Suppl. 8), 19–20.

Gellerich, F. N., Gizatullina, Z., Gainutdinov, T., Muth, K., Seppet, E., Orynbayeva, Z., et al. (2013). The control of brain mitochondrial energization by cytosolic calcium: The mitochondrial gas pedal. *IUBMB Life*, *65*, 180–190.

Gerencser, A. A., Chinopoulos, C., Birket, M. J., Jastroch, M., Vitelli, C., Nicholls, D. G., et al. (2012). Quantitative measurement of mitochondrial membrane potential in cultured cells: Calcium-induced de- and hyperpolarization of neuronal mitochondria. *The Journal of Physiology*, *590*, 2845–2871.

Giulivi, C., Boveris, A., & Cadenas, E. (1995). Hydroxyl radical generation during mitochondrial electron transfer and the formation of 8-hydroxydesoxyguanosine in mitochondrial DNA. *Archives of Biochemistry and Biophysics*, *316*, 909–916.

Gordon, G. R., Choi, H. B., Rungta, R. L., Ellis-Davies, G. C., & MacVicar, B. A. (2008). Brain metabolism dictates the polarity of astrocyte control over arterioles. *Nature*, *456*, 745–749.

Gordon, G. R., Mulligan, S. J., & MacVicar, B. A. (2007). Astrocyte control of the cerebrovasculature. *Glia*, *55*, 1214–1221.

Guo, X., Popadin, K. Y., Markuzon, N., Orlov, Y. L., Kraytsberg, Y., Krishnan, K. J., et al. (2010). Repeats, longevity and the sources of mtDNA deletions: Evidence from 'deletional spectra'. *Trends in Genetics*, *26*, 340–343.

Hajos, N., Ellender, T. J., Zemankovics, R., Mann, E. O., Exley, R., Cragg, S. J., et al. (2009). Maintaining network activity in submerged hippocampal slices: Importance of oxygen supply. *The European Journal of Neuroscience*, *29*, 319–327.

Hamilton, N. B., Attwell, D., & Hall, C. N. (2010 May 21). Pericyte-mediated regulation of capillary diameter: A component of neurovascular coupling in health and disease. *Front Neuroenergetics*, *2*. pii: 5. http://dx.doi.org/10.3389/fnene.2010.00005. eCollection 2010.

Haydon, P. G., & Carmignoto, G. (2006). Astrocyte control of synaptic transmission and neurovascular coupling. *Physiological Reviews*, *86*, 1009–1031.

Hertz, L., Xu, J., Song, D., Yan, E., Gu, L., & Peng, L. (2013). Astrocytic and neuronal accumulation of elevated extracellular K(+) with a 2/3 K(+)/Na(+) flux ratio-consequences for energy metabolism, osmolarity and higher brain function. *Frontiers in Computational Neuroscience*, 7, 114.

Hoffmeyer, H. W., Enager, P., Thomsen, K. J., & Lauritzen, M. J. (2007). Nonlinear neurovascular coupling in rat sensory cortex by activation of transcallosal fibers. *Journal of Cerebral Blood Flow and Metabolism*, *27*, 575–587.

Hossmann, K. A. (2006). Pathophysiology and therapy of experimental stroke. *Cellular and Molecular Neurobiology*, *26*, 1057–1083.

Huchzermeyer, C., Albus, K., Gabriel, H. J., Otahal, J., Taubenberger, N., Heinemann, U., et al. (2008). Gamma oscillations and spontaneous network activity in the hippocampus are highly sensitive to decreases in pO_2 and concomitant changes in mitochondrial redox state. *The Journal of Neuroscience*, *28*, 1153–1162.

Huchzermeyer, C., Berndt, N., Holzhutter, H. G., & Kann, O. (2013). Oxygen consumption rates during three different neuronal activity states in the hippocampal CA3 network. *Journal of Cerebral Blood Flow and Metabolism*, *33*, 263–271.

Iadecola, C., & Davisson, R. L. (2008). Hypertension and cerebrovascular dysfunction. *Cell Metabolism*, 7, 476–484.

Iadecola, C., & Nedergaard, M. (2007). Glial regulation of the cerebral microvasculature. *Nature Neuroscience*, *10*, 1369–1376.

Imlay, J. A., & Linn, S. (1988). DNA damage and oxygen radical toxicity. *Science*, *240*, 1302–1309.

Isaeva, E., Hernan, A., Isaev, D., & Holmes, G. L. (2012). Thrombin facilitates seizures through activation of persistent sodium current. *Annals of Neurology*, 72, 192–198.

Ivens, S., Kaufer, D., Flores, L. P., Bechmann, I., Zumsteg, D., Tomkins, O., et al. (2007). TGF-beta receptor-mediated albumin uptake into astrocytes is involved in neocortical epileptogenesis. *Brain*, *130*, 535–547.

Jarrett, S. G., Liang, L. P., Hellier, J. L., Staley, K. J., & Patel, M. (2008). Mitochondrial DNA damage and impaired base excision repair during epileptogenesis. *Neurobiology of Disease*, *30*, 130–138.

Kaneko, K., Itoh, K., Berliner, L. J., Miyasaka, K., & Fujii, H. (2002). Consequences of nitric oxide generation in epileptic-seizure rodent models as studied by in vivo EPR. *Magnetic Resonance in Medicine*, *48*, 1051–1056.

Kann, O., Huchzermeyer, C., Kovacs, R., Wirtz, S., & Schuelke, M. (2011). Gamma oscillations in the hippocampus require high complex I gene expression and strong functional performance of mitochondria. *Brain*, *134*, 345–358.

Kann, O., & Kovacs, R. (2007). Mitochondria and neuronal activity. *American Journal of Physiology Cell Physiology*, *292*, C641–C657.

Kann, O., Kovacs, R., & Heinemann, U. (2003). Metabotropic receptor-mediated Ca^{2+} signaling elevates mitochondrial Ca^{2+} and stimulates oxidative metabolism in hippocampal slice cultures. *Journal of Neurophysiology*, *90*, 613–621.

Kann, O., Kovacs, R., Njunting, M., Behrens, C. J., Otahal, J., Lehmann, T. N., et al. (2005). Metabolic dysfunction during neuronal activation in the ex vivo hippocampus from chronic epileptic rats and humans. *Brain*, *128*, 2396–2407.

Kann, O., Taubenberger, N., Huchzermeyer, C., Papageorgiou, I. E., Benninger, F., Heinemann, U., et al. (2012). Muscarinic receptor activation determines the effects of store-operated Ca(2+)-entry on excitability and energy metabolism in pyramidal neurons. *Cell Calcium*, *51*, 40–50.

Kasischke, K. A., Vishwasrao, H. D., Fisher, P. J., Zipfel, W. R., & Webb, W. W. (2004). Neural activity triggers neuronal oxidative metabolism followed by astrocytic glycolysis. *Science*, *305*, 99–103.

Kelleher, R. J., & Soiza, R. L. (2013). Evidence of endothelial dysfunction in the development of Alzheimer's disease: Is Alzheimer's a vascular disorder? *American Journal of Cardiovascular Disease*, *3*, 197–226.

Khatri, R., McKinney, A. M., Swenson, B., & Janardhan, V. (2012). Blood-brain barrier, reperfusion injury, and hemorrhagic transformation in acute ischemic stroke. *Neurology*, *79*, S52–S57.

Kniesel, U., & Wolburg, H. (2000). Tight junctions of the blood-brain barrier. *Cellular and Molecular Neurobiology*, *20*, 57–76.

Kong, Q., Takahashi, K., Schulte, D., Stouffer, N., Lin, Y., & Lin, C. L. (2012). Increased glial glutamate transporter EAAT2 expression reduces epileptogenic processes following pilocarpine-induced status epilepticus. *Neurobiology of Disease*, *47*, 145–154.

Kovacs, R., Gutierrez, R., Kivi, A., Schuchmann, S., Gabriel, S., & Heinemann, U. (1999). Acute cell damage after low Mg^{2+}-induced epileptiform activity in organotypic hippocampal slice cultures. *Neuroreport*, *10*, 207–213.

Kovacs, R., & Heinemann, U. (2014). Models in research of pharmacoresistant epilepsy: Present and future in development of antiepileptic drugs. *Current Medicinal Chemistry*, *21*, 689–703.

Kovacs, R., Heinemann, U., & Steinhauser, C. (2012). Mechanisms underlying blood-brain barrier dysfunction in brain pathology and epileptogenesis: Role of astroglia. *Epilepsia*, *53*(Suppl. 6), 53–59.

Kovacs, R., Kardos, J., Heinemann, U., & Kann, O. (2005). Mitochondrial calcium ion and membrane potential transients follow the pattern of epileptiform discharges in hippocampal slice cultures. *The Journal of Neuroscience*, *25*, 4260–4269.

Kovacs, R., Rabanus, A., Otahal, J., Patzak, A., Kardos, J., Albus, K., et al. (2009). Endogenous nitric oxide is a key promoting factor for initiation of seizure-like events in hippocampal and entorhinal cortex slices. *The Journal of Neuroscience*, *29*, 8565–8577.

Kovacs, R., Schuchmann, S., Gabriel, S., Kann, O., Kardos, J., & Heinemann, U. (2002). Free radical-mediated cell damage after experimental status epilepticus in hippocampal slice cultures. *Journal of Neurophysiology*, *88*, 2909–2918.

Kovacs, R., Schuchmann, S., Gabriel, S., Kardos, J., & Heinemann, U. (2001). Ca^{2+} signalling and changes of mitochondrial function during low-Mg^{2+}-induced epileptiform activity in organotypic hippocampal slice cultures. *The European Journal of Neuroscience*, *13*, 1311–1319.

Kovacs, R., Szilagyi, N., Barabas, P., Heinemann, U., & Kardos, J. (2000). Low-[Mg^{2+}]-induced Ca^{2+} fluctuations in organotypic hippocampal slice cultures. *Neuroreport*, *11*, 2107–2111.

Kraytsberg, Y., Kudryavtseva, E., McKee, A. C., Geula, C., Kowall, N. W., & Khrapko, K. (2006). Mitochondrial DNA deletions are abundant and cause functional impairment in aged human substantia nigra neurons. *Nature Genetics*, *38*, 518–520.

Kudin, A. P., Bimpong-Buta, N. Y., Vielhaber, S., Elger, C. E., & Kunz, W. S. (2004). Characterization of superoxide-producing sites in isolated brain mitochondria. *The Journal of Biological Chemistry*, *279*, 4127–4135.

Kudin, A. P., Kudina, T. A., Seyfried, J., Vielhaber, S., Beck, H., Elger, C. E., et al. (2002). Seizure-dependent modulation of mitochondrial oxidative phosphorylation in rat hippocampus. *The European Journal of Neuroscience*, *15*, 1105–1114.

Kudin, A. P., Zsurka, G., Elger, C. E., & Kunz, W. S. (2009). Mitochondrial involvement in temporal lobe epilepsy. *Experimental Neurology*, *218*, 326–332.

Kunz, W. S., Kudin, A. P., Vielhaber, S., Blumcke, I., Zuschratter, W., Schramm, J., et al. (2000). Mitochondrial complex I deficiency in the epileptic focus of patients with temporal lobe epilepsy. *Annals of Neurology*, *48*, 766–773.

Kussmaul, L., & Hirst, J. (2006). The mechanism of superoxide production by NADH:ubiquinone oxidoreductase (complex I) from bovine heart mitochondria. *Proceedings of the National Academy of Sciences of the United States of America*, *103*, 7607–7612.

Kwan, P., Schachter, S. C., & Brodie, M. J. (2011). Drug-resistant epilepsy. *The New England Journal of Medicine*, *365*, 919–926.

Lee, K. R., Drury, I., Vitarbo, E., & Hoff, J. T. (1997). Seizures induced by intracerebral injection of thrombin: A model of intracerebral hemorrhage. *Journal of Neurosurgery*, *87*, 73–78.

Li, X., Breteler, M. M., de Bruyne, M. C., Meinardi, H., Hauser, W. A., & Hofman, A. (1997). Vascular determinants of epilepsy: The Rotterdam study. *Epilepsia*, *38*, 1216–1220.

Liang, L. P., Ho, Y. S., & Patel, M. (2000). Mitochondrial superoxide production in kainate-induced hippocampal damage. *Neuroscience*, *101*, 563–570.

Liang, L. P., & Patel, M. (2004). Mitochondrial oxidative stress and increased seizure susceptibility in Sod2(−/+) mice. *Free Radical Biology & Medicine*, *36*, 542–554.

Liebner, S., Czupalla, C. J., & Wolburg, H. (2011). Current concepts of blood-brain barrier development. *The International Journal of Developmental Biology*, *55*, 467–476.

Liotta, A., Rōsner, J., Huchzermeyer, C., Wojtowicz, A., Kann, O., Schmitz, D., et al. (2012). Energy demand of synaptic transmission at the hippocampal Schaffer-collateral synapse. *Journal of Cerebral Blood Flow and Metabolism*, *32*, 2076–2083.

Löscher, W., & Potschka, H. (2005). Drug resistance in brain diseases and the role of drug efflux transporters. *Nature Reviews Neuroscience*, *6*, 591–602.

Luo, W., Wang, Y., & Reiser, G. (2007). Protease-activated receptors in the brain: Receptor expression, activation, and functions in neurodegeneration and neuroprotection. *Brain Research Reviews*, *56*, 331–345.

Luthi, A., Van der Putten, H., Botteri, F. M., Mansuy, I. M., Meins, M., Frey, U., et al. (1997). Endogenous serine protease inhibitor modulates epileptic activity and hippocampal long-term potentiation. *The Journal of Neuroscience*, *17*, 4688–4699.

Lux, H. D., Heinemann, U., & Dietzel, I. (1986). Ionic changes and alterations in the size of the extracellular space during epileptic activity. *Advances in Neurology*, *44*, 619–639.

Ma, H., Zhao, M., & Schwartz, T. H. (2013). Dynamic neurovascular coupling and uncoupling during ictal onset, propagation, and termination revealed by simultaneous in vivo optical imaging of neural activity and local blood volume. *Cerebral Cortex*, *23*, 885–899.

Maggio, N., Blatt, I., Vlachos, A., Tanne, D., Chapman, J., & Segal, M. (2013). Treating seizures and epilepsy with anticoagulants? *Frontiers in Cellular Neuroscience*, 7, 19.

Maggio, N., Cavaliere, C., Papa, M., Blatt, I., Chapman, J., & Segal, M. (2013). Thrombin regulation of synaptic transmission: Implications for seizure onset. *Neurobiology of Disease*, *50*, 171–178.

Maggio, N., Itsekson, Z., Dominissini, D., Blatt, I., Amariglio, N., Rechavi, G., et al. (2013). Thrombin regulation of synaptic plasticity: Implications for physiology and pathology. *Experimental Neurology*, *247*, 595–604.

Maggio, N., Shavit, E., Chapman, J., & Segal, M. (2008). Thrombin induces long-term potentiation of reactivity to afferent stimulation and facilitates epileptic seizures in rat hippocampal slices: Toward understanding the functional consequences of cerebrovascular insults. *The Journal of Neuroscience*, *28*, 732–736.

Malinska, D., Kulawiak, B., Kudin, A. P., Kovacs, R., Huchzermeyer, C., Kann, O., et al. (2010). Complex III-dependent superoxide production of brain mitochondria contributes to seizure-related ROS formation. *Biochimica et Biophysica Acta*, *1797*, 1163–1170.

Marchi, N., Angelov, L., Masaryk, T., Fazio, V., Granata, T., Hernandez, N., et al. (2007). Seizure-promoting effect of blood-brain barrier disruption. *Epilepsia*, *48*, 732–742.

Marchi, N., Granata, T., & Janigro, D. (2014). Inflammatory pathways of seizure disorders. *Trends in Neurosciences*, *37*(2), 55–65.

Marchi, N., & Lerner-Natoli, M. (2013). Cerebrovascular remodeling and epilepsy. *The Neuroscientist*, *19*, 304–312.

Mathern, G. W., Mendoza, D., Lozada, A., Pretorius, J. K., Dehnes, Y., Danbolt, N. C., et al. (1999). Hippocampal GABA and glutamate transporter immunoreactivity in patients with temporal lobe epilepsy. *Neurology*, *52*, 453–472.

Mayevsky, A., & Chance, B. (1982). Intracellular oxidation-reduction state measured in situ by a multichannel fiber-optic surface fluorometer. *Science*, *217*, 537–540.

Meldrum, B. S. (1993). Excitotoxicity and selective neuronal loss in epilepsy. *Brain Pathology*, *3*, 405–412.

Michikawa, Y., Mazzucchelli, F., Bresolin, N., Scarlato, G., & Attardi, G. (1999). Aging-dependent large accumulation of point mutations in the human mtDNA control region for replication. *Science*, *286*, 774–779.

Mueller, S. G., Trabesinger, A. H., Boesiger, P., & Wieser, H. G. (2001). Brain glutathione levels in patients with epilepsy measured by in vivo (1)H-MRS. *Neurology*, *57*, 1422–1427.

Murray, J., Taylor, S. W., Zhang, B., Ghosh, S. S., & Capaldi, R. A. (2003). Oxidative damage to mitochondrial complex I due to peroxynitrite: Identification of reactive tyrosines by mass spectrometry. *The Journal of Biological Chemistry*, *278*, 37223–37230.

Nadal, A., Fuentes, E., Pastor, J., & McNaughton, P. A. (1995). Plasma albumin is a potent trigger of calcium signals and DNA synthesis in astrocytes. *Proceedings of the National Academy of Sciences of the United States of America*, *92*, 1426–1430.

Nadal, A., Sul, J. Y., Valdeolmillos, M., & McNaughton, P. A. (1998). Albumin elicits calcium signals from astrocytes in brain slices from neonatal rat cortex. *The Journal of Physiology*, *509*(Pt 3), 711–716.

Nekhaeva, E., Bodyak, N. D., Kraytsberg, Y., McGrath, S. B., Van Orsouw, N. J., Pluzhnikov, A., et al. (2002). Clonally expanded mtDNA point mutations are abundant in individual cells of human tissues. *Proceedings of the National Academy of Sciences of the United States of America*, *99*, 5521–5526.

Neuhaus, J. F., Baris, O. R., Hess, S., Moser, N., Schroder, H., Chinta, S. J., et al. (2014). Catecholamine metabolism drives generation of mitochondrial DNA deletions in dopaminergic neurons. *Brain*, *137*, 354–365.

Nicholls, D. G., & Budd, S. L. (2000). Mitochondria and neuronal survival. *Physiological Reviews*, *80*, 315–360.

Nico, B., & Ribatti, D. (2012). Morphofunctional aspects of the blood-brain barrier. *Current Drug Metabolism*, *13*, 50–60.

Nitsch, C., & Klatzo, I. (1983). Regional patterns of blood-brain barrier breakdown during epileptiform seizures induced by various convulsive agents. *Journal of the Neurological Sciences*, *59*, 305–322.

Otáhal, J., Štěpaník, Z., Kaczmarska, A., Maršík, F., Brož, Z., & Otáhal, S. (2007). Simulation of cerebrospinal fluid transport. *Advances in Engineering Software*, *38*, 802–809.

Ott, M., Gogvadze, V., Orrenius, S., & Zhivotovsky, B. (2007). Mitochondria, oxidative stress and cell death. *Apoptosis*, *12*, 913–922.

Pan, X., Liu, J., Nguyen, T., Liu, C., Sun, J., Teng, Y., et al. (2013). The physiological role of mitochondrial calcium revealed by mice lacking the mitochondrial calcium uniporter. *Nature Cell Biology*, *15*, 1464–1472.

Parihar, M. S., Parihar, A., Villamena, F. A., Vaccaro, P. S., & Ghafourifar, P. (2008). Inactivation of mitochondrial respiratory chain complex I leads mitochondrial nitric oxide synthase to become pro-oxidative. *Biochemical and Biophysical Research Communications*, *367*, 761–767.

Patel, M. (2004). Mitochondrial dysfunction and oxidative stress: Cause and consequence of epileptic seizures. *Free Radical Biology & Medicine*, *37*, 1951–1962.

Patel, M., Liang, L. P., Hou, H., Williams, B. B., Kmiec, M., Swartz, H. M., et al. (2008). Seizure-induced formation of isofurans: Novel products of lipid peroxidation whose formation is positively modulated by oxygen tension. *Journal of Neurochemistry*, *104*, 264–270.

Patel, M., Liang, L. P., & Roberts, L. J. (2001). Enhanced hippocampal F2-isoprostane formation following kainate-induced seizures. *Journal of Neurochemistry*, *79*, 1065–1069.

Patel, K. S., Zhao, M., Ma, H., & Schwartz, T. H. (2013). Imaging preictal hemodynamic changes in neocortical epilepsy. *Neurosurgical Focus*, *34*, E10.

Pearce, L. L., Kanai, A. J., Epperly, M. W., & Peterson, J. (2005). Nitrosative stress results in irreversible inhibition of purified mitochondrial complexes I and III without modification of cofactors. *Nitric Oxide*, *13*, 254–263.

Peppiatt, C. M., Howarth, C., Mobbs, P., & Attwell, D. (2006). Bidirectional control of CNS capillary diameter by pericytes. *Nature*, *443*, 700–704.

Peterson, S. L., Morrow, D., Liu, S., & Liu, K. J. (2002). Hydroethidine detection of superoxide production during the lithium-pilocarpine model of status epilepticus. *Epilepsy Research*, *49*, 226–238.

Potschka, H. (2010). Modulating P-glycoprotein regulation: Future perspectives for pharmacoresistant epilepsies? *Epilepsia*, *51*, 1333–1347.

Radi, R., Cassina, A., Hodara, R., Quijano, C., & Castro, L. (2002). Peroxynitrite reactions and formation in mitochondria. *Free Radical Biology & Medicine*, *33*, 1451–1464.

Rigau, V., Morin, M., Rousset, M. C., de Bock, F., Lebrun, A., Coubes, P., et al. (2007). Angiogenesis is associated with blood-brain barrier permeability in temporal lobe epilepsy. *Brain*, *130*, 1942–1956.

Riobo, N. A., Clementi, E., Melani, M., Boveris, A., Cadenas, E., Moncada, S., et al. (2001). Nitric oxide inhibits mitochondrial NADH:ubiquinone reductase activity through peroxynitrite formation. *The Biochemical Journal*, *359*, 139–145.

Risher, W. C., & Eroglu, C. (2012). Thrombospondins as key regulators of synaptogenesis in the central nervous system. *Matrix Biology*, *31*, 170–177.

Robinson, K. M., Janes, M. S., & Beckman, J. S. (2008). The selective detection of mitochondrial superoxide by live cell imaging. *Nature Protocols*, *3*, 941–947.

Rosenberg, G. A. (2012). Neurological diseases in relation to the blood-brain barrier. *Journal of Cerebral Blood Flow and Metabolism*, *32*, 1139–1151.
Rowley, S., & Patel, M. (2013). Mitochondrial involvement and oxidative stress in temporal lobe epilepsy. *Free Radical Biology & Medicine*, *62*, 121–131.
Ruhrberg, C., & Bautch, V. L. (2013). Neurovascular development and links to disease. *Cellular and Molecular Life Sciences*, *70*, 1675–1684.
Ryan, K., Backos, D. S., Reigan, P., & Patel, M. (2012). Post-translational oxidative modification and inactivation of mitochondrial complex I in epileptogenesis. *The Journal of Neuroscience*, *32*, 11250–11258.
Safiulina, D., & Kaasik, A. (2013). Energetic and dynamic: How mitochondria meet neuronal energy demands. *PLoS Biology*, *11*, e1001755.
Schevon, C. A., Weiss, S. A., McKhann, G., Jr., Goodman, R. R., Yuste, R., Emerson, R. G., et al. (2012). Evidence of an inhibitory restraint of seizure activity in humans. *Nature Communications*, *3*, 1060.
Schoknecht, K., & Shalev, H. (2012). Blood-brain barrier dysfunction in brain diseases: Clinical experience. *Epilepsia*, *53*(Suppl. 6), 7–13.
Schuchmann, S., Albrecht, D., Heinemann, U., & von Bohlen und Halbach, O. (2002). Nitric oxide modulates low-Mg^{2+}-induced epileptiform activity in rat hippocampal-entorhinal cortex slices. *Neurobiology of Disease*, *11*, 96–105.
Schuchmann, S., Kovacs, R., Kann, O., Heinemann, U., & Buchheim, K. (2001). Monitoring NAD(P)H autofluorescence to assess mitochondrial metabolic functions in rat hippocampal-entorhinal cortex slices. *Brain Research. Brain Research Protocols*, *7*, 267–276.
Schuchmann, S., Lückermann, M., Kulik, A., Heinemann, U., & Ballanyi, K. (2000). Ca(2+)- and metabolism-related changes of mitochondrial potential in voltage-clamped CA1 pyramidal neurons in situ. *Journal of Neurophysiology*, *83*, 1710–1721.
Shuttleworth, C. W. (2010). Use of NAD(P)H and flavoprotein autofluorescence transients to probe neuron and astrocyte responses to synaptic activation. *Neurochemistry International*, *56*, 379–386.
Simantov, R., Crispino, M., Hoe, W., Broutman, G., Tocco, G., Rothstein, J. D., et al. (1999). Changes in expression of neuronal and glial glutamate transporters in rat hippocampus following kainate-induced seizure activity. *Brain Research Molecular Brain Research*, *65*, 112–123.
Singh, I. N., Sullivan, P. G., & Hall, E. D. (2007). Peroxynitrite-mediated oxidative damage to brain mitochondria: Protective effects of peroxynitrite scavengers. *Journal of Neuroscience Research*, *85*, 2216–2223.
Sipos, I., Tretter, L., & Adam-Vizi, V. (2003). The production of reactive oxygen species in intact isolated nerve terminals is independent of the mitochondrial membrane potential. *Neurochemical Research*, *28*, 1575–1581.
Sleven, H., Gibbs, J. E., Heales, S., Thom, M., & Cock, H. R. (2006). Depletion of reduced glutathione precedes inactivation of mitochondrial enzymes following limbic status epilepticus in the rat hippocampus. *Neurochemistry International*, *48*, 75–82.
Smith, R. A., Hartley, R. C., Cocheme, H. M., & Murphy, M. P. (2012). Mitochondrial pharmacology. *Trends in Pharmacological Sciences*, *33*, 341–352.
Srivastava, S., & Moraes, C. T. (2005). Double-strand breaks of mouse muscle mtDNA promote large deletions similar to multiple mtDNA deletions in humans. *Human Molecular Genetics*, *14*, 893–902.
Stanimirovic, D. B., & Friedman, A. (2012). Pathophysiology of the neurovascular unit: Disease cause or consequence? *Journal of Cerebral Blood Flow and Metabolism*, *32*, 1207–1221.
Stepien, K. M., Tomaszewski, M., Tomaszewska, J., & Czuczwar, S. J. (2012). The multidrug transporter P-glycoprotein in pharmacoresistance to antiepileptic drugs. *Pharmacological Reports*, *64*, 1011–1019.

Suzuki, M., Ogawa, A., Sakurai, Y., Nishino, A., Venohara, K., Mizoi, K., et al. (1992). Thrombin activity in cerebrospinal fluid after subarachnoid hemorrhage. *Stroke, 23*, 1181–1182.

Szewczyk, A., Jarmuszkiewicz, W., & Kunz, W. S. (2009). Mitochondrial potassium channels. *IUBMB Life, 61*, 134–143.

Tanaka, K., Watase, K., Manabe, T., Yamada, K., Watanabe, M., Takahashi, K., et al. (1997). Epilepsy and exacerbation of brain injury in mice lacking the glutamate transporter GLT-1. *Science, 276*, 1699–1702.

Territo, P. R., Mootha, V. K., French, S. A., & Balaban, R. S. (2000). Ca(2+) activation of heart mitochondrial oxidative phosphorylation: Role of the F(0)/F(1)-ATPase. *American Journal of Physiology. Cell Physiology, 278*, C423–C435.

Trotti, D., Danbolt, N. C., & Volterra, A. (1998). Glutamate transporters are oxidant-vulnerable: A molecular link between oxidative and excitotoxic neurodegeneration? *Trends in Pharmacological Sciences, 19*, 328–334.

Vezzani, A., Friedman, A., & Dingledine, R. J. (2013). The role of inflammation in epileptogenesis. *Neuropharmacology, 69*, 16–24.

Vielhaber, S., Niessen, H. G., Debska-Vielhaber, G., Kudin, A. P., Wellmer, J., Kaufmann, J., et al. (2008). Subfield-specific loss of hippocampal N-acetyl aspartate in temporal lobe epilepsy. *Epilepsia, 49*, 40–50.

Vielhaber, S., Von Oertzen, J. H., Kudin, A. F., Schoenfeld, A., Menzel, C., Biersack, H. J., et al. (2003). Correlation of hippocampal glucose oxidation capacity and interictal FDG-PET in temporal lobe epilepsy. *Epilepsia, 44*, 193–199.

Waldbaum, S., Liang, L. P., & Patel, M. (2010). Persistent impairment of mitochondrial and tissue redox status during lithium-pilocarpine-induced epileptogenesis. *Journal of Neurochemistry, 115*, 1172–1182.

Waldbaum, S., & Patel, M. (2010). Mitochondrial dysfunction and oxidative stress: A contributing link to acquired epilepsy? *Journal of Bioenergetics and Biomembranes, 42*, 449–455.

Wallace, D. C., Fan, W. W., & Procaccio, V. (2010). Mitochondrial energetics and therapeutics. *Annual Review of Pathology, 5*, 297–348.

Wiesner, R. J., Zsurka, G., & Kunz, W. S. (2006). Mitochondrial DNA damage and the aging process: Facts and imaginations. *Free Radical Research, 40*, 1284–1294.

Williams, G. S., Boyman, L., Chikando, A. C., Khairallah, R. J., & Lederer, W. J. (2013). Mitochondrial calcium uptake. *Proceedings of the National Academy of Sciences of the United States of America, 110*, 10479–10486.

Winkler, M. K., Chassidim, Y., Lublinsky, S., Revankar, G. S., Major, S., Kang, E. J., et al. (2012). Impaired neurovascular coupling to ictal epileptic activity and spreading depolarization in a patient with subarachnoid hemorrhage: Possible link to blood-brain barrier dysfunction. *Epilepsia, 53*(Suppl. 6), 22–30.

Yamamoto, T., Maruyama, W., Kato, Y., Yi, H., Shamoto-Nagai, M., Tanaka, M., et al. (2002). Selective nitration of mitochondrial complex I by peroxynitrite: Involvement in mitochondria dysfunction and cell death of dopaminergic SH-SY5Y cells. *Journal of Neural Transmission, 109*, 1–13.

Zhan, C., & Lu, W. (2012). The blood-brain/tumor barriers: Challenges and chances for malignant gliomas targeted drug delivery. *Current Pharmaceutical Biotechnology, 13*, 2380–2387.

Zhao, M., Ma, H., Suh, M., & Schwartz, T. H. (2009). Spatiotemporal dynamics of perfusion and oximetry during ictal discharges in the rat neocortex. *The Journal of Neuroscience, 29*, 2814–2823.

Zhao, M., Nguyen, J., Ma, H., Nishimura, N., Schaffer, C. B., & Schwartz, T. H. (2011). Preictal and ictal neurovascular and metabolic coupling surrounding a seizure focus. *The Journal of Neuroscience, 31*, 13292–13300.

Zhao, M., Suh, M., Ma, H., Perry, C., Geneslaw, A., & Schwartz, T. H. (2007). Focal increases in perfusion and decreases in hemoglobin oxygenation precede seizure onset in spontaneous human epilepsy. *Epilepsia, 48*, 2059–2067.

Zimmermann, A., Domoki, F., & Bari, F. (2008). Seizure-induced alterations in cerebrovascular function in the neonate. *Developmental Neuroscience, 30*, 293–305.

Zlokovic, B. V. (2008). The blood-brain barrier in health and chronic neurodegenerative disorders. *Neuron, 57*, 178–201.

CHAPTER TEN

Modern Techniques of Epileptic Focus Localization

Lukas Martinkovic*, Hrvoje Hecimovic†, Vlastimil Sulc*, Radek Marecek‡,§, Petr Marusic*,1

*Department of Neurology, Charles University in Prague, 2nd Faculty of Medicine, Motol University Hospital, Prague, Czech Republic
†Zagreb Epilepsy Center, Department of Neurology, University Hospital, Zagreb, Croatia
‡CEITEC—Central European Institute of Technology, Masaryk University, Brno, Czech Republic
§Brno Epilepsy Center, Department of Neurology, St. Anne's University Hospital and Faculty of Medicine, Masaryk University, Brno, Czech Republic
[1]Corresponding author: e-mail address: petr.marusic@fnmotol.cz

Contents

1. Introduction 246
2. Structural Imaging, Volumetrics, and Morphometric Analysis 247
   2.1 Image preprocessing 248
   2.2 MRI volumetry 248
   2.3 Voxel-based morphometry 250
   2.4 Surface-based morphometry 251
   2.5 Cortical morphometry 251
   2.6 Diffusion tensor imaging 252
3. Functional Neuroimaging in Epilepsies Using PET 253
   3.1 Interictal metabolic changes (FDG PET) 254
   3.2 Flumazenil PET 258
   3.3 Other tracers 259
4. Ictal Perfusion 260
   4.1 Preictal, ictal, and postictal changes (pathophysiology, hemodynamics) 260
   4.2 SPECT in epileptic seizures 261
5. Irritative Zone as Assessed by BOLD fMRI 265
   5.1 Principles 266
   5.2 Technical aspects 267
   5.3 Methodological aspects 267
   5.4 Clinical implications 269
6. Conclusions 271
Acknowledgments 271
References 271

International Review of Neurobiology, Volume 114
ISSN 0074-7742
http://dx.doi.org/10.1016/B978-0-12-418693-4.00010-8

Abstract

A clear concept of epileptic zones remains of high clinical relevance in presurgical evaluation of refractory epilepsy patients and in resection planning. Recent advances in understanding how each of the epileptic zones is functionally organized strengthened the importance of the network concept. It has been shown that neuronal networks underlying the individual epileptic zone may involve multiple brain structures with complex interactions between them. The network concept has impact not only for better understanding of pathophysiology of partial epilepsy but also for clinical practice, particularly for epilepsy surgery.

This review examines recent reports on the use of advanced imaging techniques which enable to map the epileptic zones and their structural and functional organization. Magnetic resonance postprocessing substantially improved the accuracy in detection of the epileptogenic lesions. The seizure-onset zone is primarily determined by electrophysiology but can also be localized using single photon emission computed tomography. The functional deficit zone is commonly assessed by a number of tests including methods of functional neuroimaging (positron emission tomography) which can delineate hypometabolic cortical areas and subcortical structures. Hemodynamic fluctuations associated with interictal epileptiform discharges can be detected by novel functional magnetic resonance technique which is nowadays widely used for the irritative zone localization. These techniques open new prospect for epilepsy surgery in patients who were previously considered as not suitable candidates of surgical treatment.

1. INTRODUCTION

A number of imaging techniques are capable to provide new localizing information on the ictal processes and epileptogenic zone localization. The zone concept of focal epilepsy has been implemented especially in the context of presurgical patient evaluation with the aim to identify the epileptogenic zone—the minimum amount of cortex that must be resected, inactivated, or completely disconnected, to produce seizure freedom (Luders, Najm, Nair, Widdess-Walsh, & Bingman, 2006). The location and the extent of the epileptogenic zone are deduced from the location and the extent of five brain areas identified during the presurgical evaluation: (1) the irritative zone that generates interictal spikes; (2) the seizure onset zone that initiates clinical seizures; (3) the symptomatogenic zone producing ictal symptoms; (4) the epileptogenic lesion—macroscopic lesion which is causative of the epileptic seizures because the lesion itself is epileptogenic (e.g., cortical dysplasia) or by secondary hyperexcitability of adjacent cortex; and (5) the functional deficit zone, area of cortex that is not functioning normally in the interictal period. The recognition of this set of regions responsible for seizures and interictal changes with interregional interaction already implied a network concept of epilepsy. A concept of human epilepsies as disorders of large neuronal networks was expressively presented more than 10 years ago and defined as "functionally and anatomically connected, bilaterally represented set(s) of cortical and subcortical brain structures and regions in which activity in any part affects activity in all the others" (Spencer, 2002).

This review examines neuroimaging methods used to identify some of the above-mentioned zones, methods that also brought evidence about the importance of network approach to epilepsy. Epileptogenic lesion is currently defined by anatomical imaging such as high-resolution magnetic resonance imaging (MRI). Patients with MRI-negative epilepsies have a lower chance of having surgery than those with MRI-demonstrated lesions and, if so, less chance of becoming seizure free. The MRI-negative patients for whom the surgery was successful demonstrate that the process of multimodal presurgical evaluation may lead to good outcomes (Bien et al., 2009).

Novel imaging modalities have been improved or developed to provide information about the localization of epileptic zones. The methods currently used to identify the different zones have variable degrees of error, and the knowledge about their methodology and limitations is crucial when using them for presurgical evaluation.

2. STRUCTURAL IMAGING, VOLUMETRICS, AND MORPHOMETRIC ANALYSIS

MRI is an essential method for identifying an epileptogenic lesion, i.e., structural abnormality underlying focal epilepsy, and up to 80% of patients with recurrent seizures have such a lesion evident on MRI (Salmenpera & Duncan, 2005). The sensitivity of MRI for detecting a structural abnormality depends on the pathological substrate, applied techniques, and last but not least the experience of the interpreting physician. Conventional MRI does not reliably capture epileptogenic abnormalities such as hippocampal sclerosis (HS), cortical malformations, or vascular changes; however, detection has improved with noninvasive high-resolution MRI (Von Oertzen et al., 2002).

Higher magnetic field (3 T or more), in combination with the use of phased array coils instead of a conventional quadrature coil, has resulted in further increase in image signal- and contrast-to-noise ratios (Wiggins et al., 2006). In the majority of reported cases, images obtained at 3 T allowed for a more detailed and complete delineation and characterization of structural changes in the hippocampus (atrophy and gliosis) or the neocortex (dysplasias) than those obtained at 1.5 T (Knake et al., 2005). Gadolinium does not improve the sensitivity of MRI in patients with epilepsy, but may be useful to characterize cerebral lesions associated with breakdown of the blood–brain barrier (Salmenpera & Duncan, 2005).

2.1. Image preprocessing

According to last investigations by Bonn group, an epilepsy-specific MRI protocol should include "the essential six": (1) 3D T1 1 mm isotropic three-dimensional scan in commissura anterior–commissura posterior angulation (best to differentiate gray matter (GM) and white matter (WM), available to morphometric MRI analysis); (2 + 3) fluid attenuated inversion recovery (FLAIR) sequence in axial and coronal plane, slice thickness lower than 3 mm, perpendicular to hippocampal axis; (4 + 5) T2/STIR (short tau inversion recovery) in axial and coronal plane, slice thickness lower than 3 mm, perpendicular to axis of hippocampus (this cut-plane angulation is superior to determinate pathology in mesial temporal lobe structures); (6) hemosiderin and calcification-sensitive sequences (hemo-calc). After the performance of these sequences, 99.4% of all lesions were detected except for subtle pial angiomatosis (Wellmer et al., 2013).

Advanced MRI analysis techniques can quantify the extent of GM loss and can also display the surface contours and spatial patterns of GM loss in hippocampus and other brain structures. GM loss extends beyond the affected hippocampus and includes cortical GM and WM alterations in temporal lobe and often extratemporal lobe structures (Memarian, Thompson, Engel, & Staba, 2013). In the following parts, an overview of modern imaging techniques with their usage and impact on detection an epileptogenic zone is given.

2.2. MRI volumetry

Quantitative assessment of brain volumes, obtained through volumetric MRI, has been increasingly applied in recent years to a wide range of neurologic conditions due to advances in computational technology. The clinical use of MRI-derived brain volumetry will be framed in different contexts: (1) supporting disease diagnosis, (2) understanding mechanisms and tracking clinical progression of disease, and (3) monitoring treatment effects (Giorgio & De Stefano, 2013). The prototype of manual volumetry was manual segmentation of subcortical structures, e.g., amygdala and hippocampus. Past manual techniques have used 2D measures such as linear parameters. More recently, (manual) segmentation in the three planes of the space, outlining regions of interest, has been performed. Advanced MRI and computational techniques permitted to use semiautomated, requiring some level of user intervention, or fully automated procedures. Both have demonstrated accuracy and reproducibility and have been extensively used in a number of clinical studies, especially for the volumetric assessment of the whole brain and cerebral tissue types, e.g., WM and GM. Fully automated computational methods are favorable because they accelerate analyses and allow the assessment of large datasets. Volumetric MRI can be essential in the diagnosis and management of drug-resistant epilepsy, allowing the detection of subtle abnormalities of brain tissue that are difficult or impossible to reveal on visual inspection (Bernasconi, Bernasconi, Bernhardt, & Schrader, 2011).

In temporal lobe epilepsy (TLE), the most common epilepsy in adults, MRI volumetry shows asymmetry of the hippocampal volume ipsilateral to the seizure focus with a sensitivity up to 95% (Vattipally & Bronen, 2004). A number of volumetric studies found that several brain regions were atrophic in TLE patients compared with normal subjects. Atrophy was found in not only structures of the medial temporal lobe (hippocampus, parahippocampal gyrus, and entorhinal cortex), but also bilaterally in extratemporal lobe structures, especially thalamus and parietal lobe (Keller & Roberts, 2008). In patients with right-sided mesial TLE, a reduction of GM concentration was observed in the right hippocampus and parahippocampal gyrus, left and right hemispheres of the cerebellum, bilateral thalamus, right caudate nucleus, right temporal isocortex, right temporo-occipital region, left and right middle frontal gyri, dorsal midbrain, and left parieto-occipital region. Contrary, in patients with left-sided mesial TLE, reduction of GM was found in the left hippocampus and parahippocampal gyrus, left insula, left and right cingulate gyri, left and right frontal opercula, left and right thalami, left and right parieto-occipital regions, left and right cerebellum, midbrain, and bilateral caudate nuclei (Bonilha et al., 2004). Reduction of the temporopolar WM and GM volumes ipsilateral to seizure onset was also demonstrated (Coste et al., 2002).

Atrophy in TLE is generally viewed as a consequence of seizure frequency, although edema resolution rather than true structural damage may actually be responsible for apparent volume decrease in the hippocampus of these patients (Kuzniecky & Jackson, 2005, chapter 4). In longitudinal MRI studies of patients with TLE and HS, the progression of brain atrophy in the GM seems to be associated with poorer seizure control and a longer duration of epilepsy. By contrast, an increase of GM volume can be related to the presence of developmental malformations causing reduced contrast between GM and WM on MR images (Ho, Kuzniecky, Gilliam, Faught, & Morawetz, 1998).

MRI volumetry had been an initial method to assess structural abnormalities before advanced morphometric methods were developed. Its advantage is still in precise measurement of anatomically difficult to define structures, e.g., amygdala.

2.3. Voxel-based morphometry

Voxel-based morphometry (VBM) provides an automated quantitative analysis of the distribution of GM and WM and may be applied to individuals and to groups of patients (Salmenpera & Duncan, 2005). These techniques may facilitate the detection and visualization of focal cortical dysplasia (FCD) and other potentially epileptogenic cortical malformations by highlighting structural alterations (Huppertz et al., 2005).

Whole-brain digital structural images are compared on the basis of properties corresponding with groups of voxels, and the general premise is to establish voxel-for-voxel correspondence across subjects through nonlinear registration of multiple subjects brain images to a standard anatomical template (Memarian et al., 2013). Once the images are aligned into the same coordinate space, one can perform voxel-wise statistical comparisons of GM and WM volume or concentration across different subject groups (Clarkson et al., 2011). The standard VBM process typically involves four steps: (1) spatial normalization (applying a nonlinear registration to each subject's T1-weighted MR image); (2) tissue segmentation (GM, WM, CSF based on the intensity in the image); (3) spatial smoothing (intensity in each voxel is a local weighted average generally expressed as GM, WM, CSF concentration); and (4) statistical analysis–group comparisons or correlations with covariates of interest (Ashburner & Friston, 2000).

Studies of mesial TLE using automated VBM analysis consistently found GM loss in the hippocampus ipsilateral to the site of seizure onset in patients with MRI evidence of unilateral HS (Keller et al., 2002). VBM-based GM loss has also been detected in the amygdala, entorhinal cortex, and parahippocampal gyri (Keller, Wilke, Wieshmann, Sluming, & Roberts,

2004). GM reductions extend beyond the mesial temporal lobe ipsilateral to seizure onset and often include other limbic structures, such as the thalamus, cingulate gyrus, and orbitofrontal cortex, as well as areas of temporal and parietal neocortex, striatum, and cerebellum (Keller & Roberts, 2008). VBM was also applied to detect structural abnormalities related to MRI-visible FCD on T1-weighted MRI in individual patients; studies reported increases in GM concentration that colocalized with the lesion in 63–86% of cases (Kassubek, Huppertz, Spreer, & Schulze-Bonhage, 2002).

2.4. Surface-based morphometry

Morphometric measures in surface-based morphometry (SBM) are derived from geometric models of the cortical surface or surface-based models of subcortical structures, such as the hippocampus. The techniques are used to reconstruct and analyze surfaces that represent structural boundaries within the brain. The major steps involved in SBM of the brain are preprocessing (image registration, intensity normalization, smoothing, brain extraction, and segmentation), surface extraction, inflation or parameterization, and surface mapping (Fischl & Dale, 2000). Several implementations of SBM have been made with the FreeSurfer software (Harvard Martinos Center for Biomedical Imaging, MA, USA), which is a set of automated tools for the reconstruction of the brain's cortical surface from structural MRI data.

Patient studies of mesial TLE with HS found reduced ipsilateral hippocampal GM volumes correlated with cortical GM thinning in the temporal lobe (Mueller et al., 2009). Cortical GM was on average 5–15% thinner in bilateral frontal (prominently within pre- and para-central gyri, operculum) and temporal lobes (Heschl's gyrus) of patients with mesial TLE compared with controls (McDonald et al., 2008).

In summary, SBM is well suited for detecting neocortical GM abnormalities because images are aligned with respect to neocortical folding patterns, and this improves neocortical registration compared with aligning on the basis of voxel intensity (Scanlon et al., 2011). SBM methods also have a somewhat higher reproducibility than VBM and can be more robust in longitudinal studies that measure changes in GM thickness over time (Clarkson et al., 2011).

2.5. Cortical morphometry

Cortical morphometry is a technique related to SBM concerning evaluation of neocortical GM thickness in patients with mesial TLE and HS. It involves cortical parameterization, matching cortical features across individuals, and statistical comparisons with a map of group differences in features such as cortical thickness or GM concentrations. The first step includes creating geometrical models of the cortical surface (image segmentation, registration, and 3D reconstruction). Matching cortical features across individuals is performed through neuroanatomical labeling (marking sulcal lines) and warping one brain surface onto another. Statistical testing is performed to investigate the effect of various factors (for example, disease, aging, treatment, and genetics, among others) on various cortical attributes, such as GM thickness or gyrification (Thompson et al., 2004).

In mesial TLE, GM thickness was reduced by up to 30% and reductions in cortical complexity were found in patients compared with controls (Lin et al., 2007). In another study, significant GM loss was observed bilaterally in frontal areas (pole, operculum, and orbitofrontal cortex), and lateral temporal and occipital regions independent of the hemisphere of seizure onset (Hogan et al., 2008).

Cortical morphometry is ideal for capturing small differences in GM thickness in patients without obvious pathology on MRI, although measures of volumetric GM loss can be sensitive to the geometrical calibration of the MRI scanner (Scanlon et al., 2011).

2.6. Diffusion tensor imaging

Diffusion tensor imaging (DTI) is a development of diffusion-weighted imaging, which allows better delineation of the magnitude of diffusion. Parameters derived from DTI include mean diffusivity, which reflects the overall rate of water diffusion in a voxel, and fractional anisotropy (FA), which measures how directionally constrained the diffusion process is. FA is an indirect measure of WM (fiber) density or myelination levels, and by following the principal directions of water diffusion, tract tracing methods enable to follow neural pathways throughout the brain (Memarian et al., 2013). Reduced FA in the external capsule and corpus callosum (Arfanakis et al., 2002; Concha, Beaulieu, Collins, & Gross, 2009), the uncinate fasciculus (Diehl et al., 2008), the cingulum bundle, as well as in several frontal and posterior fiber tracts including the inferior and superior longitudinal fascicles (Ahmadi et al., 2009) were reported in patients with TLE as compared to control subjects, while WM abnormalities in the fornix appear to be exclusive to TLE patients with HS (Concha et al., 2009). The effect size of mean diffusion alterations in TLE seems to decrease as a function of anatomical distance to the temporal lobe, suggesting colocalization of these changes with the seizure focus (Concha, Kim, Bernasconi, Bernhardt, & Bernasconi, 2012). A study performed in children with TLE demonstrated diffusion abnormalities in ipsilateral arcuate and uncinate fascicles which were associated with duration of epilepsy (Govindan, Makki, Sundaram, Juhász, & Chugani, 2008); however, other study performing DTI of external capsule and corpus callosum in TLE adult patients failed to demonstrate the correlation between duration of epilepsy and diffusion indices (Arfanakis et al., 2002).

These findings strongly suggest that in mesial TLE, patients' tract abnormalities are not restricted to the ipsilateral hippocampus and temporal lobe WM, but rather involve an extensive bilateral network. The most intuitive hypothesis regarding the cause of the WM abnormalities is that they are a secondary effect of ongoing seizure activity representing downstream axonal degeneration (Gross, 2011). Spatiotemporal dynamics of these abnormalities in the course of the disease are not yet fully understood.

3. FUNCTIONAL NEUROIMAGING IN EPILEPSIES USING PET

Since the early 1970s, positron emission tomography (PET) has been used first for research purposes, and then over the last two decades, it is accepted as a standard in clinical practice. Today, PET scanning is recognized as a leading application in neuroimaging and functional diagnostics, and when combined with either the CT or MRI, it additionally provides relevant information about brain morphology. It also gives important data in patients in whom MRI of brain structures is normal. The advantage of the PET is that it provides both qualitative and quantitative assessment of cortical dysfunction. In some patients, visual assessment of PET may have limited value due to a smaller or minimal cortical dysfunction, and in this situation, we can rely on quantitative PET analysis (Henry, Engel, & Mazziotta, 1993).

Many studies of brain dysfunction in epilepsy have been performed using PET with objectives to measure brain metabolism or blood flow changes (Fink et al., 1996). Some groups, based on PET data, proposed certain models of cortical dysfunction to better define cortical neural networks. For example, researchers suggested that glucose hypometabolism in the prefrontal areas, anterior cingulate, and striatum and an increase in the amygdala glucose metabolism are responsible for symptoms of apathy, attention deficits, executive function impairments, and psychomotor slowing that are commonly observed in patients with epilepsy. Glucose hypermetabolism in other regions, such as hypothalamus, insula, and brainstem appears to be responsible for symptoms of disturbed sleep, appetite, and libido. There is no consensus concerning a relationship between blood flow changes and glucose metabolism abnormalities in patients with epilepsy. More recent data from PET studies using radiolabeled isotopes other than glucose suggested more specific pathways of neurotransmission. This evidence shows more specific areas of brain dysfunction that some researchers associate with specific clinical symptoms. Various other neuroimaging modalities focused on the WM changes and more detailed analysis of the neural pathways, which offers promising new insights.

Indication for PET assessment has been expanding rapidly, and PET is now an important tool in the diagnosis and management of medically refractory epilepsies. The advantage of PET is that it is an imaging technique which allows noninvasive qualitative and quantitative examination of different biochemical and functional processes.

3.1. Interictal metabolic changes (FDG PET)

Today, a range of positron isotopes are available: ^{18}F (combined with fluorodeoxyglucose) is the most commonly used, with its good characteristics, such as higher spatial resolution, longer half-life that permits its synthetic procedures, and longer activity that is needed for its transport between a cyclotron and the PET machine. Fluorodeoxyglucose (FDG) presents an isotope and modified glucose molecule that enters neurons by the same pathway as glucose, usually via the GLUT-3 transporter. It is phosphorylated to FDG-6-phosphate, but not further. Thus, with ^{18}F-FDG application, there is a normal and "abnormal" brain metabolic activity. In order to minimize the normal metabolic activity before imaging, patients are asked not to eat in the morning prior to injection, so they become hypoglycemic and the uptake of ^{18}F-FDG is high. There is also increased activity in the urinary system as ^{18}F-FDG is not absorbed into the renal tubular cells, unlike normal glucose. There is also some ^{18}F-FDG activity in the liver due to glycolysis and some uptake in myocardiocytes. For most of these applications, the visual inspection of the ^{18}F-FDG uptake provides sufficient clinical information, and if not, a quantitative imaging analysis, which is developing rapidly as a routine software application, is utilized.

3.1.1 Pathophysiology

^{18}F-FDG PET measures glucose metabolism related to the synaptic and neuronal activity of the brain tissue. In persons with epilepsy, interictal

^{18}F-FDG PET typically shows reduced radiotracer uptake or hypometabolism in the epileptogenic region (Willmann, Wennberg, May, Woermann, & Pohlmann-Eden, 2007). A variety of mechanisms can cause this hypometabolism, including neuronal loss, diaschisis, or reduction in synaptic density. A number of studies suggested that brain glucose hypometabolism in persons with epilepsy may be associated with the time of onset, duration, frequency, and severity of their seizures (Lee & Salamon, 2009). Glucose hypometabolism is usually found in approximately 20% of children with new onset epilepsy and in about 80% of adults with longer duration of epilepsy (Henry, 2010). It is shown that increased seizure frequency and longer duration of epilepsy may lead to larger hypometabolic areas, and reduced seizure frequency may be associated with a smaller area of the hypometabolic cortex. Glucose hypometabolism in ipsilateral thalamus can be seen in cases of frontal and temporal epilepsy. Mesial TLE is commonly associated with dorsal thalamus hypometabolism. Contralateral cerebellar hypometabolism can be seen in patients with seizures of frontal, temporal, or parietal lobe origin. In some patients, hypometabolism is seen in the contralateral "mirror" region and may represent a dysfunction of these regions promoted by reduced connectivity in the corpus callosum or via the fornix in case of seizures originating in the hippocampus. More extensive areas of hypometabolism in adults with longer duration of epilepsy likely involve seizure propagation areas or damaging effect of the seizures on connected brain regions. Another contribution to the brain glucose metabolism is the use of antiepileptic drugs. The rates of glucose metabolism, irrespective of seizure frequency, can be reduced not only by some of these drugs, primarily barbiturates, but also by valproate, phenytoin, and carbamazepine.

3.1.2 Localization and extent of hypometabolism in TLE

^{18}F-FDG PET usually shows a large area of hypometabolism extending beyond the epileptogenic region. It is used for lateralization and more general localization of the seizure focus. It is also useful for evaluation of functional status of the brain in patients with an epileptogenic lesion that is considered for surgical resection. Majority of ^{18}F-FDG PET results in TLE were collected during standard presurgical evaluation (Henry, 2010). These results show that ipsilateral temporal lobe hypometabolism corresponds to seizure focus in about 86% of patients. In most of these patients, MRI demonstrated structural abnormalities. ^{18}F-FDG PET was concordant with EEG in 75% of TLE patients and with MRI in 69% patients. Current data suggest that interictal hypometabolism does not correlate with neuronal loss, but rather with cerebral dysfunction, loss of synapses, and poorly regulated electrical activity in dysfunctional cortex. Hypometabolic area is usually larger than structural abnormality, and it probably presents a dysfunction of a neural network (Lamusuo et al., 2001). ^{18}F-FDG PET in patients with HS shows more focal hypometabolism in ipsilateral temporal lobe, but also shows more widespread hypometabolism, in contralateral temporal lobe, in bilateral frontal and parietal areas, ipsilateral thalamus, and in contralateral cerebellum.

^{18}F-FDG PET is also useful in defining mesial versus lateral TLE. Patients with mesial TLE have most pronounced hypometabolism in mesial structures, but also in the lateral temporal lobe, whereas patients with lateral TLE have most significant hypometabolism in the lateral temporal cortex. Also, patients with HS have much more reduced and more focal hypometabolism than the patients with lateral TLE. The degree of hippocampal volume asymmetry does not correlate with the level of mesial or lateral temporal hypometabolism (Hajek, Antonini, Leenders, & Wieser, 1993). Similar finding suggests that a degree of neuronal loss in the hippocampus on histopathology does not correlate with the ^{18}F-FDG PET hypometabolism. Thus, glucose hypometabolism is probably a result of ictal and interictal spread, pointing to a broader epileptic network.

Even in patients without HS, ^{18}F-FDG PET is lateralizing. In 15–30% of patients with TLE, MRI is normal and ^{18}F-FDG PET may help in surgical decision making (Fig. 10.1). In up to 87% of these patients, ^{18}F-FDG PET can identify hypometabolism in ipsilateral temporal lobe and, if this is concordant with the EEG findings, reduces need for intracranial electrode placement.

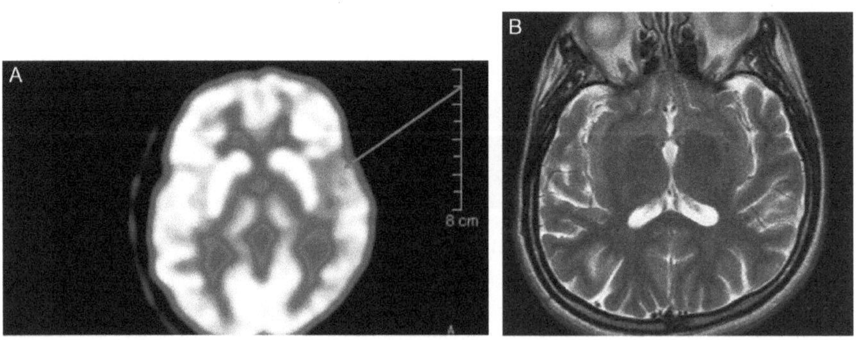

Figure 10.1 ^{18}F-FDG PET showing hypometabolism in the left temporal region that is concordant with the epileptic area defined by EEG (A). Brain MRI was normal (B).

3.1.3 Clinical implications in TLE

Results from a meta-analysis of all publications from 1992 to 2006 show that the presence of ipsilateral glucose hypometabolism provides better surgical outcome in 86% of TLE patients following temporal lobectomy. In the patients with TLE and normal MRI, ipsilateral hypometabolism also predicted good surgical outcome in about 77% of TLE patients. However, almost half of TLE patients with bitemporal hypometabolism have independent bilateral foci, and surgical outcome is much worse, only about 50% of patients have a good outcome (Shandal, Veenstra, Behen, Sundaram, & Chugani, 2012). ^{18}F-FDG PET is also very helpful in evaluation of patients with dual pathology, either with mesial and neocortical temporal foci or if a secondary noncontiguous epileptic focus is suspected.

^{18}F-FDG PET can also provide some other important prognostic information. In patients with unilateral TLE, the presence of bitemporal glucose hypometabolism is associated with lower memory performance and prefrontal hypometabolism is associated with impaired executive function. Similarly, associated hypometabolism of the ipsilateral insular cortex may correlate with emotional or somesthetic symptoms (Bouilleret et al., 2002).

3.1.4 Clinical implications in extratemporal lobe epilepsy

In patients with frontal lobe epilepsy, sensitivity of ^{18}F-FDG PET to detect an epileptic region is about 45–92% (generally it is about 55%) (Henry, 2010). Despite this, ^{18}F-FDG PET can provide additional information about the epileptic focus in almost two-thirds of patients with frontal lobe epilepsy and help in surgical decision making. Some studies suggest that ^{18}F-FDG PET alone may be the basis for surgical decision in almost 17% of patients. Poor localization using ^{18}F-FDG PET has been reported in patients with parietal or occipital lobe epilepsy.

Incidence of extratemporal epilepsy is higher in children. Most of them have normal MRI findings, and ^{18}F-FDG PET has an important role in clinical management and presurgical evaluation in the pediatric population. ^{18}F-FDG PET is particularly useful in cortical dysplasia and heterotopias, which also presents more commonly in pediatric epilepsy population. Other beneficial applications of ^{18}F-FDG PET include patients in whom available MR imaging is negative or if only a limited number of lesions are suspected to be epileptogenic out of several detected lesions, such as in cavernomas or tuberous sclerosis patients.

3.1.5 Statistical parametric mapping

Statistical parametric mapping (SPM) is a voxel-based analysis technique used in epilepsy patients (Merlet, Ryvlin, et al., 2004). This is a statistical technique for assessment of differences in the brain activity recorded during functional neuroimaging experiments using PET or fMRI. Parametric statistical models use a general linear model to describe variability in the data in terms of experimental and confounding effects. Hypotheses expressed in terms of the model parameters are assessed at each voxel with univariate statistics. It objectively defines focal abnormalities that correspond to seizure foci. SPM has greater than 75% sensitivity and specificity in both lesional and nonlesional brains. General localization value of qualitative assessment and SPM analysis is comparable, but the SPM is particularly useful in finding epileptic areas in mesial structures that may be missed on visual examination. Seizures arising from these cortical regions, e.g., parasagittal area, can show a fast bilateral synchrony on the ictal electroencephalography, thus missing potential candidates for surgical resection. SPM analysis appears to be a particularly useful approach for assessing medially located seizure foci, and its result may optimize intracranial electrode coverage (Merlet, Ryvlin, et al., 2004). Use of SPM analysis has several other advantages over basic qualitative assessment. SPM could help in lateralizing epileptic focus, even in patients with bilateral structural (MRI) or functional (PET) abnormalities. Most researchers reported that SPM analysis is equally good in both left and right hemispheres, and that visual assessment has bias in detecting left hemisphere ^{18}F-FDG hypometabolism. The reason may be that visual analysis is more dependent on our perception and detection of left or right hemispheric asymmetries. There is also a physiological mild ^{18}F-FDG uptake hypometabolic region in the left inferior frontal gyrus that is detectable by visual assessment but corrected by the SPM analysis. Thus, visual and SPM analyses appear to be complementary and the SPM is useful in increasing chances of the ^{18}F-FDG PET for detection of epileptic focus.

3.2. Flumazenil PET

Another PET tracer used in clinical application for detecting epileptic brain regions is ^{11}C-flumazenil. This compound binds to subunits of the $GABA_A$ benzodiazepine receptor. ^{11}C-flumazenil binding in the medial temporal lobe is more specific than the ^{18}F-FDG PET uptake, and it offers superior visualization of an epileptic focus in the mesial temporal structures. This is the main reason that ^{11}C-flumazenil PET is preferred over ^{18}F-FDG PET for use in mesial TLE, showing more focally decreased tracer binding in the hippocampus (Juhasz et al., 2001).

3.3. Other tracers

A number of other radioligands that bind to translocator proteins, opioid receptors, histamine receptors, *N*-methyl-D-aspartate receptors, serotonin (5-HT) receptors, and acetylcholine receptors have been developed for clinical research in epilepsy.

3.3.1 5-HT_{1A} receptor antagonists

Binding of the 5-HT_{1A} receptor antagonists (^{18}F-FCWAY, ^{18}F-MPPF) in TLE patients is reduced in epileptic region (Merlet, Ostrowsky, et al., 2004), more in mesial than in lateral structures, and it shows reduced receptor density or an increase of endogenous serotonin that binds to this receptor. Binding in the hippocampal area is reduced in patients with hippocampal atrophy but not in those with normal hippocampal volumes. The binding is reduced not only in the epileptic focus but also in the seizure propagation area, as recorded by intracranial invasive EEG monitoring.

3.3.2 Tryptophan metabolism

Another radioligand, ^{11}C-α-methyl-L-tryptophan (α-MTrp), which does not bind to a receptor but measures tryptophan metabolism, appears to be a promising candidate (Natsume et al., 2003). α-MTrp, an artificial amino acid, is an analog of L-Trp. α-MTrp is converted under *in vivo* conditions to 5-hydroxy-α-methyl-L-tryptophan, which is also a substrate for aromatic amino acid decarboxylase. This yields α-methyl-serotonin (α-M5-HT). α-M5-HT is not a substrate for monoamine oxidase and, as such, remains in the brain for quite a long time after being formed from α-MTrp. In addition, α-MTrp is not incorporated into brain proteins. Studies have shown that there is an excellent concordance between the localization of endogenous 5-HT and radioactively labeled α-MTrp, as well as tryptophan and radioactively labeled α-MTrp. Interestingly, PET performed with this radioligand can show increased tryptophan uptake in the epileptogenic cortex interictally, making visual detection easier than with those tracers whose uptake or binding is usually reduced in the interictal phase. The tracer shows increased uptake, particularly in patients with cortical dysplasia type IIB, who usually have a good surgical outcome. This is an important finding because it indicates that there is an *in vivo* imaging technique that can predict pathologic subtype and may provide presurgical prognostic information.

3.3.3 *Inflammatory changes*

Translocator protein (TSPO), which is a marker of inflammation, has been measured *in vivo* using PET labeled with the ^{11}C-(*R*)-PK11195, but results have not been consistent. Increased uptake of radioactivity after injection of ^{11}C-(*R*)-PK11195 was reported in two cases of Rasmussen encephalitis and one case of intractable seizures due to encephalopathy of unknown origin but not confirmed in three patients with TLE and HS (Banati et al., 1999). Unfortunately, ^{11}C-(*R*)-PK11195 may lack sufficient sensitivity to detect increased TSPO in TLE. A novel compound ^{11}C-PBR28 is an improved PET radioligand for TSPO with an increased sensitivity to detect TSPO changes in TLE. ^{11}C-PBR28 (*N*-acetyl-*N*-(2-^{11}C-methoxybenzyl)-2-phenoxy-5-pyridinamine) is a marker of activated microglia-mediated neuroinflammation, in the hippocampus, parahippocampal gyrus, amygdala, and fusiform gyrus ipsilateral to the epileptic focus in patients with TLE (Hirvonen et al., 2012). There is still a controversy whether observed neuroinflammatory changes are the cause or the effect of seizures.

4. ICTAL PERFUSION

4.1. Preictal, ictal, and postictal changes (pathophysiology, hemodynamics)

The relationship between preictal and ictal state is complicated, as several studies have reported preictal changes on EEG and other diagnostic modalities scales (Bragin et al., 2002; Worrell et al., 2008). So-called microseizures observed on microelectrodes with a high sampling rate are thought to arise from submillimeter scale neuronal assemblies in epileptogenic zone (Stead et al., 2010). As the energy is spent on ion channel opening, increase in neuronal firing raises focal metabolic demands for oxygen and glucose which triggers increased blood perfusion via the mechanism of neurovascular coupling (Arthurs, Donovan, Spiegelhalter, Pickard, & Boniface, 2007). PET, single-photon emission computed tomography (SPECT), and fMRI studies usually indicate that relative increase in cerebral blood flow (CBF) more than meets regional metabolic need, resulting in relative hyper-oxygenation of the region (Jackson, Connelly, Cross, Gordon, & Gadian, 1994). On the other hand, studies with oxygen-sensitive electrodes, near infrared spectroscopy, or optical recording of intrinsic signals suggest that regional metabolic demands are not met (Adelson et al., 1999; Geneslaw, Zhao, Ma, & Schwartz, 2011).

More recently, it has also been observed that hemodynamic changes are present before seizure often even without concurrent changes observed on electrophysiology (Schwartz, Hong, Bagshaw, Chauvel, & Bénar, 2011). It has also been demonstrated that, at least for visual stimulation, the hemodynamic changes are correlated to the local field potential (>30 Hz) and not to neuronal spiking (Logothetis, Pauls, Augath, Trinath, & Oeltermann, 2001). The importance of gamma band in the neurovascular coupling has been confirmed by several human and animal studies (Lachaux et al., 2007; Niessing et al., 2005). Glial cells, namely astrocytes, have been also shown to play an important role in neurovascular coupling (Figley & Stroman, 2011).

However, it is also possible that vascular changes are not preictal but rather ictal phenomenon and the equipment used in these combined intracranial EEG/CBF studies is not sensitive enough to capture subtle electrophysiological changes occurring in much smaller spatial and temporal domains (Bragin et al., 2002; Stead et al., 2010; Worrell et al., 2008). Another possibility is that local hemodynamic response is caused by changes in local environment and glial cells such as astrocytes (Lauritzen, 2001).

The whole field of those preictal and early ictal changes is complicated by the fact that also interictal paroxysmal changes may or may not increase focal metabolic demand and cause hemodynamic changes and the electrophysiological changes are due to their small temporal and spatial scale difficult to observe on standard scalp EEG.

4.2. SPECT in epileptic seizures

In clinical settings, the most knowledge about ictal perfusion changes on macroscopic scale is gathered from ictal SPECT injection, which is able to provide a snapshot of brain perfusion during seizure. The tracer is administered intravenously, is rapidly taken up with an estimated peak brain uptake of 30 s (Andersen, 1989; Kaminska et al., 2003), and is not redistributed which enables data acquisition after patient recovers from the seizure. The interictal SPECT is usually performed for subtraction analysis but can occasionally show decreased cerebral perfusion in the region of suspected epileptogenic focus.

Various seizure-related factors affect the localization values of ictal SPECT. Early ictal injection and injection before or without seizure generalization increase the yield of ictal SPECT. Total seizure duration, etiology of epilepsy, its type, and especially propagation pattern can also influence the results, necessitating the importance of availability of these factors for correct interpretation of ictal SPECT. The localized ictal SPECT had been reported to be more frequent in a group with localized ictal EEG patterns. SPECT tracer distribution is also dependent on the type of the compound used.

Two most commonly used traces are ^{99m}Tc-hexamethylenepropylene amine (^{99m}Tc-HMPAO, Ceretec) and ^{99m}Tc-ethyl cysteinate dimer (^{99m}Tc-ECD, Neurolite). ^{99m}Tc-ECD is retained in the brain after an enzymatic conversion to ionized acid compounds and ^{99m}Tc-HMPAO after conversion to a nondiffusible hydrophilic compound after cell uptake. These different mechanisms of brain retention could explain the differences in cerebral distribution of the two tracers (Asenbaum, Brucke, Pirker, Pietrzyk, & Podreka, 1998). ^{99m}Tc-ECD is stable for 6–8 h, and the stabilized form of ^{99m}Tc-HMPAO for 4–6 h.

4.2.1 Propagation pattern

Because of the low temporal resolution of SPECT due to the combination of time-to-injection and time-to-uptake delays, ictal SPECT usually shows characteristic ictal hyperperfusion pattern, often containing both the ictal onset zone and propagation with a multilobular hourglass appearance (Van Paesschen, Dupont, Sunaert, Goffin, & Van Laere, 2007).

In a group SPECT study of patients with refractory TLE and complex partial seizures (CPSs) the ipsilateral temporal lobe, middle frontal and precentral gyrus, both occipital lobes, and parts of the contralateral postcentral gyrus showed ictal hyperperfusion. The frontal lobe, contralateral posterior cerebellum, and ipsilateral precuneus showed hypoperfusion (Van Paesschen, Dupont, Van Driel, Van Billoen, & Maes, 2003). In mesial temporal TLE, an initial hyperperfusion of the whole temporal lobe had been described followed by hypoperfusion of lateral temporal structures, and from 2 to 15 min postictally, there was hypoperfusion of the whole temporal lobe (Duncan, Patterson, Roberts, Hadley, & Bone, 1993; Newton et al., 1995).

Other studies also showed ictal ipsilateral or contralateral cerebellar hyperperfusion (Bohnen, O'Brien, Mullan, & So, 1998; Shin, Hong, Tae, Seo, & Kim, 2001). Ipsilateral thalamic (Spencer, 2002) and ipsilateral or contralateral basal ganglia hyperperfusion has been reported (Shin et al., 2001; Spencer, 2002). The often found bi-occipital hyperperfusion could be explained by relative occipital hypoactivity during interictal injection which usually takes place in dimly lit environment sometimes with eyes closed. In cases when the occipital hyperperfusion is dominant, the hyperperfusion focus however may indicate the presence of occipitotemporal epilepsy (Aykut-Bingol & Spencer, 1999). In a study in TLE with CPS and with later injection times, hyperperfusion changes were found in the basal ganglia, thalamus, and brainstem, suggesting that these structures are involved during later stages of seizure (Blumenfeld et al., 2003).

Both animal and human studies have implicated that the cerebellum has seizure-inhibiting potential. During seizure generalization, cerebellar hyperperfusion is present mainly in the region of the superior medial cerebellum and deep cerebellar nuclei. Early after seizure termination, the increase in the superior midline cerebellum becomes more pronounced which later shifts into lateral cerebellar hemispheres with midline regions less involved. The amount of cerebellar hyperperfusion was not shown to be related to total seizure duration, duration of clonic phase, or severity of the tonic contractions (Blumenfeld et al., 2009).

In a study of generalized tonic–clonic seizures during electroconvulsive therapy (ECT), bilateral cerebellar and parietotemporal hyperperfusion was observed during bilateral and right-sided ECT. Bilateral frontal hyperperfusion was present during bilateral ECT and right frontotemporal hyperperfusion during right-sided ECT. Relative sparing of many brain regions during seizures suggested that specific networks might be more important than others in these "generalized" seizures. A group analysis of secondarily generalized tonic–clonic seizures revealed consistent subcortical increases, which were correlated with decreased CBF in the frontoparietal association cortex. These regions closely resembled the default mode network areas postulated to play an important role in normal attention and consciousness (Raichle et al., 2001). The abnormal activity in frontoparietal association cortices during secondarily generalized seizures and CPS may be the neural substrate of impaired consciousness. Focal seizure activity produces abnormal polyspike discharge in the temporal lobe which is then propagated to subcortical structures particularly to GABAergic inhibitory neurons in structures such as the lateral septal nuclei, anterior hypothalamic ventrolateral preoptic area, thalamic reticular nucleus, habenula, substantia nigra pars reticulata, ventral pallidum, and cerebellar Purkinje cells (Blumenfeld, 2012). This spread of seizure-related changes disrupts subcortical arousal systems in the upper brainstem, thalamus, hypothalamus, and basal forebrain, which prevents normal activation of the cortex and leads to decreased function of the default mode areas of the frontoparietal association cortex. These widespread network changes are not seen in simple partial seizures with spared consciousness, where changes seen on SPECT are confined mainly to the temporal lobe of seizure onset.

4.2.2 Subtraction (SISCOM vs. STATISCOM)

Subtraction ictal SPECT coregistered to MRI (SISCOM) has been proved to be clinically useful in epilepsy surgery evaluation. SISCOM compares patient ictal scan with interictal to produce subtraction image that is later coregistered to and visualized on patient's MRI. Both ictal and interictal scans are intensity normalized to adjust for differences in administered dose. The resulting images typically demonstrated areas of regional hyperperfusion, reflecting increased neuronal activity in parts of the brain involved in seizure activity (Fig. 10.2). This method improved localization of 39% when using side-by-side visual inspection to 88% when using SISCOM (O'Brien et al., 1998).

The standard subtraction is not able to determine whether observed changes are not physiological, i.e., how much do those changes vary from expected random variance between two studies. In order to account for this, a database of healthy controls can be used in SPM analysis to model the random variance. Patient's ictal and interictal scans are normalized for total brain counts, and nonlinear spatial normalization is applied to transform SPECT scans into the same reference space. Contrasts are then set in SPM to detect relative hyperperfusion and hypoperfusion. It has been reported that using a control group, it is possible to achieve higher sensitivity than by using the subtraction alone (Kazemi et al., 2010; McNally et al., 2005).

4.2.3 Clinical implications

Correlations to surgical outcome suggest that SISCOM provides additional information to MRI or electrophysiologic findings and resection of the area

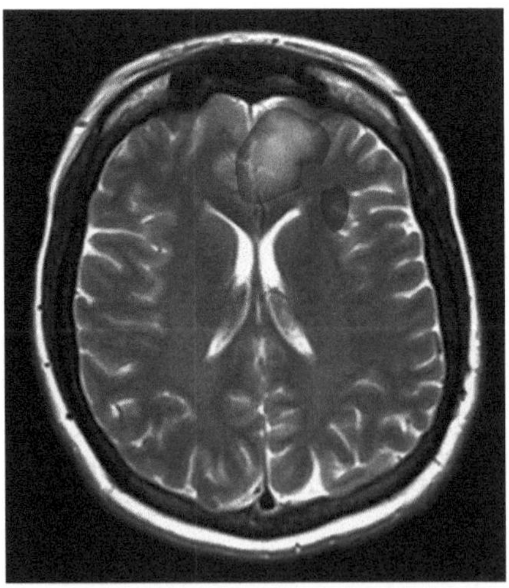

Figure 10.2 SISCOM at the MRI-negative patient with refractory epilepsy and hypermotor seizures. Ictal hyperperfusion in the left prefrontal region. The SPECT result was used to guide intracranial electrodes placement. *Courtesy of Ing. Jan Sanda.*

of increased perfusion is associated with better surgical outcome (O'Brien et al., 1999). SISCOM has been described as particularly useful in identifying the seizure-onset zone in FCDs (Dupont et al., 2006).

In patients with TLE and normal MRI, a favorable outcome was associated with SISCOM abnormity coregistered to the resection site (Bell et al., 2009). Also in nonlesional and extratemporal epilepsies, SISCOM is a highly valuable diagnostic tool to localize the seizure-onset zone and can be used to guide intracranial electrodes placement and improve surgical outcome (Von Oertzen et al., 2011).

The interpretation of ictal perfusion changes should be done in context of all available diagnostic modalities—especially in extratemporal lobe epilepsy. In practice, using different thresholds for SPECT analysis can help to see propagation patterns (Newey et al., 2013; Van Paesschen et al., 2007).

Early ictal SPECT tracer injection is critical to avoid patterns of hyperperfusion due to seizure propagation (Van Paesschen et al., 2007). A study on secondarily generalized tonic–clonic seizures (Varghese et al., 2009) found that in the pre-generalization and generalization phase, ictal SPECT showed more regions of CBF increases than in partial seizures without secondary generalization. This made identification of a single unambiguous region impossible 50% of time, although correct side of the seizure origin could be identified in 84% of cases. When a single unambiguous region was present, the correct localization was 80% of time. In the same study, a CBF decrease during generalization and early postictal period was present in contralateral hemisphere to side of seizure onset although other studies reported good localization of seizure onset based on postictal CBF decrease (O'Brien et al., 1999). Opposite pattern of CBF decrease in the hemisphere ipsilateral to side of onset has also been observed, although this study included only patients with partial seizures without secondary generalization (McNally et al., 2005).

Studies with nonlocalizing initial ictal SPECT showed that second ictal SPECT was able to localize epileptic focus in 62% of patients. This could be caused by significantly shorter injection times of second ictal SPECT. The localized group showed better surgical outcome compared to the nonlocalized group (Lee et al., 2011).

5. IRRITATIVE ZONE AS ASSESSED BY BOLD fMRI

The fluctuations in brain perfusion and hemodynamics mirror ongoing brain neural activity. In epileptic patients, the neural activity is obviously strongly affected during the seizures. Furthermore, the presence of interictal epileptiform discharges (IEDs) in scalp EEG shows that the neural activity and consequently perfusion and hemodynamic fluctuations are affected during interictal periods too. The hemodynamic fluctuations can be acquired with relatively high spatial resolution by Echo Planar Imaging sequences which are sensitive to blood oxygenation level-dependent (BOLD) changes in perfusion. These data provide unique opportunities for localization of IED source (or sources) with the millimeter accuracy. In fact, this has motivated development of techniques for simultaneous EEG and functional MRI (EEG–fMRI). This method combines the information from the two modalities and is nowadays widely used in presurgical patient evaluation being complementary to the traditional methods for focus localization, e.g., scalp EEG description or more recently proposed IED Source Imaging (Michel et al., 2004; Vulliemoz et al., 2009).

5.1. Principles

The basic idea how to use EEG-fMRI data in the task of epileptogenic focus localization is to identify brain region (or regions) where the hemodynamic fluctuations are in close time correlation with occurrence of IED measured on the scalp, mostly by estimating IED-informed general linear model (GLM) of the fMRI time series. The whole procedure begins with acquiring simultaneous resting-state EEG–fMRI data. During the scanning, the subject is instructed to lie still with eyes closed or with some visual fixation, without any external stimulation. Then, the EEG is corrected for imaging and pulse artifacts and subjected to epileptiform events' timing identification. These events (categorized according to IED type if reasonable) are considered as a kind of intrinsic stimulation and are convolved with chosen model of hemodynamic response function (HRF) to create model of epileptiform hemodynamic fluctuations. The GLM is estimated for each fMRI data voxel, and the resulting statistical parametric map shows brain regions with hemodynamic fluctuations significantly related to IED (Al-Asmi et al., 2003; Bénar et al., 2006; Chaudhary et al., 2012; Gholipour, Moeller, Pittau, Dubeau, & Gotman, 2011; Hamandi et al., 2008; Jacobs et al., 2009; Kobayashi et al., 2006; Lemieux et al., 2008; Luo et al., 2010; Pittau, Dubeau, & Gotman, 2012; Rathakrishnan, Moeller, Levan, Dubeau, & Gotman, 2010; Seeck et al., 1998; Thornton et al., 2010; Van Houdt et al., 2013) (Fig. 10.3).

5.2. Technical aspects

The basic feature of EEG–fMRI is MR-compatible EEG system. It contains cap connected with short cable to amplifier which converts the analog signal to its digital representation and via optical cable sends the data into the recording device. The amplifier usually allows recording of additional physiological signals, e.g., ECG, respiration, and EMG. In particular, the acquisition of ECG signal is important for accurate pulse artifact removal from EEG. The majority of commercially available EEG–fMRI systems work with the sampling rate around 5 kHz and allow for registering signals from 30 to 256 electrodes.

EEG signal acquired in the MR scanner is highly contaminated by artifacts related predominantly to MR scan acquisition (imaging artifact, Fig. 10.3A), blood pulsation (pulse artifact, Fig. 10.3B), and movements. The methodology used to remove these artifacts is described elsewhere (Mulert & Lemieux, 2010).

5.3. Methodological aspects

Accurate IED identification is crucial for results' validity. Some MR-specific artifacts may resemble epileptiform activity, and therefore, the identification should be done with great carefulness. Here, the comparison with IED found in clinical EEG recordings can be of great help and is generally recommended.

From the statistical point of view, the power of the GLM statistic depends on the number of events (IEDs) recorded and identified. It was shown that too few events cause the decrease of GLM estimation efficiency (Van Houdt et al., 2010) and dramatically reduce sensitivity of the method. Moreover, in approximately 30% of recordings, there is no IED detected at all, despite the fact that clinical EEG contained interictal epileptiform activity (Al-Asmi et al., 2003; Kobayashi et al., 2006; Pittau et al., 2012; Salek-Haddadi et al., 2006).

The sensitivity and validity of GLM depend on prior assumptions made about the shape and timing of hemodynamic response to IED. The first studies assumed that the hemodynamic response to IED is similar to HRF observed in healthy people as a response to simple sensory input (also known as a canonical HRF) (Al-Asmi et al., 2003; Seeck et al., 1998). Later studies show that the hemodynamic response may differ among patients and even among brain regions (Lu et al, 2006) and suggest using models which allow for variability in HRF, e.g., using several canonical HRF peakings at the

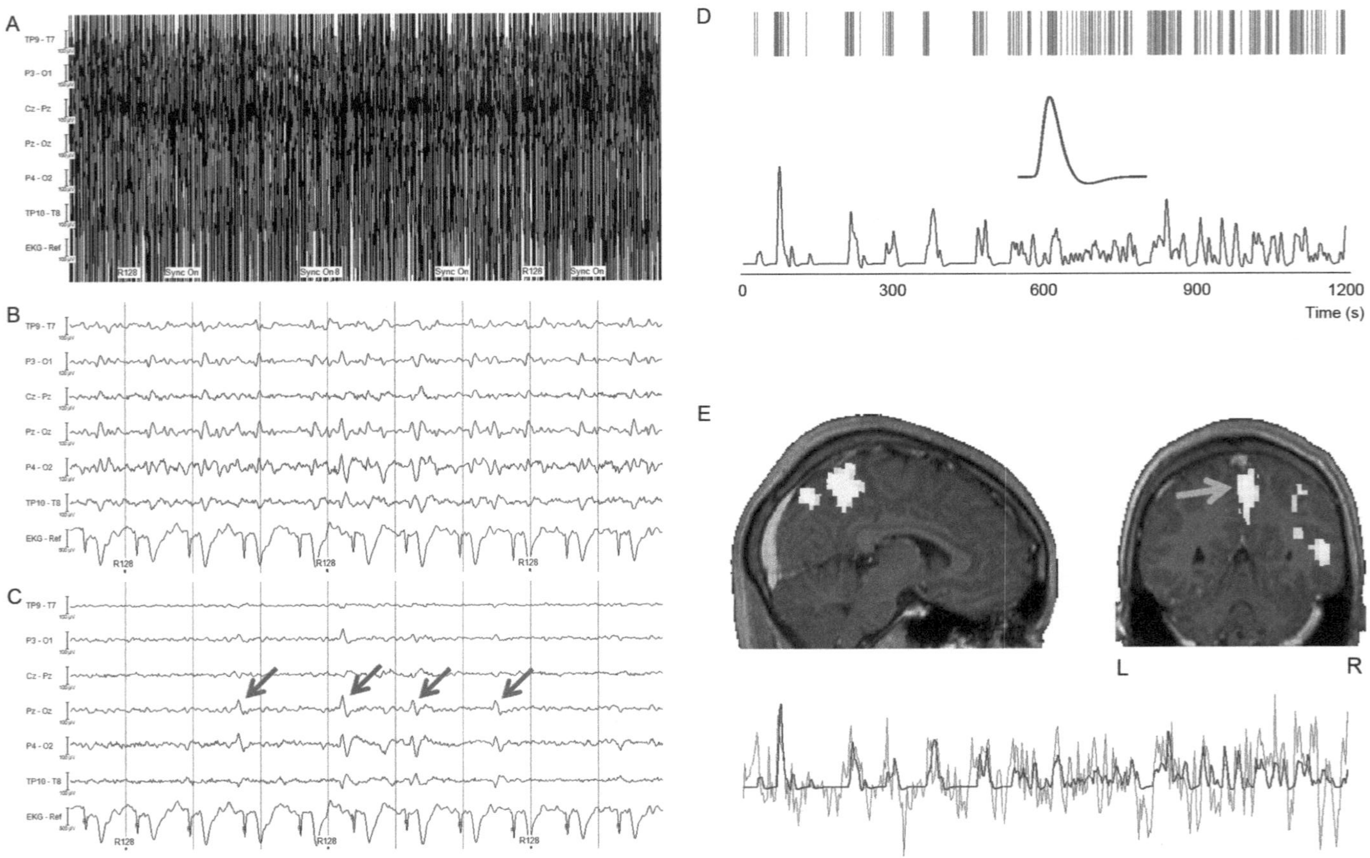

Figure 10.3 The schema of EEG–fMRI application for focus localization. The EEG measured in MR has to be corrected for imaging (A) and pulse (B) artifacts. Corrected EEG is subjected to epileptiform events identification (C). The time series of these events (top part of D) is convolved with hemodynamic response function (single curve in D) to create a model of hemodynamic response to epileptiform activity (bottom part of D) which is compared with fMRI time series from each voxel (bottom part of E, grey line) using general linear model. The resulting parametric map shows brain regions with hemodynamic fluctuations significantly related to epileptiform activity (top part of E).

range of 9 s before to 9 s after IED (Bagshaw et al., 2004; Bénar et al., 2006; Hawco, Bagshaw, Lu, Dubeau, & Gotman, 2007; Jacobs et al., 2009) or by using some basis set to allow for certain degree of shape variability. These adjustments of HRF may be appropriate especially in cases with the hemodynamic system strongly affected by structural or ischemic lesions (Krainik, Hund-Georgiadis, Zysset, & von Cramon, 2005). Other work concludes that in focal epilepsy, the hemodynamic response resembles the canonical one and the regions with altered HRF are usually distant from presumed focus, possibly showing false positives (Lemieux et al., 2008).

Both EEG and positive BOLD signal are known to be generated by field potentials (Logothetis et al., 2001), and it is assumed that the regions with positive BOLD response (activations) represent sources and propagations of IED. Nevertheless, there is a common experience of finding also negative BOLD responses whose interpretation is less straightforward and was shown to be less concordant with epileptogenic zone when compared to positive responses (Salek-Haddadi et al., 2006). From the pathophysiological point of view, there are several proposed interpretations of negative responses. In some cases, the regions with negative response resemble resting-state networks which are known to decrease its activity during cognitive or executive load induced by external or intrinsic stimuli. Hence, the IED can play a role of pathological intrinsic stimuli (Laufs et al., 2007). Other interpretation assumes that some of IEDs have anti-ictogenic role (Walker, Chaudhary, & Lemieux, 2010) and cause the inhibition in local and distant regions.

5.4. Clinical implications

According to zone concept of epilepsy, the IEDs are generated in the irritative zone which is assumed to be corresponding to regions with IED-related BOLD responses (Rosenow & Lüders, 2001).

The sensitivity of the method is *a priori* limited as there is considerable portion of patients who have either nonactive EEG in MR (i.e., no identifiable IED) or no IED-related activations/deactivations. In the work of Salek-Haddadi et al. (2006), 25 patients of 59 had nonactive EEG, and in 11 patients from the remaining 34, there were no activations/deactivations detected, yielding 61% patients from the whole group with no result. In more recent study, Pittau et al. (2012) found nonactive EEG in 10 patients of 43. In all others, the activations or deactivations were found, probably due to adjustments in HRF modeling, yielding only 23% patients with no result.

Interesting and promising way how to treat nonactive EEG recordings was published by Grouiller et al. (2011) who suggest using clinical EEG to find typical spatial topography of IED. This topography is then correlated with spatial topography in each time frame of EEG measured in MR, and resulting time series of correlation coefficients, convolved with HRF, is used in GLM as a model of interictal epileptiform activity. The lack of visible IED in EEG has also motivated research in the field of employing data-driven methods for focus localization (Jann et al., 2008; Lopes, Lina, Fahoum, & Gotman, 2012; Rodionov et al., 2007), though the interpretation of data-driven methods results is generally less straightforward. In cases with IED-related activation or deactivation, the sensitivity is considered in terms of concordance with epileptogenic zone. This issue was handled by Pittau et al. (2012) who examined 33 patients with activation or deactivation and found that in 29 cases, there were BOLD responses with localization concordant to scalp EEG spike field. Moreover, in 12 of 14 patients, the concordance of localization was further validated either by examination with intracerebral electrodes or by MRI lesion. Other study (Van Houdt et al., 2013) compared localization of BOLD responses with electrocorticographic (ECoG) data and surgical outcome. The authors showed for all 16 patients that there was at least one area with IED-related BOLD response concordant to the interictally active ECoG and concluded that the IED-related BOLD responses are spatially accurate at the level of brain lobes .

Often, the BOLD response to IED contains multiple regions. This is in most studies interpreted as a consequence of IED propagation establishing possible evidence for epilepsy network concept (Laufs, 2012). With respect to focus localization, this always brings up the question which one is the most relevant to epileptogenic zone.

Another very important issue when considering EEG–fMRI as a clinical tool for focus localization is reproducibility of the results. This was studied in a recent work (Gholipour et al., 2011), where 15 patients were scanned either twice in 3 T MR (7 patients) or once in 1.5 T MR and once in 3 T MR (8 patients). The study shows reasonable reproducibility of results in 12 patients. Moreover, in 4 patients, the 3 T scan showed better results than 1.5 T scan in terms of BOLD response significance.

To conclude, despite many limitations and problems which have emerged during almost two decades of using the IED-informed EEG–fMRI in presurgical patient evaluation, the method has been established in many epilepsy centers, in particular as an additional investigation when planning intracranial EEG examination.

6. CONCLUSIONS

Introduction of novel imaging modalities and sophisticated techniques of data analysis brought new insights to our understanding of epileptic seizures and epilepsy. These techniques increase the information yield of presurgical evaluation, improve the localization of each of the epileptic zones, and provide more detailed information about the spatiotemporal distribution of ictal changes and the interictal altered activity. Moreover, studies that utilized these techniques demonstrated that it is important to apply the network concepts to the epileptic zone and also to accept the existence of functional heterogeneity within each of these zones.

ACKNOWLEDGMENTS

This work has been supported by Grants GAUK 334 911, IGA MZ CR NT/11536-5, and NT/14489-3.

REFERENCES

Adelson, P. D., Nemoto, E., Scheuer, M., Painter, M., Morgan, J., & Yonas, H. (1999). Noninvasive continuous monitoring of cerebral oxygenation periictally using near-infrared spectroscopy: A preliminary report. *Epilepsia, 40*(11), 1484–1489.

Ahmadi, M. E., Hagler, D. J., McDonald, C. R., Tecoma, E. S., Iragui, V. J., Dale, A. M., et al. (2009). Side matters: Diffusion tensor imaging tractography in left and right temporal lobe epilepsy. *American Journal of Neuroradiology, 30*(9), 1740–1747.

Al-Asmi, A., Bénar, C.-G., Gross, D. W., Khani, Y. A., Andermann, F., Pike, B., et al. (2003). fMRI activation in continuous and spike-triggered EEG-fMRI studies of epileptic spikes. *Epilepsia, 44*(10), 1328–1339.

Andersen, A. R. (1989). 99mTc-D, L-hexamethylene-propyleneamine oxime (99mTc-HMPAO): Basic kinetic studies of a tracer of cerebral blood flow. *Cerebrovascular and Brain Metabolism Reviews, 1*(4), 288.

Arfanakis, K., Hermann, B. P., Rogers, B. P., Carew, J. D., Seidenberg, M., & Meyerand, M. E. (2002). Diffusion tensor MRI in temporal lobe epilepsy. *Magnetic Resonance Imaging, 20*(7), 511–519.

Arthurs, O. J., Donovan, T., Spiegelhalter, D. J., Pickard, J. D., & Boniface, S. J. (2007). Intracortically distributed neurovascular coupling relationships within and between human somatosensory cortices. *Cerebral Cortex, 17*(3), 661–668.

Asenbaum, S., Brucke, T., Pirker, W., Pietrzyk, U., & Podreka, I. (1998). Imaging of cerebral blood flow with technetium-99m-HMPAO and technetium-99m-ECD: A comparison. *Journal of Nuclear Medicine, 39*(4), 613–618.

Ashburner, J., & Friston, K. J. (2000). Voxel-based morphometry—The methods. *NeuroImage, 11*(6), 805–821.

Aykut-Bingol, C., & Spencer, S. S. (1999). Nontumoral occipitotemporal epilepsy: Localizing findings and surgical outcome. *Annals of Neurology, 46*(6), 894–900.

Bagshaw, A. P., Aghakhani, Y., Bénar, C.-G., Kobayashi, E., Hawco, C., Dubeau, F., et al. (2004). EEG-fMRI of focal epileptic spikes: Analysis with multiple haemodynamic functions and comparison with gadolinium-enhanced MR angiograms. *Human Brain Mapping, 22*(3), 179–192.

Banati, R. B., Goerres, G. W., Myers, R., Gunn, R. N., Turkheimer, F. E., Kreutzberg, G. W., et al. (1999). [11C](R)-PK11195 positron emission tomography imaging of activated microglia in vivo in Rasmussen's encephalitis. *Neurology, 53*(9), 2199–2203.

Bell, M. L., Rao, S., So, E. L., Trenerry, M., Kazemi, N., Matt Stead, S., et al. (2009). Epilepsy surgery outcomes in temporal lobe epilepsy with a normal MRI. *Epilepsia, 50*(9), 2053–2060.

Bénar, C.-G., Grova, C., Kobayashi, E., Bagshaw, A. P., Aghakhani, Y., Dubeau, F., et al. (2006). EEG-fMRI of epileptic spikes: Concordance with EEG source localization and intracranial EEG. *NeuroImage, 30*(4), 1161–1170.

Bernasconi, A., Bernasconi, N., Bernhardt, B. C., & Schrader, D. (2011). Advances in MRI for 'cryptogenic' epilepsies. *Nature Reviews Neurology, 7*(2), 99–108.

Bien, C. G., Szinay, M., Wagner, J., Clusmann, H., Becker, A. J., & Urbach, H. (2009). Characteristics and surgical outcomes of patients with refractory magnetic resonance imaging-negative epilepsies. *Archives of Neurology, 66*(12), 1491.

Blumenfeld, H. (2012). Impaired consciousness in epilepsy. *The Lancet Neurology, 11*(9), 814–826.

Blumenfeld, H., Varghese, G. I., Purcaro, M. J., Motelow, J. E., Enev, M., McNally, K. A., et al. (2009). Cortical and subcortical networks in human secondarily generalized tonic-clonic seizures. *Brain, 132*(4), 999–1012.

Blumenfeld, H., Westerveld, M., Ostroff, R. B., Vanderhill, S. D., Freeman, J., Necochea, A., et al. (2003). Selective frontal, parietal, and temporal networks in generalized seizures. *NeuroImage, 19*(4), 1556–1566.

Bohnen, N. I., O'Brien, T. J., Mullan, B. P., & So, E. L. (1998). Cerebellar changes in partial seizures: Clinical correlations of quantitative SPECT and MRI analysis. *Epilepsia, 39*(6), 640–650.

Bonilha, L., Rorden, C., Castellano, G., Pereira, F., Rio, P. A., Cendes, F., et al. (2004). Voxel-based morphometry reveals gray matter network atrophy in refractory medial temporal lobe epilepsy. *Archives of Neurology, 61*(9), 1379–1384.

Bouilleret, V., Dupont, S., Spelle, L., Baulac, M., Samson, Y., & Semah, F. (2002). Insular cortex involvement in mesiotemporal lobe epilepsy: A positron emission tomography study. *Annals of Neurology, 51*(2), 202–208.

Bragin, A., Wilson, C. L., Staba, R. J., Reddick, M., Fried, I., & Engel, J. (2002). Interictal high-frequency oscillations (80-500 Hz) in the human epileptic brain: Entorhinal cortex. *Annals of Neurology, 52*(4), 407–415.

Chaudhary, U. J., Rodionov, R., Carmichael, D. W., Thornton, R. C., Duncan, J. S., & Lemieux, L. (2012). Improving the sensitivity of EEG-fMRI studies of epileptic activity by modelling eye blinks, swallowing and other video-EEG detected physiological confounds. *NeuroImage, 61*(4), 1383–1393.

Clarkson, M. J., Cardoso, M. J., Ridgway, G. R., Modat, M., Leung, K. K., Rohrer, J. D., et al. (2011). A comparison of voxel and surface based cortical thickness estimation methods. *NeuroImage, 57*(3), 856–865.

Concha, L., Beaulieu, C., Collins, D. L., & Gross, D. W. (2009). White-matter diffusion abnormalities in temporal-lobe epilepsy with and without mesial temporal sclerosis. *Journal of Neurology, Neurosurgery & Psychiatry, 80*(3), 312–319.

Concha, L., Kim, H., Bernasconi, A., Bernhardt, B. C., & Bernasconi, N. (2012). Spatial patterns of water diffusion along white matter tracts in temporal lobe epilepsy. *Neurology, 79*(5), 455–462.

Coste, S., Ryvlin, P., Hermier, M., Ostrowsky, K., Adeleine, P., Froment, J. C., et al. (2002). Temporopolar changes in temporal lobe epilepsy: A quantitative MRI-based study. *Neurology, 59*(6), 855–861.

Diehl, B., Busch, R. M., Duncan, J. S., Piao, Z., Tkach, J., & Lüders, H. O. (2008). Abnormalities in diffusion tensor imaging of the uncinate fasciculus relate to reduced memory in temporal lobe epilepsy. *Epilepsia, 49*(8), 1409–1418.

Duncan, R., Patterson, J., Roberts, R., Hadley, D. M., & Bone, I. (1993). Ictal/postictal SPECT in the pre-surgical localisation of complex partial seizures. *Journal of Neurology, Neurosurgery & Psychiatry*, *56*(2), 141–148.

Dupont, P., Van Paesschen, W., Palmini, A., Ambayi, R., Van Loon, J., Goffin, J., et al. (2006). Ictal perfusion patterns associated with single MRI-visible focal dysplastic lesions: Implications for the noninvasive delineation of the epileptogenic zone. *Epilepsia*, *47*(9), 1550–1557.

Figley, C. R., & Stroman, P. W. (2011). The role (s) of astrocytes and astrocyte activity in neurometabolism, neurovascular coupling, and the production of functional neuroimaging signals. *European Journal of Neuroscience*, *33*(4), 577–588.

Fink, G. R., Pawlik, G., Stefan, H., Pietrzyk, U., Wienhard, K., & Heiss, W. D. (1996). Temporal lobe epilepsy: Evidence for interictal uncoupling of blood flow and glucose metabolism in temporomesial structures. *Journal of the Neurological Sciences*, *137*(1), 28–34.

Fischl, B., & Dale, A. M. (2000). Measuring the thickness of the human cerebral cortex from magnetic resonance images. *Proceedings of the National Academy of Sciences of the United States of America*, *97*(20), 11050–11055.

Geneslaw, A. S., Zhao, M., Ma, H., & Schwartz, T. H. (2011). Tissue hypoxia correlates with intensity of interictal spikes. *Journal of Cerebral Blood Flow & Metabolism*, *31*(6), 1394–1402.

Gholipour, T., Moeller, F., Pittau, F., Dubeau, F., & Gotman, J. (2011). Reproducibility of interictal EEG-fMRI results in patients with epilepsy. *Epilepsia*, *52*(3), 433–442.

Giorgio, A., & De Stefano, N. (2013). Clinical use of brain volumetry. *Journal of Magnetic Resonance Imaging*, *37*(1), 1–14.

Govindan, R. M., Makki, M. I., Sundaram, S. K., Juhász, C., & Chugani, H. T. (2008). Diffusion tensor analysis of temporal and extra-temporal lobe tracts in temporal lobe epilepsy. *Epilepsy Research*, *80*(1), 30–41.

Gross, D. W. (2011). Diffusion tensor imaging in temporal lobe epilepsy. *Epilepsia*, *52*(s4), 32–34.

Grouiller, F., Thornton, R. C., Groening, K., Spinelli, L., Duncan, J. S., Schaller, K., et al. (2011). With or without spikes: Localization of focal epileptic activity by simultaneous electroencephalography and functional magnetic resonance imaging. *Brain*, *134*(10), 2867–2886.

Hajek, M., Antonini, A., Leenders, K. L., & Wieser, H. G. (1993). Mesiobasal versus lateral temporal lobe epilepsy: Metabolic differences in the temporal lobe shown by interictal 18F-FDG positron emission tomography. *Neurology*, *43*(1 Part 1), 79–86.

Hamandi, K., Laufs, H., Nöth, U., Carmichael, D. W., Duncan, J. S., & Lemieux, L. (2008). BOLD and perfusion changes during epileptic generalised spike wave activity. *NeuroImage*, *39*(2), 608–618.

Hawco, C. S., Bagshaw, A. P., Lu, Y., Dubeau, F., & Gotman, J. (2007). BOLD changes occur prior to epileptic spikes seen on scalp EEG. *NeuroImage*, *35*(4), 1450–1458.

Henry, T. R. (2010). Neuroimaging in epilepsy. In H. T. Chugani (Ed.), *Positron emission tomography: Glucose metabolism studies in temporal lobe epilepsy* (pp. 122–141). New York: Oxford University Press.

Henry, T. R., Engel, J., Jr., & Mazziotta, J. C. (1993). Clinical evaluation of interictal fluorine-18-fluorodeoxyglucose PET in partial epilepsy. *Journal of Nuclear Medicine: Official Publication, Society of Nuclear Medicine*, *34*(11), 1892–1898.

Hirvonen, J., Kreisl, W. C., Fujita, M., Dustin, I., Khan, O., Appel, S., et al. (2012). Increased in vivo expression of an inflammatory marker in temporal lobe epilepsy. *Journal of Nuclear Medicine*, *53*(2), 234–240.

Ho, S. S., Kuzniecky, R. I., Gilliam, F., Faught, E., & Morawetz, R. (1998). Temporal lobe developmental malformations and epilepsy: Dual pathology and bilateral hippocampal abnormalities. *Neurology*, *50*(3), 748–754.

Hogan, R. E., Carne, R. P., Kilpatrick, C. J., Cook, M. J., Patel, A., King, L., et al. (2008). Hippocampal deformation mapping in MRI negative PET positive temporal lobe epilepsy. *Journal of Neurology, Neurosurgery & Psychiatry*, *79*(6), 636–640.

Huppertz, H.-J., Grimm, C., Fauser, S., Kassubek, J., Mader, I., Hochmuth, A., et al. (2005). Enhanced visualization of blurred gray-white matter junctions in focal cortical dysplasia by voxel-based 3D MRI analysis. *Epilepsy Research*, *67*(1), 35–50.

Jackson, G. D., Connelly, A., Cross, J. H., Gordon, I., & Gadian, D. G. (1994). Functional magnetic resonance imaging of focal seizures. *Neurology*, *44*(5), 850–856.

Jacobs, J., LeVan, P., Moeller, F., Boor, R., Stephani, U., Gotman, J., et al. (2009). Hemodynamic changes preceding the interictal EEG spike in patients with focal epilepsy investigated using simultaneous EEG-fMRI. *NeuroImage*, *45*(4), 1220–1231.

Jann, K., Wiest, R., Hauf, M., Meyer, K., Boesch, C., Mathis, J., et al. (2008). BOLD correlates of continuously fluctuating epileptic activity isolated by independent component analysis. *NeuroImage*, *42*(2), 635–648.

Juhasz, C., Chugani, D., Muzik, O., Shah, A., Shah, J., Watson, C., et al. (2001). Relationship of flumazenil and glucose PET abnormalities to neocortical epilepsy surgery outcome. *Neurology*, *56*(12), 1650–1658.

Kaminska, A., Chiron, C., Ville, D., Dellatolas, G., Hollo, A., Cieuta, C., et al. (2003). Ictal SPECT in children with epilepsy: Comparison with intracranial EEG and relation to postsurgical outcome. *Brain*, *126*(1), 248–260.

Kassubek, J., Huppertz, H.-J., Spreer, J., & Schulze-Bonhage, A. (2002). Detection and localization of focal cortical dysplasia by voxel-based 3-D MRI analysis. *Epilepsia*, *43*(6), 596–602.

Kazemi, N. J., Worrell, G. A., Stead, S. M., Brinkmann, B. H., Mullan, B. P., O'Brien, T. J., et al. (2010). Ictal SPECT statistical parametric mapping in temporal lobe epilepsy surgery. *Neurology*, *74*(1), 70–76.

Keller, S. S., Mackay, C. E., Barrick, T. R., Wieshmann, U. C., Howard, M. A., & Roberts, N. (2002). Voxel-based morphometric comparison of hippocampal and extrahippocampal abnormalities in patients with left and right hippocampal atrophy. *NeuroImage*, *16*(1), 23–31.

Keller, S. S., & Roberts, N. (2008). Voxel-based morphometry of temporal lobe epilepsy: An introduction and review of the literature. *Epilepsia*, *49*(5), 741–757.

Keller, S. S., Wilke, M., Wieshmann, U. C., Sluming, V. A., & Roberts, N. (2004). Comparison of standard and optimized voxel-based morphometry for analysis of brain changes associated with temporal lobe epilepsy. *NeuroImage*, *23*(3), 860–868.

Knake, S., Triantafyllou, C., Wald, L. L., Wiggins, G., Kirk, G. P., Larsson, P. G., et al. (2005). 3T phased array MRI improves the presurgical evaluation in focal epilepsies: A prospective study. *Neurology*, *65*(7), 1026–1031.

Kobayashi, E., Bagshaw, A. P., Bénar, C.-G., Aghakhani, Y., Andermann, F., Dubeau, F., et al. (2006). Temporal and extratemporal BOLD responses to temporal lobe interictal spikes. *Epilepsia*, *47*(2), 343–354.

Krainik, A., Hund-Georgiadis, M., Zysset, S., & von Cramon, D. Y. (2005). Regional impairment of cerebrovascular reactivity and BOLD signal in adults after stroke. *Stroke*, *36*(6), 1146–1152.

Kuzniecky, R. I., & Jackson, G. D. (2005). *Magnetic resonance in epilepsy: Neuroimaging techniques* (2nd ed.). San Diego: Elsevier Academic Press.

Lachaux, J.-P., Fonlupt, P., Kahane, P., Minotti, L., Hoffmann, D., Bertrand, O., et al. (2007). Relationship between task-related gamma oscillations and BOLD signal: New insights from combined fMRI and intracranial EEG. *Human Brain Mapping*, *28*(12), 1368–1375.

Lamusuo, S., Jutila, L., Ylinen, A., Kalviainen, R., Mervaala, E., Haaparanta, M., et al. (2001). [18F] FDG-PET reveals temporal hypometabolism in patients with temporal lobe epilepsy even when quantitative MRI and histopathological analysis show only mild hippocampal damage. *Archives of Neurology*, *58*(6), 933.

Laufs, H. (2012). Functional imaging of seizures and epilepsy: evolution from zones to networks. *Current Opinion in Neurology, 25,* 194–200. http://dx.doi.org/10.1097/WCO.0b013e3283515db9.

Laufs, H., Hamandi, K., Salek-Haddadi, A., Kleinschmidt, A. K., Duncan, J. S., & Lemieux, L. (2007). Temporal lobe interictal epileptic discharges affect cerebral activity in "default mode" brain regions. *Human Brain Mapping, 28*(10), 1023–1032.

Lauritzen, M. (2001). Relationship of spikes, synaptic activity, and local changes of cerebral blood flow. *Journal of Cerebral Blood Flow & Metabolism, 21*(12), 1367–1383.

Lee, J. Y., Joo, E. Y., Park, H. S., Song, P., Young Byun, S., Seo, D. W., et al. (2011). Repeated ictal SPECT in partial epilepsy patients: SISCOM analysis. *Epilepsia, 52*(12), 2249–2256.

Lee, K. K., & Salamon, N. (2009). [18F] Fluorodeoxyglucose-positron-emission tomography and MR imaging coregistration for presurgical evaluation of medically refractory epilepsy. *American Journal of Neuroradiology, 30*(10), 1811–1816.

Lemieux, L., Laufs, H., Carmichael, D., Paul, J. S., Walker, M. C., & Duncan, J. S. (2008). Noncanonical spike-related BOLD responses in focal epilepsy. *Human Brain Mapping, 29*(3), 329–345.

Lin, J. J., Salamon, N., Lee, A. D., Dutton, R. A., Geaga, J. A., Hayashi, K. M., et al. (2007). Reduced neocortical thickness and complexity mapped in mesial temporal lobe epilepsy with hippocampal sclerosis. *Cerebral Cortex, 17*(9), 2007–2018.

Logothetis, N. K., Pauls, J., Augath, M., Trinath, T., & Oeltermann, A. (2001). Neurophysiological investigation of the basis of the fMRI signal. *Nature, 412*(6843), 150–157.

Lopes, R., Lina, J.-M., Fahoum, F., & Gotman, J. (2012). Detection of epileptic activity in fMRI without recording the EEG. *NeuroImage, 60*(3), 1867–1879.

Lu, Y., Bagshaw, A. P., Grova, C., Kobayashi, E., Dubeau, F., & Gotman, J. (2006). Using voxel-specific hemodynamic response function in EEG-fMRI data analysis. *NeuroImage, 32*(1), 238–247.

Luders, H. O., Najm, I., Nair, D., Widdess-Walsh, P., & Bingman, W. (2006). The epileptogenic zone: General principles. *Epileptic Disorders, 8,* S1.

Luo, C., Yao, Z., Li, Q., Lei, X., Zhou, D., Qin, Y., et al. (2010). Imaging foci of epileptic discharges from simultaneous EEG and fMRI using the canonical HRF. *Epilepsy Research, 91*(2), 133–142.

McDonald, C. R., Hagler, D. J., Ahmadi, M. E., Tecoma, E., Iragui, V., Gharapetian, L., et al. (2008). Regional neocortical thinning in mesial temporal lobe epilepsy. *Epilepsia, 49*(5), 794–803.

McNally, K. A., Paige, A. L., Varghese, G., Zhang, H., Novotny, E. J., Spencer, S. S., et al. (2005). Localizing value of ictal-interictal SPECT analyzed by SPM (ISAS). *Epilepsia, 46*(9), 1450–1464.

Memarian, N., Thompson, P. M., Engel, J., & Staba, R. J. (2013). Quantitative analysis of structural neuroimaging of mesial temporal lobe epilepsy. *Imaging in Medicine, 5*(3), 219–235.

Merlet, I., Ostrowsky, K., Costes, N., Ryvlin, P., Isnard, J., Faillenot, I., et al. (2004). 5-HT1A receptor binding and intracerebral activity in temporal lobe epilepsy: An [18F] MPPF-PET study. *Brain, 127*(4), 900–913.

Merlet, I., Ryvlin, P., Costes, N., Dufournel, D., Isnard, J., Faillenot, I., et al. (2004). Statistical parametric mapping of 5-HT1A receptor binding in temporal lobe epilepsy with hippocampal ictal onset on intracranial EEG. *NeuroImage, 22*(2), 886–896.

Michel, C. M., Lantz, G., Spinelli, L., de Peralta, R. G., Landis, T., & Seeck, M. (2004). 128-Channel EEG source imaging in epilepsy: Clinical yield and localization precision. *Journal of Clinical Neurophysiology, 21*(2), 71–83.

Mueller, S. G., Laxer, K. D., Barakos, J., Cheong, I., Garcia, P., & Weiner, M. W. (2009). Widespread neocortical abnormalities in temporal lobe epilepsy with and without mesial sclerosis. *NeuroImage, 46*(2), 353–359.

Mulert, C., & Lemieux, L. (2010). *EEG-fMRI: Physiological basis, technique, and applications*. Berlin: Springer-Verlag.

Natsume, J., Kumakura, Y., Bernasconi, N., Soucy, J.-P., Nakai, A., Rosa, P., et al. (2003). alpha-[11C] methyl-L-tryptophan and glucose metabolism in patients with temporal lobe epilepsy. *Neurology, 60*(5), 756–761.

Newey, C. R., Wong, C., Irene Wang, Z., Chen, X., Wu, G., & Alexopoulos, A. V. (2013). Optimizing SPECT SISCOM analysis to localize seizure-onset zone by using varying z scores. *Epilepsia, 54*(5), 793–800.

Newton, M. R., Berkovic, S. F., Austin, M. C., Rowe, C. C., McKay, W. J., & Bladin, P. F. (1995). SPECT in the localisation of extratemporal and temporal seizure foci. *Journal of Neurology, Neurosurgery & Psychiatry, 59*(1), 26–30.

Niessing, J., Ebisch, B., Schmidt, K. E., Niessing, M., Singer, W., & Galuske, R. A. (2005). Hemodynamic signals correlate tightly with synchronized gamma oscillations. *Science, 309*(5736), 948–951.

O'Brien, T. J., So, E. L., Mullan, B. P., Hauser, M. F., Brinkmann, B. H., Bohnen, N. I., et al. (1998). Subtraction ictal SPECT co-registered to MRI improves clinical usefulness of SPECT in localizing the surgical seizure focus. *Neurology, 50*(2), 445–454.

O'Brien, T. J., So, E. L., Mullan, B. P., Hauser, M. F., Brinkmann, B. H., Jack, C. R., et al. (1999). Subtraction SPECT co-registered to MRI improves postictal SPECT localization of seizure foci. *Neurology, 52*(1), 137–146.

Pittau, F., Dubeau, F., & Gotman, J. (2012). Contribution of EEG/fMRI to the definition of the epileptic focus. *Neurology, 78*(19), 1479–1487.

Raichle, M. E., MacLeod, A. M., Snyder, A. Z., Powers, W. J., Gusnard, D. A., & Shulman, G. L. (2001). A default mode of brain function. *Proceedings of the National Academy of Sciences of the United States of America, 98*(2), 676–682.

Rathakrishnan, R., Moeller, F., Levan, P., Dubeau, F., & Gotman, J. (2010). BOLD signal changes preceding negative responses in EEG-fMRI in patients with focal epilepsy. *Epilepsia, 51*(9), 1837–1845.

Rodionov, R., De Martino, F., Laufs, H., Carmichael, D. W., Formisano, E., Walker, M., et al. (2007). Independent component analysis of interictal fMRI in focal epilepsy: Comparison with general linear model-based EEG-correlated fMRI. *NeuroImage, 38*(3), 488–500.

Rosenow, F., & Lüders, H. (2001). Presurgical evaluation of epilepsy. *Brain, 124*(9), 1683–1700.

Salek-Haddadi, A., Diehl, B., Hamandi, K., Merschhemke, M., Liston, A., Friston, K., et al. (2006). Hemodynamic correlates of epileptiform discharges: An EEG-fMRI study of 63 patients with focal epilepsy. *Brain Research, 1088*(1), 148–166.

Salmenpera, T. M., & Duncan, J. S. (2005). Imaging in epilepsy. *Journal of Neurology, Neurosurgery & Psychiatry, 76*(Suppl. 3), iii2–iii10.

Scanlon, C., Mueller, S. G., Tosun, D., Cheong, I., Garcia, P., Barakos, J., et al. (2011). Impact of methodologic choice for automatic detection of different aspects of brain atrophy by using temporal lobe epilepsy as a model. *American Journal of Neuroradiology, 32*(9), 1669–1676.

Schwartz, T. H., Hong, S.-B., Bagshaw, A. P., Chauvel, P., & Bénar, C.-G. (2011). Preictal changes in cerebral haemodynamics: Review of findings and insights from intracerebral EEG. *Epilepsy Research, 97*(3), 252–266.

Seeck, M., Lazeyras, F., Michel, C., Blanke, O., Gericke, C., Ives, J., et al. (1998). Noninvasive epileptic focus localization using EEG-triggered functional MRI and electromagnetic tomography. *Electroencephalography and Clinical Neurophysiology, 106*(6), 508–512.

Shandal, V., Veenstra, A. L., Behen, M., Sundaram, S., & Chugani, H. (2012). Long-term outcome in children with intractable epilepsy showing bilateral diffuse cortical glucose

hypometabolism pattern on positron emission tomography. *Journal of Child Neurology, 27*(1), 39–45.
Shin, W. C., Hong, S. B., Tae, W. S., Seo, D. W., & Kim, S. E. (2001). Ictal hyperperfusion of cerebellum and basal ganglia in temporal lobe epilepsy: SPECT subtraction with MRI coregistration. *Journal of Nuclear Medicine, 42*(6), 853–858.
Spencer, S. S. (2002). Neural networks in human epilepsy: Evidence of and implications for treatment. *Epilepsia, 43*(3), 219–227.
Stead, M., Bower, M., Brinkmann, B. H., Lee, K., Marsh, W. R., Meyer, F. B., et al. (2010). Microseizures and the spatiotemporal scales of human partial epilepsy. *Brain, 133*(9), 2789–2797.
Thompson, P. M., Hayashi, K. M., de Zubicaray, G. I., Janke, A. L., Rose, S. E., Semple, J., et al. (2004). Mapping hippocampal and ventricular change in Alzheimer disease. *NeuroImage, 22*(4), 1754–1766.
Thornton, R., Laufs, H., Rodionov, R., Cannadathu, S., Carmichael, D. W., Vulliemoz, S., et al. (2010). EEG correlated functional MRI and postoperative outcome in focal epilepsy. *Journal of Neurology, Neurosurgery & Psychiatry, 81*(8), 922–927.
Van Houdt, P. J., de Munck, J. C., Leijten, F. S., Huiskamp, G. J., Colon, A. J., Boon, P. A., et al. (2013). EEG-fMRI correlation patterns in the presurgical evaluation of focal epilepsy: A comparison with electrocorticographic data and surgical outcome measures. *NeuroImage, 75*, 238–248.
Van Houdt, P. J., de Munck, J. C., Zijlmans, M., Huiskamp, G., Leijten, F. S., Boon, P. A., et al. (2010). Comparison of analytical strategies for EEG-correlated fMRI data in patients with epilepsy. *Magnetic Resonance Imaging, 28*(8), 1078–1086.
Van Paesschen, W., Dupont, P., Sunaert, S., Goffin, K., & Van Laere, K. (2007). The use of SPECT and PET in routine clinical practice in epilepsy. *Current Opinion in Neurology, 20*(2), 194–202.
Van Paesschen, W., Dupont, P., Van Driel, G., Van Billoen, H., & Maes, A. (2003). SPECT perfusion changes during complex partial seizures in patients with hippocampal sclerosis. *Brain, 126*(5), 1103–1111.
Varghese, G. I., Purcaro, M. J., Motelow, J. E., Enev, M., McNally, K., Levin, A. R., et al. (2009). Clinical use of ictal SPECT in secondarily generalized tonic-clonic seizures. *Brain, 132*(8), 2102–2113.
Vattipally, V. R., & Bronen, R. A. (2004). MR imaging of epilepsy: Strategies for successful interpretation. *Neuroimaging Clinics of North America, 14*(3), 349–372.
Von Oertzen, J., Mormann, F., Urbach, H., Reichmann, K., Koenig, R., Clusmann, H., et al. (2011). Prospective use of subtraction ictal SPECT coregistered to MRI (SISCOM) in presurgical evaluation of epilepsy. *Epilepsia, 52*(12), 2239–2248.
Von Oertzen, J., Urbach, H., Jungbluth, S., Kurthen, M., Reuber, M., Fernandez, G., et al. (2002). Standard magnetic resonance imaging is inadequate for patients with refractory focal epilepsy. *Journal of Neurology, Neurosurgery & Psychiatry, 73*(6), 643–647.
Vulliemoz, S., Thornton, R., Rodionov, R., Carmichael, D. W., Guye, M., Lhatoo, S., et al. (2009). The spatio-temporal mapping of epileptic networks: Combination of EEG-fMRI and EEG source imaging. *NeuroImage, 46*(3), 834–843.
Walker, M., Chaudhary, U., & Lemieux, L. (2010). EEG-fMRI. In C. Mulert & L. Lemieux (Eds.), *EEG-fMRI in adults with focal epilepsy* (pp. 309–332). Berlin: Springer.
Wellmer, J., Quesada, C. M., Rothe, L., Elger, C. E., Bien, C. G., & Urbach, H. (2013). Proposal for a magnetic resonance imaging protocol for the detection of epileptogenic lesions at early outpatient stages. *Epilepsia, 54*(11), 1977–1987.
Wiggins, G., Triantafyllou, C., Potthast, A., Reykowski, A., Nittka, M., & Wald, L. L. (2006). 32-Channel 3 Tesla receive-only phased-array head coil with soccer-ball element geometry. *Magnetic Resonance in Medicine, 56*(1), 216–223.

Willmann, O., Wennberg, R., May, T., Woermann, F. G., & Pohlmann-Eden, B. (2007). The contribution of 18F-FDG PET in preoperative epilepsy surgery evaluation for patients with temporal lobe epilepsy: A meta-analysis. *Seizure*, *16*(6), 509–520.
Worrell, G. A., Gardner, A. B., Stead, S. M., Hu, S., Goerss, S., Cascino, G. J., et al. (2008). High-frequency oscillations in human temporal lobe: Simultaneous microwire and clinical macroelectrode recordings. *Brain*, *131*(4), 928–937.

CHAPTER ELEVEN

From Treatment to Cure: Stopping Seizures, Preventing Seizures, and Reducing Brain Propensity to Seize

Ivan Pavlov[1], **Stephanie Schorge**[1]
Department of Clinical and Experimental Epilepsy, UCL Institute of Neurology, London, United Kingdom
[1]Correspondence to either IP (i.pavlov@ucl.ac.uk) or SS (s.schorge@ucl.ac.uk)

Contents

1. From Anticonvulsant Therapy to Disease-Modifying Treatment 280
2. Seizure Prevention: Toward Personalized Medicine to Stop Seizures from Happening 280
3. Seizure Prevention: Stopping Epileptogenesis 282
4. Reducing Potential Adverse Effects by Treating Seizures in Real Time 283
5. Real-Time Seizure Treatments: What to Deliver? 283
6. Optogenetics 284
7. Chemical Genetics 287
8. What to Hit: Are Interneurons a Good Target for Interventions? 288
9. Gene Therapy: Introducing Genes to Selected Neurons 290
10. Treatments to Restore Excitability: Is there a Possibility of a Cure? 291
11. How Far to Cure(s)? 293

Acknowledgments 294
References 294

Abstract

Current antiepileptic treatments are aimed at stopping or preventing seizures rather than curing epilepsy. Here, we discuss how recent advances in genetics and neurophysiological research may not only help us better understand the pathophysiology of different epilepsies, but may also drive the development of novel approaches, including treatments that can modify the underlying disease. Although these new therapeutic strategies are still in the very early stages, rapid progress in this direction means that the ability to stop or even to reverse changes caused by epileptogenesis may not be that far away.

International Review of Neurobiology, Volume 114
ISSN 0074-7742
http://dx.doi.org/10.1016/B978-0-12-418693-4.00011-X

1. FROM ANTICONVULSANT THERAPY TO DISEASE-MODIFYING TREATMENT

Presently, most treatments for epilepsy are designed to stop seizures, not to cure epilepsy. This focus on seizures is partly a consequence of the historical blurring between seizures as a symptom and the underlying disease that increases the likelihood of having spontaneous seizures. The focus on seizures is also partly because the rational development of treatments is impossible without understanding what the underlying changes that lead to epilepsy are. Yet, even seizures themselves, mechanisms of their initiation, maintenance, and termination, remain poorly understood. However, advances in genetics have now identified a number of proteins, mutations in which are closely associated with an increased risk of developing epilepsy and in rare cases are even sufficient to cause seizures. Biophysical studies in epileptic tissue and in models of epilepsy are revealing how individual cell types and altered ionic currents in these cells can contribute to seizure generation. Perhaps, it is time to consider shifting our strategy for treatment: from simply trying to suppress seizures to repairing the changes that underlie epilepsy. This change in strategy requires definitions that make clear the difference, and publications are beginning to make explicit the difference between anticonvulsants (which stop seizures) and antiepileptics (which target the changes underlying high propensity of the network to generate seizures). Ultimately, the treatment to which researchers and clinicians should aspire is one that removes the propensity to seize, without side effects on other aspects of brain function. This includes the potential to restore some damaged tissue in patients who have already experienced uncontrolled seizures.

This chapter will not provide a comprehensive review of refinements of existing therapies, but instead will be focusing on novel approaches, especially those which are aimed at transitioning from anticonvulsant treatment strategies to true antiepileptic therapeutics. From symptom alleviation to curing the underlying tendency to seize and even the restoration of lost or damaged tissue—where is treatment of epilepsy going?

2. SEIZURE PREVENTION: TOWARD PERSONALIZED MEDICINE TO STOP SEIZURES FROM HAPPENING

At present, there are some specific types of epilepsy which respond so well to specific treatments that seizures are suppressed (though the underlying disease may still reemerge if treatment was stopped). In some cases, current medications work with exceptional success and can prevent seizures until the underlying tendency to seize disappears. For instance, childhood absence epilepsy (CAE) responds well to several current drugs and often resolves entirely as children pass through puberty (Matricardi, Verrotti, Chiarelli, Cerminara, & Curatolo, 2014). This is in contrast to the many epilepsy syndromes that show various degrees of drug resistance. Up to 30% of patients with temporal lobe epilepsy (TLE) do not respond to present drugs, and even when seizures are stopped by medications, patients can be concerned about side effects, for example, a choice between Topirimate and Sodium Valproate may involve choosing between potential deleterious cognitive effects and weight gain (Meador, Loring, Hulihan, Kamin, & Karim, 2003). There are indications that drug response may be linked to genetic polymorphisms (e.g. Tate et al., 2005) but it is likely the more important refinement would be to map out how more specific types of epilepsy respond to different drugs.

Therefore, while major efforts in current pharmacotherapy development are aimed at the identification of novel drug targets and creation of substances with fewer side effects, there is also interest in determining which patients will benefit most from different treatments. It is possible that as the genetic causes of different epilepsies are determined, new treatments may be designed to target the underlying causes. For example, in patients with early onset absence epilepsy, approximately 10% may have mutations in the glucose transporter gene SLC2A1. Identifying these patients by sequencing suggests an effective treatment with a ketogenic diet (Suls et al., 2009), while avoiding unnecessary trials of the diet in patients with a different etiology. In the case of SLC2A1, the link between glucose transport and ketogenic diet is mechanistically satisfying although at present still indirect (as alternative mechanisms are still being proposed for this diet).

While whole-genome analysis in all patients is still impractical, rapid development of genome sequencing methodology and its continuous cost reduction will likely introduce this procedure to the routine clinical practice in the future. The ILAE has recently produced a first report of types of epilepsy most likely to benefit from genetic testing and highlights how ability to refine treatment to fit the genetic basis is central for justifying treatments (Ottman et al., 2010). These are just the first steps toward personalized medicine where the treatment approach will be individually tailored according to the established genetic underpinnings of the epilepsy syndrome.

Identification of the genetic abnormality, however, does not avoid the challenge of finding curative medication that will fully restore the function. A special exception, where the line between anticonvulsant and antiepileptic is blurred is Rapamycin, which may play a disease-modifying role in tuberous sclerosis (Zeng, Xu, Gutmann, & Wong, 2008). An excellent recent review of drug development in epilepsy highlights how innovatively designed research and clinical trials may be key to improving how seizures are treated, and for moving toward treatments that can cure (i.e., disease-modifying treatments) (Löscher, Klitgaard, Twyman, & Schmidt, 2013).

3. SEIZURE PREVENTION: STOPPING EPILEPTOGENESIS

An alternative approach which is potentially of great clinical value for nongenetic epilepsies would be to predict who is going to develop epilepsy, so that treatments may be delivered before the disease is established. Trauma, stroke, and status epilepticus are all insults which may trigger epileptogenesis, a poorly understood process which after a latent, seizure-free period, transforms the brain into one capable of generating spontaneous seizures. However, as it is not yet known what the key change may be, it is not yet possible to stop the process. Moreover, as most patients with insults do not go on to develop spontaneous seizures, it is important to determine ways of detecting who is at risk so that the smallest number of patients at highest risk may be subjected to potentially invasive treatments. The ideal biomarker would be highly predictive and easily obtained. Full characteristics of ideal biomarkers are recently reviewed in Engel et al. (2013).

As well as electroencephalographic (EEG) changes, current top candidates include changes revealed in magnetic resonance imaging (Parekh et al., 2010), microRNAs in serum (Gorter et al., 2014; Henshall, 2013), or noninvasive measures of inflammation in the brain (Vezzani & Friedman, 2011). Ideally, a biomarker would allow selection of an effective treatment that could prevent the disease from developing, thereby protecting patients from the damage incurred from seizures (Echegoyen, Armstrong, Morgan, & Soltesz, 2009). Protecting patients from unnecessary treatment is also important: recent animal studies have shown that even though some drugs can reduce hippocampal damage after an insult, they may also increase mortality, and in spite of the hippocampal preservation, none of the so far tested drugs improved cognitive performance (Halbsgut, Fahim, Kapoor, Hong, & Friedman, 2013).

4. REDUCING POTENTIAL ADVERSE EFFECTS BY TREATING SEIZURES IN REAL TIME

For patients who have already developed chronic epilepsy, it is too late to block epileptogenesis. Many patients take drugs, which may have side effects, and which must be taken continuously, while seizures are only intermittent. Extended exposure to drugs which alter neuronal excitability may not only hamper normal network function in the brain but also may cause a variety of homeostatic changes in expression of membrane channels (Turrigiano, 2011) that can have further unwanted side effects. A series of studies are developing strategies to administer treatment in a temporally controlled fashion. Ideally, these strategies would detect that a seizure is imminent (or just begun) and deliver a fast onset treatment that stops the seizure (or better yet, prevents it from starting). This is a two-part process which requires development of seizure detection/prediction algorithms and a strategy for fast intervention.

Seizure prediction is a fast-moving field. A first in human study recently demonstrated that in 11 of 15 patients implanting an EEG monitoring device was tolerated and allowed for generation of algorithms that in several cases were able to reliably predict that a seizure was imminent (Cook et al., 2013). This proof-of-principle study opens the door for many different algorithms being developed to analyse EEG recordings (Litt & Echauz, 2002; Litt et al., 2001). A slightly less elegant, but possibly more efficient approach is to detect seizures and deliver treatment only after they have started. The advantage is that EEG changes during seizures may be more reliably detected, but a disadvantage is that the patient may experience seizure onset.

5. REAL-TIME SEIZURE TREATMENTS: WHAT TO DELIVER?

The second step is to develop a way of delivering treatment in response to the predicted seizure. Several groups are tackling this aspect in parallel with the prediction strategy. Approaches range from implanted pumps to devices delivering electrical or light stimuli. However, as noted in a recent review, these devices are not designed to cure epilepsy, but to stop seizures, and as such they operate as anticonvulsants rather than antiepileptics (Fisher, 2012). Nonetheless, in case of refractory epilepsy, this approach may be much less invasive and less risky than surgical resection of the epileptic focus.

The first generation of devices utilized direct electrical stimulation via an implant and were open-loop systems that delivered intermittent stimulation. Deep brain stimulation is probably best established in Parkinson's disease (Wichmann & Delong, 2006), but can also be used to prevent seizures (Pereira, Green, Stacey, & Aziz, 2012). A growing number of possibilities for stimulation in refractory epilepsy are being explored. These include stimulation of the vagus nerve, the trigeminal nerve, thalamus, and hypothalamus (recently reviewed in: Bergey, 2013; Fisher, 2011, 2012; Pereira et al., 2012). In addition, a recent clinical trial has suggested stimulating the hippocampus itself may be safe and effective (Cukiert, Cukiert, Burattini, & Lima, 2014). More advanced, closed-loop systems are also being designed. They feature a stimulator coupled to a device that constantly monitors brain activity and so a stimulus is delivered only when required (Heck et al., 2014; Morrell, 2011). If seizure prediction advances, a clear goal would be to move toward seizure prevention.

Unfortunately, in most cases, the mechanisms linking direct stimulation to seizure modification are unknown. A disadvantage of direct stimulation is its lack of specificity. All neurons in the range of the device will be activated, including both excitatory and inhibitory cells. An approach that would allow specific stimulation or suppression of only the cells implicated in ictogenesis is an attractive alternative. Gene therapy offers the potential to express almost any protein or RNA in specific subsets of cells. At present, there are two approaches being developed to use gene therapy to make subsets of neurons amenable to targeted stimulation. Both involve expression of synthetic genes and make targeted neurons either sensitive to photostimulation (optogenetics) or to synthetic ligands (chemogenetics). Optogenetics are by far the most studied, and we begin with a review of the field that demonstrates what is possible, and what is being learned by this directed approach to stopping seizures.

6. OPTOGENETICS

The optogenetic approach combines genetic manipulations of brain cells and optical methods of light delivery and control. It has rapidly spread as a more selective alternative to electrical stimulation traditionally used to activate neurons in neurophysiology. Unlike direct stimulation, optogenetics also offers an ability to suppress neuronal firing. The essence of this methodology is introduction into the cells of interest genes that encode light-sensitive membrane proteins, opsins. Opsin molecules can be either membrane channels or ion pumps that respond to light. When activated, they cause changes in the membrane potential either exciting or inhibiting cells that express them. In the first paper demonstrating that neurons can be made responsive to light stimulation, neuronal cultures were transfected by a blue light-activated algal cation channel channelrhodopsin-2 (ChR2) using lentivirus vector (Boyden, Zhang, Bamberg, Nagel, & Deisseroth, 2005). This breakthrough study achieved rapid activation of neurons simply by illuminating them with brief pulses of light (450–490 nm). Photocurrents in the transduced neurons can be used to reliably evoke subthreshold depolarizations, elicit individual action potentials, and drive trains of spikes at various frequencies. A similar approach was later used to transduce neurons with an inhibitory opsin *Natronomonas* halorhodopsin (NpHR). NpHR is a light-activated chloride pump which hyperpolarizes neuronal membrane upon photostimulation with flashes of yellow light (Zhang et al., 2007). Creation of new opsin molecules with defined kinetic properties has considerably improved response times and reliability of opsin activation, and allows transduced neurons to be switched on and off with millisecond timescale precision both *in vitro* and *in vivo*.

Akin to other methods of gene therapy, the main advantage of the optogenetic approach is the ability to control neuronal activity in a subset of well-defined neurons without directly affecting other cells. This cell type-specific manipulation holds a lot of promise as a potential treatment approach for a wide range of neurological disorders. Indeed, several recent studies have provided the proof-of-principle for the concept that optogenetics can be used to suppress seizures. The first study of that kind assumed a bold approach by asking whether generation of epileptic activity can be prevented by the light stimulation-induced silencing of excitatory neurons transduced with NpHR (Tønnesen, Sørensen, Deisseroth, Lundberg, & Kokaia, 2009). Here, the transgene expression of NpHR in hippocampal neurons was driven by CaMKIIα promoter that has relatively high specificity for glutamatergic primary neurons. Epileptiform activity, in this case induced by an electrical train stimulation in hippocampal organotypic cultures, was considerably suppressed by continuous photoactivation-induced hyperpolarization of transduced neurons. The potential translational perspective of this strategy has been further corroborated in the *in vivo* studies using rat model of focal cortical epilepsy induced by tetanus toxin injection (Wykes et al., 2012) and in the mouse model of TLE induced by unilateral kainic acid injection into the dorsal hippocampus (Krook-Magnuson, Armstrong, Oijala, & Soltesz, 2013). In both studies, photostimulation was performed *in vivo* through an implanted optical fiber. The latter study also implemented a closed-loop system with an online detection of the pathological activity, and so spontaneously occurring seizures were suppressed shortly after their initiation (Krook-Magnuson et al., 2013). Although about 57% of seizures stopped within 1 s of pulse delivery near the site of kainic acid injection, it is not clear whether longer pulses may be required to control network activity in the situation with less well-defined focus of seizure generation. This raises a concern that prolonged activation of NpHR may result in loading neurons with excessive Cl^-, depolarizing shift of GABA reversal potential (E_{GABA}) and thus conversion of initially inhibitory GABAergic inputs into excitatory ones (Raimondo, Kay, Ellender, & Akerman, 2012). This scenario may be particularly relevant for the forms of TLE where Cl^- homeostasis is disrupted, and E_{GABA} already is more positive than in control neurons (Cohen, Navarro, Clemenceau, Baulac, & Miles, 2002; Jin, Huguenard, & Prince, 2005). The use of another inhibitory opsin, proton pump archaerhodopsin (Arch) may help to avoid this potential caveat (Chow et al., 2010).

An alternative strategy to transducing primary neurons with inhibitory opsins is to drive expression of the excitatory opsins in interneurons in order to increase GABAergic neurotransmission. The first reports of such an approach where hippocampal interneurons expressing ChR2 were photostimulated suggested that this may also be used to delay ongoing epileptiform bursts generated *in vitro* (Ledri, Madsen, Nikitidou, Kirik, & Kokaia, 2014) and to suppress spontaneous seizures *in vivo* (Krook-Magnuson et al., 2013).

The ability to deliver localized stimulation is an attractive feature of the optogenetic approach and seems to be ideal for treatment of focal epilepsies when only relatively small area of epileptic tissue in a well-defined focus needs to be affected. However, the effects of photostimulation may not necessarily be restricted to the local circuitry. For example, in the above mentioned study similar extend of seizure suppression was achieved with light activation of ChR2 in interneurons located in the hemisphere contralateral to the recording site (Krook-Magnuson et al., 2013).

The variety of epileptogenesis- and epilepsy-induced changes in neuronal networks responsible for seizure generation may make certain optogenetic strategies more or less efficient depending on the nature of particular alterations (e.g., the loss of specific populations of neurons (Cossart et al., 2001) or axonal sprouting (Peng et al., 2013)). Unfortunately, even for relatively established models of chronic epilepsy experimental data often suggest contrasting cellular mechanisms involved in the pathology. Thus, it may be extremely difficult to accurately predict the exact pattern of changes in neuronal connectivity in individual patients. In such cases, targeting long-range connections may be the way forward. Indeed, this strategy was successfully used to interrupt spontaneous ictal activity developed in rats after focal cortical stroke which results in maladaptive changes in the peristroke region (Paz et al., 2012). Brief light-induced inhibition of thalamocortical neurons at seizure onset using a closed-loop system in this study aborted epileptic activity.

Neither ChR2 nor inhibitory opsins NpHR and Arch seem to have any major impact on intrinsic properties of the transduced neurons, although potential risks of the deleterious effects of the long-term expression of opsins on the brain tissue still need to be carefully evaluated. Stability of the expression over time is another issue that requires extensive future testing. There also are many technical hurdles related to difficulties in increasing the efficiency of light delivery into the tissue: light scattering and low-tissue penetration may result in insufficient number of photostimulated cells, or require increased number of optic fibers to be implanted making the procedure too invasive. On the other hand, excessive increase in the power of the light source may cause local tissue heating. Therefore, despite very encouraging initial observations, a lot of theoretical and practical questions need to be addressed in order to turn the optogenetic approach into a real treatment option.

7. CHEMICAL GENETICS

Light-activated proteins have the advantage of immediacy; light may be turned on and off instantaneously. However, the disadvantage is that light does not penetrate tissue easily. The alternative is a chemical ligand that crosses the blood brain barrier easily, but may take many minutes to reach target concentrations. The ideal chemical ligand would bind only to the receptor of interest and would be orally available. A field of receptor biology has recently emerged that uses mutation and rational modular rearrangements of existing proteins to generate synthetic ion channels and receptors that can be activated by specific ligands that have no natural counterparts (Rogan & Roth, 2011). As yet, this field has received less publicity than the light-activated channels, but in the longer term the ability to design channels that are activated by otherwise biologically inert ligands may prove medically more accessible.

At present, there are two main families of G protein-coupled receptors that have been modified to respond to synthetic or designer ligands. The designer receptors exclusively activated by designer drugs (DREADDs) derived from muscarinic receptors, and activated by the small molecule clozapine-N-oxide (CNO) (Armbruster, Li, Pausch, Herlitze, & Roth, 2007), and the receptors activated solely by a synthetic ligand originally derived from opioid receptors (Coward et al., 1998). However, G protein-coupled receptors require coupling to downstream effectors in neurons in order to link ligand binding to cellular response, a more direct route is to make a membrane channel that can be gated by a synthetic ligand. This approach has recently been achieved using pores from the Cys-loop receptor superfamily and engineered ligand-binding domains (Magnus et al., 2011). The potential of expressing ion channels with designated permeability (either cation or anion selective) coupled to binding of an orally available ligand may move treatment of many diseases into a new phase. Thus, far reports of the use of these chemogenetic receptors have been limited. However, it has been recently shown that expression of DREADDs in noradrenergic neurons of the locus coeruleus can be used to modify a model of narcolepsy (Hasegawa, Yanagisawa, Sakurai, & Mieda, 2014) and expression in neurons important for energy homeostasis can modify feeding behavior in rodents (Krashes et al., 2011). Thus, this tool has promise for neurological disorders including epilepsy. Indeed, focal seizures evoked by chemoconvulsants pilocarpine and picrotoxin have been shown to be successfully inhibited by chemical-genetic suppression of excitatory neurotransmission following systemic administration of CNO to animals transduced with the silencer DREADD hM4D$_i$ (Kätzel, Nicholson, Schorge, Walker, & Kullmann, 2014). However, as patients with epilepsy may be expected to take the synthetic ligands for long durations (perhaps lifelong), it will be important to ensure they are safe, harmless, and not exorbitantly expensive to make.

It is unlikely that the speed of oral administration would allow a responsive mode of treatment, unless seizure prediction develops to a point where seizures may be accurately predicted many minutes in advance. More rapid application may be available with implantable pumps, but these have another set of possible adverse effects, as well as surgical implantation. Nor is expression of the synthetic effector protein is not likely to have an impact on the underlying disease. As with optogenetics and other devices, the synthetic channels and receptors would only be serving to enhance treatment of the symptoms—the seizures. However, they do open an entirely new vista for these treatments.

8. WHAT TO HIT: ARE INTERNEURONS A GOOD TARGET FOR INTERVENTIONS?

Once an opto- or chemogenetic approach has been chosen, what is the most efficient way of suppressing seizures without having an adverse impact of the stimulation on the cognitive processes and motor activity? Broadly, there are two strategies: either to suppress excessive excitation, or enhance inhibitory neurotransmission. On the one hand, the first strategy runs the risk of completely "shutting down" brain activity in the targeted region in order to suppress seizures that have already been initiated. On the other hand, the fact that experimental disinhibition often results in epileptic activity is consistent with the principle that a lack of sufficient GABAergic inhibition contributes to pathological discharges, and so the second strategy looks appealing. However, a number of clinical and animal studies have shown that the excitatory drive often masks a massive recruitment of inhibitory neurons during ictal and interictal events (Avoli, Louvel, Pumain, & Köhling, 2005; Marchionni & Maccaferri, 2009; Schevon et al., 2012; Trevelyan, Sussillo, Watson, & Yuste, 2006). This has sparked a controversy regarding the role played by GABA-mediated neurotransmission in the generation of epileptiform activity (Avoli & de Curtis, 2011; Pavlov, Kaila, Kullmann, & Miles, 2013). While some research suggests that ictogenesis is accompanied by the failure of inhibition to contain the spread of excitation (Schevon et al., 2012; Trevelyan et al., 2006; Trevelyan, Sussillo, & Yuste, 2007), other studies argue that GABA-mediated signaling paradoxically contributes to the epileptiform discharges (Cohen et al., 2002; D'Antuono et al., 2004; Gnatkovsky, Librizzi, Trombin, & de Curtis, 2008; Köhling et al., 1998; Köhling, Vreugdenhil, Bracci, & Jefferys, 2000), and even that activation of the interneuron network alone can drive the prototypic epileptiform activity (Fujiwara-Tsukamoto et al., 2010). Indeed, interneurons could potentially facilitate aberrant discharges through several mechanisms. These include rapid synchronization of neuronal firing across larger areas by networks of interconnected interneurons and rapid changes of GABAergic signaling from inhibitory to excitatory following sustained activity of $GABA_A$ receptors (Ilie, Raimondo, & Akerman, 2012).

The situation is further complicated by the fact that interneurons constitute a highly heterogeneous population of cells characterized by diverse firing properties, projection targets, and roles in regulating network dynamics (Klausberger & Somogyi, 2008). How individual subtypes of interneurons fire at different stages of pathological activity remains largely unknown with a few notable exceptions (Fujiwara-Tsukamoto, Isomura, Kaneda, & Takada, 2004; Karlócai et al., 2014; Velazquez & Carlen, 1999; Ziburkus, Cressman, Barreto, & Schiff, 2006). Interneurons may also change their intrinsic firing depending on their membrane conductances (Pavlov et al., 2014). It is unclear whether activation and suppression of excitability of an interneuron by chemo- and optogenetic stimulation, or expression of different ion channels preserves its natural pattern of spiking.

Furthermore, since some interneurons may tend to enter the state of depolarizing block during excessive network activity (Cammarota, Losi, Chiavegato, Zonta, & Carmignoto, 2013; Karlócai et al., 2014), it may be difficult to manipulate their firing in a desired way (e.g., additional depolarization by opto- or chemogenetic stimulation in such cells may paradoxically reduce rather than increase their spiking).

Clearly, further research is needed to resolve the ambiguity of the role of GABAergic neurotransmission in ictogenesis to device appropriate therapeutic strategies based on manipulations of interneuron firing. In this sense, the potential of gene therapy to target specific subsets of neurons has outstripped the physiological understanding of the roles of different neurons in triggering and perpetuating seizures.

9. GENE THERAPY: INTRODUCING GENES TO SELECTED NEURONS

Several recent reviews highlight the promise and pitfalls of gene therapy in human disease, including epilepsy (Kullmann, Schorge, Walker, & Wykes, 2014; Noe, Sørensen, Kokaia, & Vezzani, 2012; Simonato, 2013). We focus here on how gene therapy may serve as a bridge between treatment of symptoms (i.e., with optogenetics or synthetic receptors) and treatment of the underlying disease (i.e., by expressing genes which can stop epileptogenesis or restore neuronal excitability, see below). The big caveat—for both optogenetic and chemogenetic approaches—is that the nonnative effector proteins (opsins or engineered receptors) must somehow be expressed in the target neurons. At present, the most promising approach to this in humans is to use genetically modified neurotropic viral particles to deliver DNA.

Most gene therapy requires surgical administration and can only target a small region of tissue. We have proposed (Wykes et al., 2012), this makes many current gene therapy approaches ideal for focal epilepsy, and, in particular, drug-resistant neocortical focal epilepsy, for which there may not be any treatment option. However, the number of viral delivery packages is presently growing rapidly, with new recombinant adeno-associated virus (AAV) vectors being developed that can cross the blood brain barrier, and transduce neurons widely through the CNS in rodents (Stoica, Ahmed, Gao, & Sena-Esteves, 2013; Zhang et al., 2011). These new particles open the possibility of widely expressing virtually any protein or nucleic acid sequence in the brain—pending the ability to fit the transcript in the viral capsule. However, epilepsy is often described as an imbalance of excitation and inhibition, and a treatment to alter neuronal excitability may misfire if the wrong subset of neurons is hit (see also a discussion on targeting interneurons). Thus, one of the primary concerns in gene therapy is that the viral particle either selectively targets specific cell types or that the payload is only activated in the desired cell types.

Cell-type-specific gene expression is controlled by promoters and enhancers. Several gene therapy approaches are exploiting specific promoters to direct the gene therapy payload to specific subsets of neurons, typically excitatory (glutamatergic, often using CaMKIIα promoters) or inhibitory (GAD67 promoters). Increasingly refined promoters (parvalbumin, somatostatin, etc.) allow further selectivity. Although at present, viral targeting is still limited to few rather broad cell classes that consist of functionally different cell types (e.g., expression driven by somatostatin promoter in the hippocampus will target oriens-lacunosum/moleculare and bistratified interneurons), there is rapid movement in the field of generating means to achieve higher level of cell specificity. Many endogenous specific promoters do not give robust enough expression of the gene therapy payload to generate sufficient gene expression in the desired cells. To overcome this hurdle, endogenous promoters can be given a boost. For example, to send a potassium leak channel to hippocampal neurons, the powerful (but nonspecific) CMV promoter was used to enhance the GABRA4 promoter (specific to neurons, but not so powerful)—the combined promoter allowed the therapy to reduce severity of status epilepticus in mice (Dey et al., 2014).

The next step may be synthetic biology. If the receptors being expressed can be built from stray parts, why not the promoters that drive them? Admittedly, promoter engineering is still largely confined to single-celled organisms, but a recent online database, PromoterCAD, mines databases of sequences and gene expression data to facilitate design of synthetic promoters for plants and mice (Nishikata et al., 2014).

Given the possibility of expressing a wide-variety of proteins or nucleic acids in epileptic brains, and the potential of controlling where they are expressed, is it possible to move beyond stopping seizures and toward reversing epileptogenesis—should we be working toward a cure?

10. TREATMENTS TO RESTORE EXCITABILITY: IS THERE A POSSIBILITY OF A CURE?

There are several strategies for removing the tendency to seize from epileptic brains. This can be done by expressing ion channels that stop neurons from firing excessively in seizures. We recently used overexpression of a shaker potassium channel to reduce the excitability of a small number of pyramidal neurons in induced epileptic foci and demonstrated that this can be sufficient to reduce EEG-detected seizures in rodents (Wykes et al., 2012). Similar approach is to increase expression of a potassium leak channel (Dey et al., 2014), which would also reduce cell excitability. These interventions are aimed at stopping the hyperexcitable phenotype that develops during epileptogenesis, and which is thought to contribute to triggering spontaneous seizures in chronic models. Additionally, related "single-gene" approaches include overexpressing the peptide NPY as an anticonvulsant (Noe et al., 2008, 2010; Woldbye et al., 2010) or somatostatin (Zafar, King, & Carney, 2012). A different strategy is to administer cells, rather than viruses, that secrete substances capable of inhibiting seizures. Galanin- and GABA-producing cell transplants have been used to reduce induced convulsions (Handreck, Backofen-Wehrhahn, Bröer, Löscher, & Gernert, 2014; Nikitidou et al., 2014). However, altering the level of only one protein to reduce excitability is not the same as restoring an epileptic network to a normal one. The exceptions may be the rare cases where epilepsy can be attributed to a single gene—such as in Dravet syndrome—in those cases restoring the expression of the gene, if given early enough and able to hit enough neurons, may provide a cure (unfortunately, the NaV1.1 transcript implicated in Dravet syndrome is likely too large for AAVs).

More often, alterations of gene expression in epilepsy are too widespread with changes in inflammatory pathways, peptide signaling, microRNA, and many other pathways. So ideally, one would want to target the processes that drive these changes, the master regulators of epileptogenesis. Candidates high on the regulatory chain include transcriptional repressor neuron-restrictive silencer factor (McClelland et al., 2011) and microRNAs (Gorter et al., 2006, 2014; Henshall, 2013; Jimenez-Mateos et al., 2012). But gene therapy manipulations that modify many genes have a correspondingly high risk for off-target effects. A recent review has highlighted some of the challenges of determining what happens during epileptogenesis—or indeed what epileptogenesis is (Sloviter & Bumanglag, 2013).

In cases where neurons are already damaged or destroyed by seizures or by an original insult, the problem of identifying which genes are central may be skirted by taking a black box approach and administering neuronal progenitors—letting them restore impaired neural network. Such cell therapy has shown impressive promise in models of epilepsy. Neuronal progenitors from the fetal brain that develop mostly in GABAergic neurons have been demonstrated to integrate well into the adult brain, though early studies revealed limited migration from the injection site (Hattiangady, Rao, & Shetty, 2008; Waldau, Hattiangady, Kuruba, & Shetty, 2010). Nevertheless, they reduced seizures in epileptic rats. Remarkably, early postnatal grafts of precursor cells that develop into GABAergic interneurons could even considerably reduce generalized seizures in mice lacking Kv1.1 potassium channel when transplanted in neonates (Baraban et al., 2009). More recently, transplantation of inhibitory GABAergic progenitors into the hippocampus was shown not only to reduce EEG-recorded seizures in a rodent model of chronic epilepsy, but also to restore normal behavior and spatial learning deficits (Hunt, Girskis, Rubenstein, Alvarez-Buylla, & Baraban, 2013). Abundant supplies of neural progenitors can also be generated from embryonic stem cells which when grafted into dentate gyrus of adult mice with chronic epilepsy could be successfully developed into GABAergic interneurons (Maisano et al., 2012). The approach may provide an opportunity to restore normal brain function even after extensive epilepsy-induced damage of the tissue, something which provides an advantage compared to gene therapy.

The ability to reprogram adult cells (e.g., fibroblasts) into so-called induced pluripotent stem cells has opened further possibilities for the regenerative cell-based therapy (Takahashi & Yamanaka, 2006), and it is the matter of time when the possibilities of this new technology to restore epileptic tissue will be explored experimentally.

It is worth noting that epilepsy may be associated with changes in neurogenesis. Initially increased after the precipitating insult in models of TLE, it becomes suppressed in chronic epilepsy (Bengzon et al., 1997; Hattiangady, Rao, & Shetty, 2004; Parent et al., 1997). The impact of seizures on newly generated neurons, however, remains unknown. It cannot be excluded that pathological network activity may cause highly plastic newly generated neurons to abnormally integrate into existing circuitry. Therefore, the danger that stem cells transplanted into epileptic tissue will develop in neurons supporting rather than suppressing seizures should not be overlooked (Kuruba, Hattiangady, & Shetty, 2009).

It is too early to judge to what extent the original network should be intact for successful recovery and whether the transplant itself will not turn into the source of pathological activity as in the case of one of the early studies (Buzsáki, Bayardo, Miles, Wong, & Gage, 1989).

11. HOW FAR TO CURE(S)?

It is most likely that different types of epilepsy will respond to different treatments. Some (such as CAE, mentioned above) may already be controlled well enough to negate more invasive approaches. Other types of epilepsy, such as focal neocortical epilepsy, are much more likely to justify gene therapy or device implantation. However, to advance these treatments from palliative reduction of seizures to true treatment of the underlying diseases, researchers will have to refocus on the underlying changes, and treatments which reduce the underlying changes that lead to spontaneous seizures. This will likely require a transition from EEG-based reporting of ongoing seizures to biomarker-based prediction of epileptogenesis. Ultimately, if treatments truly restore the epileptic network, they will not only stop seizures, but will also alleviate comorbidities, such as depression and learning difficulties, which are hugely important to patient welfare, and improving patient welfare is the real goal behind all these treatments.

ACKNOWLEDGMENTS

The authors are supported by fellowships from The Worshipful Company of Pewterers (I. P.), and The Royal Society (S. S.).

REFERENCES

Armbruster, B. N., Li, X., Pausch, M. H., Herlitze, S., & Roth, B. L. (2007). Evolving the lock to fit the key to create a family of G protein-coupled receptors potently activated by an inert ligand. *Proceedings of the National Academy of Sciences of the United States of America*, *104*, 5163–5168.

Avoli, M., & de Curtis, M. (2011). GABAergic synchronization in the limbic system and its role in the generation of epileptiform activity. *Progress in Neurobiology*, *95*, 104–132.

Avoli, M., Louvel, J., Pumain, R., & Köhling, R. (2005). Cellular and molecular mechanisms of epilepsy in the human brain. *Progress in Neurobiology*, *77*, 166–200.

Baraban, S. C., Southwell, D. G., Estrada, R. C., Jones, D. L., Sebe, J. Y., Alfaro-Cervello, C., et al. (2009). Reduction of seizures by transplantation of cortical GABAergic interneuron precursors into Kv1.1 mutant mice. *Proceedings of the National Academy of Sciences of the United States of America*, *106*, 15472–15477.

Bengzon, J., Kokaia, Z., Elmér, E., Nanobashvili, A., Kokaia, M., & Lindvall, O. (1997). Apoptosis and proliferation of dentate gyrus neurons after single and intermittent limbic seizures. *Proceedings of the National Academy of Sciences of the United States of America*, *94*, 10432–10437.

Bergey, G. K. (2013). Neurostimulation in the treatment of epilepsy. *Experimental Neurology*, *244*, 87–95.

Boyden, E. S., Zhang, F., Bamberg, E., Nagel, G., & Deisseroth, K. (2005). Millisecond-timescale, genetically targeted optical control of neural activity. *Nature Neuroscience*, *8*, 1263–1268.

Buzsáki, G., Bayardo, F., Miles, R., Wong, R. K., & Gage, F. H. (1989). The grafted hippocampus: An epileptic focus. *Experimental Neurology*, *105*, 10–22.

Cammarota, M., Losi, G., Chiavegato, A., Zonta, M., & Carmignoto, G. (2013). Fast spiking interneuron control of seizure propagation in a cortical slice model of focal epilepsy. *Journal of Physiology*, *591*, 807–822.

Chow, B. Y., Han, X., Dobry, A. S., Qian, X., Chuong, A. S., Li, M., et al. (2010). High-performance genetically targetable optical neural silencing by light-driven proton pumps. *Nature*, *463*, 98–102.

Cohen, I., Navarro, V., Clemenceau, S., Baulac, M., & Miles, R. (2002). On the origin of interictal activity in human temporal lobe epilepsy in vitro. *Science*, *298*, 1418–1421.

Cook, M. J., O'Brien, T. J., Berkovic, S. F., Murphy, M., Morokoff, A., Fabinyi, G., et al. (2013). Prediction of seizure likelihood with a long-term, implanted seizure advisory system in patients with drug-resistant epilepsy: A first-in-man study. *Lancet Neurology*, *12*, 563–571.

Cossart, R., Dinocourt, C., Hirsch, J. C., Merchan-Perez, A., De Felipe, J., Ben-Ari, Y., et al. (2001). Dendritic but not somatic GABAergic inhibition is decreased in experimental epilepsy. *Nature Neuroscience*, *4*, 52–62.

Coward, P., Wada, H. G., Falk, M. S., Chan, S. D., Meng, F., Akil, H., et al. (1998). Controlling signaling with a specifically designed Gi-coupled receptor. *Proceedings of the National Academy of Sciences of the United States of America*, *95*, 352–357.

Cukiert, A., Cukiert, C. M., Burattini, J. A., & Lima, A. M. (2014). Seizure outcome after hippocampal deep brain stimulation in a prospective cohort of patients with refractory temporal lobe epilepsy. *Seizure*, *23*, 6–9.

D'Antuono, M., Louvel, J., Köhling, R., Mattia, D., Bernasconi, A., Olivier, A., et al. (2004). GABAA receptor-dependent synchronization leads to ictogenesis in the human dysplastic cortex. *Brain*, *127*, 1626–1640.

Dey, D., Eckle, V.-S., Vitko, I., Sullivan, K. A., Lasiecka, Z. M., Winckler, B., et al. (2014). A potassium leak channel silences hyperactive neurons and ameliorates status epilepticus. *Epilepsia*, *55*, 203–213.

Echegoyen, J., Armstrong, C., Morgan, R. J., & Soltesz, I. (2009). Single application of a CB1 receptor antagonist rapidly following head injury prevents long-term hyperexcitability in a rat model. *Epilepsy Research*, *85*, 123–127.

Engel, J., Pitkänen, A., Loeb, J. A., Dudek, F. E., Bertram, E. H., Cole, A. J., et al. (2013). Epilepsy biomarkers. *Epilepsia*, *54*(Suppl. 4), 61–69.

Fisher, R. S. (2011). Direct brain stimulation is an effective therapy for epilepsy. *Neurology*, *77*, 1220–1221.

Fisher, R. S. (2012). Therapeutic devices for epilepsy. *Annals of Neurology*, *71*, 157–168.

Fujiwara-Tsukamoto, Y., Isomura, Y., Imanishi, M., Ninomiya, T., Tsukada, M., Yanagawa, Y., et al. (2010). Prototypic seizure activity driven by mature hippocampal fast-spiking interneurons. *Journal of Neuroscience*, *30*, 13679–13689.

Fujiwara-Tsukamoto, Y., Isomura, Y., Kaneda, K., & Takada, M. (2004). Synaptic interactions between pyramidal cells and interneuron subtypes during seizure-like activity in the rat hippocampus. *Journal of Physiology*, *557*, 961–979.

Gnatkovsky, V., Librizzi, L., Trombin, F., & de Curtis, M. (2008). Fast activity at seizure onset is mediated by inhibitory circuits in the entorhinal cortex in vitro. *Annals of Neurology*, *64*, 674–686.

Gorter, J. A., Iyer, A., White, I., Colzi, A., van Vliet, E. A., Sisodiya, S., et al. (2014). Hippocampal subregion-specific microRNA expression during epileptogenesis in experimental temporal lobe epilepsy. *Neurobiology of Disease*, *62*, 508–520.

Gorter, J. A., van Vliet, E. A., Aronica, E., Breit, T., Rauwerda, H., Lopes da Silva, F. H., et al. (2006). Potential new antiepileptogenic targets indicated by microarray analysis in a rat model for temporal lobe epilepsy. *Journal of Neuroscience*, *26*, 11083–11110.

Halbsgut, L. R., Fahim, E., Kapoor, K., Hong, H., & Friedman, L. K. (2013). Certain secondary antiepileptic drugs can rescue hippocampal injury following a critical growth period despite poor anticonvulsant activity and cognitive deficits. *Epilepsy and Behavior*, *29*, 466–477.

Handreck, A., Backofen-Wehrhahn, B., Bröer, S., Löscher, W., & Gernert, M. (2014). Anticonvulsant effects by bilateral and unilateral transplantation of GABA-producing cells into the subthalamic nucleus in an acute seizure model. *Cell Transplantation, 23*, 111–132.

Hasegawa, E., Yanagisawa, M., Sakurai, T., & Mieda, M. (2014). Orexin neurons suppress narcolepsy via 2 distinct efferent pathways. *Journal of Clinical Investigation, 124*, 604–616.

Hattiangady, B., Rao, M. S., & Shetty, A. K. (2004). Chronic temporal lobe epilepsy is associated with severely declined dentate neurogenesis in the adult hippocampus. *Neurobiology of Disease, 17*, 473–490.

Hattiangady, B., Rao, M. S., & Shetty, A. K. (2008). Grafting of striatal precursor cells into hippocampus shortly after status epilepticus restrains chronic temporal lobe epilepsy. *Experimental Neurology, 212*, 468–481.

Heck, C. N., King-Stephens, D., Massey, A. D., Nair, D. R., Jobst, B. C., Barkley, G. L., et al. (2014). Two-year seizure reduction in adults with medically intractable partial onset epilepsy treated with responsive neurostimulation: Final results of the RNS System Pivotal trial. *Epilepsia, 55*, 432–441.

Henshall, D. C. (2013). MicroRNAs in the pathophysiology and treatment of status epilepticus. *Frontiers in Molecular Neuroscience, 6*, 37.

Hunt, R. F., Girskis, K. M., Rubenstein, J. L., Alvarez-Buylla, A., & Baraban, S. C. (2013). GABA progenitors grafted into the adult epileptic brain control seizures and abnormal behavior. *Nature Neuroscience, 16*, 692–697.

Ilie, A., Raimondo, J. V., & Akerman, C. J. (2012). Adenosine release during seizures attenuates GABAA receptor-mediated depolarization. *Journal of Neuroscience, 32*, 5321–5332.

Jimenez-Mateos, E. M., Engel, T., Merino-Serrais, P., McKiernan, R. C., Tanaka, K., Mouri, G., et al. (2012). Silencing microRNA-134 produces neuroprotective and prolonged seizure-suppressive effects. *Nature Medicine, 18*, 1087–1094.

Jin, X., Huguenard, J. R., & Prince, D. A. (2005). Impaired Cl^- extrusion in layer V pyramidal neurons of chronically injured epileptogenic neocortex. *Journal of Neurophysiology, 93*, 2117–2126.

Karlócai, M. R., Kohus, Z., Káli, S., Ulbert, I., Szabó, G., Máté, Z., et al. (2014). Physiological sharp wave-ripples and interictal events in vitro: What's the difference? *Brain, 137*, 463–485.

Kätzel, D., Nicholson, E., Schorge, S., Walker, M. C., & Kullmann, D. M. (2014). Chemical–genetic attenuation of focal neocortical seizures. *Nature Communications, 5*, 3847.

Klausberger, T., & Somogyi, P. (2008). Neuronal diversity and temporal dynamics: The unity of hippocampal circuit operations. *Science, 321*, 53–57.

Köhling, R., Lücke, A., Straub, H., Speckmann, E. J., Tuxhorn, I., Wolf, P., et al. (1998). Spontaneous sharp waves in human neocortical slices excised from epileptic patients. *Brain, 121*, 1073–1087.

Köhling, R., Vreugdenhil, M., Bracci, E., & Jefferys, J. G. R. (2000). Ictal epileptiform activity is facilitated by hippocampal GABA A receptor-mediated oscillations. *Journal of Neuroscience, 20*, 6820–6829.

Krashes, M. J., Koda, S., Ye, C., Rogan, S. C., Adams, A. C., Cusher, D. S., et al. (2011). Rapid, reversible activation of AgRP neurons drives feeding behavior in mice. *Journal of Clinical Investigation, 121*, 1424–1428.

Krook-Magnuson, E., Armstrong, C., Oijala, M., & Soltesz, I. (2013). On-demand optogenetic control of spontaneous seizures in temporal lobe epilepsy. *Nature Communications, 4*, 1376.

Kullmann, D. M., Schorge, S., Walker, M. C., & Wykes, R. C. (2014). Gene therapy in epilepsy - is it time for clinical trials? *Nature Reviews Neurology, 10*, 300–304.

Kuruba, R., Hattiangady, B., & Shetty, A. K. (2009). Hippocampal neurogenesis and neural stem cells in temporal lobe epilepsy. *Epilepsy and Behavior, 14*(Suppl. 1), 65–73.

Ledri, M., Madsen, M. G., Nikitidou, L., Kirik, D., & Kokaia, M. (2014). Global optogenetic activation of inhibitory interneurons during epileptiform activity. *Journal of Neuroscience, 34*, 3364–3377.

Litt, B., & Echauz, J. (2002). Prediction of epileptic seizures. *Lancet Neurology, 1*, 22–30.

Litt, B., Esteller, R., Echauz, J., D'Alessandro, M., Shor, R., Henry, T., et al. (2001). Epileptic seizures may begin hours in advance of clinical onset: A report of five patients. *Neuron, 30*, 51–64.

Löscher, W., Klitgaard, H., Twyman, R. E., & Schmidt, D. (2013). New avenues for antiepileptic drug discovery and development. *Nature Reviews Drug Discovery, 12*, 757–776.

Magnus, C. J., Lee, P. H., Atasoy, D., Su, H. H., Looger, L. L., & Sternson, S. M. (2011). Chemical and genetic engineering of selective ion channel-ligand interactions. *Science, 333*, 1292–1296.

Maisano, X., Litvina, E., Tagliatela, S., Aaron, G. B., Grabel, L. B., & Naegele, J. R. (2012). Differentiation and functional incorporation of embryonic stem cell-derived GABAergic interneurons in the dentate gyrus of mice with temporal lobe epilepsy. *Journal of Neuroscience, 32*, 46–61.

Marchionni, I., & Maccaferri, G. (2009). Quantitative dynamics and spatial profile of perisomatic GABAergic input during epileptiform synchronization in the CA1 hippocampus. *Journal of Physiology, 587*, 5691–5708.

Matricardi, S., Verrotti, A., Chiarelli, F., Cerminara, C., & Curatolo, P. (2014). Current advances in childhood absence epilepsy. *Pediatric Neurology, 50*, 205–212.

McClelland, S., Flynn, C., Dubé, C., Richichi, C., Zha, Q., Ghestem, A., et al. (2011). Neuron-restrictive silencer factor-mediated hyperpolarization-activated cyclic nucleotide gated channelopathy in experimental temporal lobe epilepsy. *Annals of Neurology, 70*, 454–464.

Meador, K. J., Loring, D. W., Hulihan, J. F., Kamin, M., & Karim, R. (2003). Differential cognitive and behavioral effects of topiramate and valproate. *Neurology, 60*, 1483–1488.

Morrell, M. J. (2011). Responsive cortical stimulation for the treatment of medically intractable partial epilepsy. *Neurology, 77*, 1295–1304.

Nikitidou, L., Torp, M., Fjord-Larsen, L., Kusk, P., Wahlberg, L. U., & Kokaia, M. (2014). Encapsulated galanin-producing cells attenuate focal epileptic seizures in the hippocampus. *Epilepsia, 55*, 167–174.

Nishikata, K., Cox, R. S., Shimoyama, S., Yoshida, Y., Matsui, M., Makita, Y., et al. (2014). Database construction for PromoterCAD: Synthetic promoter design for mammals and plants. *ACS Synthetic Biology, 3*, 192–196.

Noe, F., Pool, A.-H., Nissinen, J., Gobbi, M., Bland, R., Rizzi, M., et al. (2008). Neuropeptide Y gene therapy decreases chronic spontaneous seizures in a rat model of temporal lobe epilepsy. *Brain, 131*, 1506–1515.

Noe, F. M., Sørensen, A. T., Kokaia, M., & Vezzani, A. (2012). Gene therapy of focal onset epilepsy using adeno-associated virus vector-mediated overexpression of neuropeptide Y. In J. L. Noebels, M. Avoli, M. A. Rogawski, R. W. Olsen, & A. V. Delgado-Escueta (Eds.), *Jasper's basic mechanisms of the epilepsies* (4th ed.), Bethesda (MD): National Center for Biotechnology Information (US).

Noe, F., Vaghi, V., Balducci, C., Fitzsimons, H., Bland, R., Zardoni, D., et al. (2010). Anticonvulsant effects and behavioural outcomes of rAAV serotype 1 vector-mediated neuropeptide Y overexpression in rat hippocampus. *Gene Therapy, 17*, 643–652.

Ottman, R., Hirose, S., Jain, S., Lerche, H., Lopes-Cendes, I., Noebels, J. L., et al. (2010). Genetic testing in the epilepsies—report of the ILAE Genetics Commission. *Epilepsia, 51*, 655–670.

Parekh, M. B., Carney, P. R., Sepulveda, H., Norman, W., King, M., & Mareci, T. H. (2010). Early MR diffusion and relaxation changes in the parahippocampal gyrus precede the onset of spontaneous seizures in an animal model of chronic limbic epilepsy. *Experimental Neurology, 224*, 258–270.

Parent, J. M., Yu, T. W., Leibowitz, R. T., Geschwind, D. H., Sloviter, R. S., & Lowenstein, D. H. (1997). Dentate granule cell neurogenesis is increased by seizures and contributes to aberrant network reorganization in the adult rat hippocampus. *Journal of Neuroscience, 17*, 3727–3738.

Pavlov, I., Kaila, K., Kullmann, D. M., & Miles, R. (2013). Cortical inhibition, pH and cell excitability in epilepsy: What are optimal targets for antiepileptic interventions? *Journal of Physiology, 591*, 765–774.

Pavlov, I., Savtchenko, L. P., Song, I., Koo, J., Pimashkin, A., & Rusakov, D. A. (2014). Tonic GABA A conductance bidirectionally controls interneuron firing pattern and synchronization in the CA3 hippocampal network. *Proceedings of the National Academy of Sciences of the United States of America, 111*, 504–509.

Paz, J. T., Davidson, T. J., Frechette, E. S., Delord, B., Parada, I., Peng, K., et al. (2012). Closed-loop optogenetic control of thalamus as a tool for interrupting seizures after cortical injury. *Nature Neuroscience, 16*, 64–70.

Peng, Z., Zhang, N., Wei, W., Huang, C. S., Cetina, Y., Otis, T. S., et al. (2013). A reorganized GABAergic circuit in a model of Epilepsy: Evidence from optogenetic labeling and stimulation of somatostatin interneurons. *Journal of Neuroscience, 33*, 14392–14405.

Pereira, E. A. C., Green, A. L., Stacey, R. J., & Aziz, T. Z. (2012). Refractory epilepsy and deep brain stimulation. *Journal of Clinical Neurosciences, 19*, 27–33.

Raimondo, J. V., Kay, L., Ellender, T. J., & Akerman, C. J. (2012). Optogenetic silencing strategies differ in their effects on inhibitory synaptic transmission. *Nature Neuroscience, 15*, 1102–1104.

Rogan, S. C., & Roth, B. L. (2011). Remote control of neuronal signaling. *Pharmacological Reviews, 63*, 291–315.

Schevon, C. A., Weiss, S. A., McKhann, G., Goodman, R. R., Yuste, R., Emerson, R. G., et al. (2012). Evidence of an inhibitory restraint of seizure activity in humans. *Nature Communications, 3*, 1060.

Simonato, M. (2013). Gene therapy for epilepsy. *Epilepsy and Behavior.* http://dx.doi.org/10.1016/j.yebeh.2013.09.013. [Epub ahead of print].

Sloviter, R. S., & Bumanglag, A. V. (2013). Defining "epileptogenesis" and identifying "antiepileptogenic targets" in animal models of acquired temporal lobe epilepsy is not as simple as it might seem. *Neuropharmacology, 69*, 3–15.

Stoica, L., Ahmed, S. S., Gao, G., & Sena-Esteves, M. (2013). Gene transfer to the CNS using recombinant adeno-associated virus. *Current Protocols in Microbiology.* http://dx.doi.org/10.1002/9780471729259.mc14d05s29. Chapter 14:Unit14D.5.

Suls, A., Mullen, S. A., Weber, Y. G., Verhaert, K., Ceulemans, B., Guerrini, R., et al. (2009). Early-onset absence epilepsy caused by mutations in the glucose transporter GLUT1. *Annals of Neurology, 66*, 415–419.

Takahashi, K., & Yamanaka, S. (2006). Induction of pluripotent stem cells from mouse embryonic and adult fibroblast cultures by defined factors. *Cell, 126*, 663–676.

Tate, S. K., Depondt, C., Sisodiya, S. M., Cavalleri, G. L., Schorge, S., Soranzo, N., et al. (2005). Genetic predictors of the maximum doses patients receive during clinical use of the anti-epileptic drugs carbamazepine and phenytoin. *Proceedings of the National Academy of Sciences of the United States of America, 102*, 5507–5512.

Tønnesen, J., Sørensen, A. T., Deisseroth, K., Lundberg, C., & Kokaia, M. (2009). Optogenetic control of epileptiform activity. *Proceedings of the National Academy of Sciences of the United States of America, 106*, 12162–12167.

Trevelyan, A. J., Sussillo, D., Watson, B. O., & Yuste, R. (2006). Modular propagation of epileptiform activity: Evidence for an inhibitory veto in neocortex. *Journal of Neuroscience*, *26*, 12447–12455.

Trevelyan, A. J., Sussillo, D., & Yuste, R. (2007). Feedforward inhibition contributes to the control of epileptiform propagation speed. *Journal of Neuroscience*, *27*, 3383–3387.

Turrigiano, G. (2011). Too many cooks? Intrinsic and synaptic homeostatic mechanisms in cortical circuit refinement. *Annual Review of Neuroscience*, *34*, 89–103.

Velazquez, J. L., & Carlen, P. L. (1999). Synchronization of GABAergic interneuronal networks during seizure-like activity in the rat horizontal hippocampal slice. *European Journal of Neuroscience*, *11*, 4110–4118.

Vezzani, A., & Friedman, A. (2011). Brain inflammation as a biomarker in epilepsy. *Biomarkers in Medicine*, *5*, 607–614.

Waldau, B., Hattiangady, B., Kuruba, R., & Shetty, A. K. (2010). Medial ganglionic eminence-derived neural stem cell grafts ease spontaneous seizures and restore GDNF expression in a rat model of chronic temporal lobe epilepsy. *Stem Cells*, *28*, 1153–1164.

Wichmann, T., & Delong, M. R. (2006). Deep brain stimulation for neurologic and neuropsychiatric disorders. *Neuron*, *52*, 197–204.

Woldbye, D. P. D., Angehagen, M., Gøtzsche, C. R., Elbrønd-Bek, H., Sørensen, A. T., Christiansen, S. H., et al. (2010). Adeno-associated viral vector-induced overexpression of neuropeptide Y Y2 receptors in the hippocampus suppresses seizures. *Brain*, *133*, 2778–2788.

Wykes, R. C., Heeroma, J. H., Mantoan, L., Zheng, K., Macdonald, D. C., Deisseroth, K., et al. (2012). Optogenetic and potassium channel gene therapy in a rodent model of focal neocortical epilepsy. *Science Translational Medicine*, *4*, 161ra152.

Zafar, R., King, M. A., & Carney, P. R. (2012). Adeno associated viral vector-mediated expression of somatostatin in rat hippocampus suppresses seizure development. *Neuroscience Letters*, *509*, 87–91.

Zeng, L.-H., Xu, L., Gutmann, D. H., & Wong, M. (2008). Rapamycin prevents epilepsy in a mouse model of tuberous sclerosis complex. *Annals of Neurology*, *63*, 444–453.

Zhang, F., Wang, L.-P., Brauner, M., Liewald, J. F., Kay, K., Watzke, N., et al. (2007). Multimodal fast optical interrogation of neural circuitry. *Nature*, *446*, 633–639.

Zhang, H., Yang, B., Mu, X., Ahmed, S. S., Su, Q., He, R., et al. (2011). Several rAAV vectors efficiently cross the blood-brain barrier and transduce neurons and astrocytes in the neonatal mouse central nervous system. *Molecular Therapy*, *19*, 1440–1448.

Ziburkus, J., Cressman, J. R., Barreto, E., & Schiff, S. J. (2006). Interneuron and pyramidal cell interplay during in vitro seizure-like events. *Journal of Neurophysiology*, *95*, 3948–3954.

INDEX

Note: Page numbers followed by "*f*" indicate figures and "*t*" indicate tables.

A

Absence seizures, 135–137, 142–143, 145, 148
Acidosis, in seizure termination, 190–191
Acquired epilepsy, 10, 18–19, 157–158
Active states, of slow-wave sleep
  in deep cortical layers, 13
  termination of, 14
Adenosine, 192
Allopregnanolone, 80
Allotetrahydrodeoxycorticosterone (THDOC), 80
Ammon's horn sclerosis, 64–65
Amygdalohippocampectomy, for drug-resistant limbic epilepsy, 93–94
Androstane neurosteroids, 78–80
Anterior interamygdalian commissural connectivity, 73
Anteromesial temporal lobectomy, for drug-resistant limbic epilepsy, 93–94
Anticonvulsant treatment, 280, 282
Antiepileptic drugs, 3
Apoptosis, 39–40
Astrocytes, 173–174

B

BBB. *See* Blood brain barrier (BBB)
Biophysical models
  chloride and calcium intracellular concentrations, 127–128
  extracellular potassium concentration, 124–126
  gap junctions, 128, 129
  Hodgkin–Huxley equations, 123–124
  homeostatic synaptic plasticity, 129–130, 130*f*
  lumped models, 137–146
  network topology, 130–135, 132*f*
  posttraumatic epileptogenic tissue, 130
  power-law behavior, 146–148
  single neuron dynamics, 123
  sodium concentrations, 126
  stochastic models, 146
  thalamocortical loop, 135–137
  VFOs, 128
Blood brain barrier (BBB)
  angiogenesis, 226
  astrocytes, 224–226
  breakdown, 228–231
  cerebral vessel maturation, 226
  endothelial cells, 224–226
  neurovascular unit, 224–226, 225*f*
  pharmacoresistance, 231–232
Blood oxygenation level-dependent (BOLD) fMRI, irritative zone assessment
  clinical implications, 269–270
  interictal epileptiform discharges, 265–266
  methodological aspects, 267–269
  principles, 266
  technical aspects, 267
Brain recordings, 159–162
Bumetanide, 169–170

C

CA3-driven interictal spikes, 74
Cajal–Retzius cells, 42
^{11}C-α-methyl-L-tryptophan (α-MTrp), 259
Caspases, 39–40
Catamenial seizure exacerbations, 80–81
Cellular substrates, of ictogenesis
  GABAergic neurotransmission, 167–170
  glutamatergic neurotransmission, 165
  interneurons, 165–167
  neuronal behavior, 162–164
  pyramidal cells, 164–165
Cerebral cortex
  characteristic structure, 36
  neocortical focus
    (un)identified, 45–50
    removal of, 50–53

Cerebral cortex (*Continued*)
normal and abnormal development of
developmental excitatory–inhibitory shift of GABA, 43–45
gliogenesis and myelination, 40
neurogenesis and apoptotic cell death, 37–40
neuronal migration, 40–42
transient neurons and circuits, 42–43
prenatal and early postnatal development, 36–37, 37*f*
^{11}C-flumazenil PET, 258–259
Chemical genetics, 287–288
Childhood absence epilepsy (CAE), 280–281
Chronic epileptic tissue
metabolic dysfunction of
mitochondrial dysfunction, 217–220
oxidative stress, 217
mitochondrial genome alteration, 220–224
Conductance-based modeling approach, 123
Convulsive seizures, 10
Cortical atrophy, 105–106
Cortical malformations, epilepsy surgery for, 52–53
Cortical morphometry, 251–252
Corticothalamic neural field model, 142–143
Cytochrome *c* oxidase, 223–224, 223*f*

D

Delta oscillation, 12–15
Dentate gyrus, 121-153, 130–133
Designer receptors exclusively activated by designer drugs (DREADDs), 287–288
Developmental excitatory–inhibitory shift, of GABA, 43–45
Diffusion tensor imaging (DTI), 96–97, 252–253
Dravet syndrome, 291–292
Drug-resistant limbic epilepsy, treatment for, 93–94
DTI. *See* Diffusion tensor imaging (DTI)
Dynamic systems theory, 199–200, 201*f*

E

Electroconvulsive therapy (ECT), 263
Electroencephalographic (EEG) studies
EEG–fMRI
feature of, 267
for focus localization, 267, 268*f*
issues, 270
principle, 266
in limbic epilepsy
computer simulations, 97–98
ictal EEG onset patterns, 99–101, 100*f*, 102*f*
intracranial EEG analysis, 99
theta-band functional connectivity, 97–98
Epilepsy, 156–157
acquired, 10, 18–19
causes of, 10
definition of, 3–4, 91–92
gene expression, alterations of, 292
genetic, 10
MRI protocol, 248
neuronal plasticity, 22–25
PET (*see* Positron emission tomography (PET))
progression, 157–158
Epileptic focus, 1–2, 21
definition of, 20–21
molecular, structural, and functional alterations
animal models, 46–48
experimental studies in human tissues, 49–50
undercut cortex model
advantage of, 21–22
computational studies, 21–22
spontaneous electrographic seizures, 22, 23*f*
Epileptic tissue
blood flow regulation in, 224–228
chronic (*see* Chronic epileptic tissue)
Epileptic transformation, of astrocytes, 228
Epileptic zone, 156–157
Epileptiform synchronization
high-frequency oscillations, 69–70
ictal discharges, 70–73
interactions between limbic areas, 73–75
interictal spikes, 67–69
neurosteroids, modulation of, 78–81
scalp EEG, ictal activity on, 67
stereo-EEG method, 67
Epileptogenesis prevention, 282

Epileptogenicity index (EI), 99
Epileptogenic network, definition of, 90
Epileptogenic zone (EZ), 2
  definition, 50–51
  location of, 246–247
Epileptor, 143
Extracellular potassium concentration, 124–126
Extracellular space, 172–173

F

^{18}F-Fluorodeoxyglucose positron emission tomography (FDG PET)
  extratemporal lobe epilepsy, clinical implications in, 257
  hypometabolism, localization and extent of, 255–256
  pathophysiology, 254–255
  statistical parametric mapping, 258
  temporal lobe epilepsy, clinical implications in, 257
Fluro-deoxyglucose-PET, in limbic epilepsy, 109–110
Focal cortical dysplasia (FCD)
  types of, 92–93
  voxel-based morphometry, 250
Focal epilepsy, 1–2
  altered glial cells, 4
  motor-evoked potential, 52–53
Focal neocortical epilepsy, 293–294
Focal seizures, 1–2
Fractional anisotropy (FA), 252–253
FreeSurfer software, 251
Functional connectivity (FC)
  in temporal lobe epilepsy, 106–107
  theta-band, 97–98
  working memory, 107–108
Functional magnetic resonance imaging (fMRI), 97, 105–109

G

GABAergic neurotransmission, 167–170
Gap junctions, 128, 129
Gene therapy, 284, 290–291
Genetic epilepsy, 10
Gliogenesis, 40
Glutamatergic neurotransmission, 165
Graphs
  clustering coefficient, 199
  definition, 197–198
  eigenvector centrality, 199
  functional networks, 197–198
  at seizure onset, 198
  structural networks, 197–198

H

Hidden Markov model, 146
High-frequency oscillations (HFOs), 69–70, 101–104, 103*f*
Hippocampal atrophy, 95–96, 102*f*
Hippocampal sclerosis (HS)
  axon fiber sprouting, 130–131
  magnetic resonance spectroscopy, 157–158
  in MTLE patients, 92–93
Hippocampus, topological changes, 130–133
Hodgkin–Huxley equations, 123–124
Homeostatic synaptic plasticity (HSP), 129–130, 130*f*
5-HT_{1A} receptor antagonists, 259
Human hippocampus, deletional spectra in CA3 region, 222*f*
Human limbic system, 91–92, 92*f*
Human mesial temporal lobe epilepsy, 4–5
Hyper-synchronous seizure onset (HYP)
  CA3 region of hippocampus, 79*f*
  characterization of, 75
  fast ripple occurrence, 78
  ictal discharges, 77*f*
  in vitro brain slice preparations, 76
  vs. LVF seizure onset, 75–76
  origin of, 77–78

I

Ictal perfusion
  preictal, ictal, and postictal changes, 260–261
  SPECT (*see* Single-photon emission computed tomography (SPECT))
Ictogenesis
  astrocytes, 173–174
  cellular substrates of, 162–170
  description of, 70–71, 158
  extracellular calcium ions, 172
  extracellular potassium ions, 170–172

Ictogenesis (*Continued*)
extracellular space, 172–173
$GABA_A$ receptor signaling, 71–73
hippocampal outputs, 74–75
interictal spikes, 69
neurosteroids, 80
Interictal spikes
firing patterns of single units, 69
$GABA_A$ receptor antagonists, 67, 68*f*
K^+ channel blockers, 67–69
partial epilepsy, noninvasive diagnosis of, 66–67
Interneurons, 165–167, 288–290
cortical, 14
GABAergic interneurons, 37–38, 40–41, 292–293
Intraoperative electrocorticography (ECoG), 51–52, 73
In vitro limbic epileptiform synchronization, 4AP model of, 76

L

Light-activated proteins, 287
Limbic epilepsy, 64–65
clinical features of, 91–95
DTI studies, 96–97
electroencephalographic studies, 97–105
and epileptiform synchronization (*see* Epileptiform synchronization)
FDG-PET, 109–110
fMRI (*see* Functional magnetic resonance imaging (fMRI))
high-frequency oscillations, 101–104, 103*f*
magnetic resonance spectroscopy, 109, 111
magnetoencephalography, 98–99
MRI studies, 95–96
PET scanning, 109–110
SPECT, 109, 110–111
treatment of, 93–94
Limbic seizures
characteristic ictal manifestations, 91–92
clinical features of, 91–92, 93*t*
Low-voltage fast (LVF) seizure onset, 67
CA3 region of hippocampus, 79*f*
characterization of, 75–76
HFO occurrence, 78
vs. hyper-synchronous seizure onset, 75–76
ictal discharges, 77*f*
in vitro brain slice preparations, 76, 77*f*
zones of, 77–78
Lumped models
corticothalamic neural field model, 142–143
Epileptor, 143
excitatory interneurons, 140–142
excitatory pyramidal cells, 140–142
hypothesis of, 138
inhibitory interneurons, 140–142
phase clustering index, 140
propagation of partial seizures, 143–145

M

Magnetic resonance imaging (MRI)
epileptogenic lesion, 247
gray matter loss, 248
limbic epilepsy, 95–96
sensitivity of, 247
volumetric, 248–250
Magnetic resonance spectroscopy (MRS)
hippocampal sclerosis, 157–158
limbic epilepsy, 109, 111
Magnetoencephalography (MEG), in limbic epilepsy, 98–99
Mammalian target of rapamycin (mTOR), 39
Mesial temporal lobe epilepsy (MTLE), 90–91
anterior mesial temporal resection, 73
chronic models of, 77–78
cortical morphometry, 251–252
depth electrode recordings, 75
description of, 64–65
focal cortical dysplasia, 92–93
hippocampal outputs, 74–75
hippocampal sclerosis, 92–93
HYP seizures (*see* Hyper-synchronous seizure onset (HYP))
interictal and ictal epileptiform discharges, 65, 66*f*
irritative zone, electrophysiological localization of, 66–67
LVF seizures (*see* Low-voltage fast (LVF) seizure onset)
neocorticolimbic connectivity in, 73
voxel-based morphometry, 250–251

Mesial temporal sclerosis, 64–65
Mitochondria
dysfunction, 217–220
fission and fusion cycles, 210–212
genome alteration, 220–224
oxidative phosphorylation, impairment of, 221–223
respiratory chain, 211*f*
role in ATP synthesis, 210–212
Mitochondrial DNA (mtDNA), 210–212
damage, 220, 221–223
mutations, 220–221, 223–224
repair capacity of, 220–221
Mitochondrial permeability transition pore (mPTP), 213–214
Mossy fiber sprouting, in dentate gyrus, 131–133
Motor-evoked potential (MEP), 52–53
Multicompartment model, 123
Myelination, 40

N

Natronomonas halorhodopsin (NpHR), 284–286
Neocortical focus
(un)identified, 45–50
removal of, 50–53
Neocortical neuronal activities
during seizures, 15–17
during states of vigilance, 10–15
Neocortical structural lesion, identification of, 45, 46*f*
Neural mass models. *See* Lumped models
Neural population models. *See* Lumped models
Neurobiological research, in field of epilepsy, 2
Neurogenesis, 38–40
Neuronal migrations, 40–42
Neurosteroids, 78–81
Nonconvulsive seizures, 10
Non-REM sleep, 11–15

O

Opsins, 284–285, 286, 287
Optogenetics
advantage of, 285–286
feature of, 286
Natronomonas halorhodopsin, 284–286
opsins, 284–285, 286, 287
Oxidative stress, 217

P

Paroxysmal depolarizing shift, 17
Partial epilepsy, noninvasive diagnosis of, 66–67
Perinatal hypoxia, 38–39
P-glycoprotein (P-gp), 231–232
Positron emission tomography (PET)
advantage of, 253, 254
^{11}C-flumazenil, 258–259
^{18}F-FDG PET, 254–258
glucose hypermetabolism, 253–254
in limbic epilepsy, 109–110
Power-law behavior, 146–148
Pregnane neurosteroids, 78–80
Preictal dynamics, 159–162
Programmed cell death, 39–40
Pyramidal cells, 37–38, 164–165

R

Rapamycin, 282
Recombinant adeno-associated virus vectors, 290–291
Refractory epilepsy, 283, 284
REM sleep, 11
Retinal waves, 38

S

Seizure focus, 20–21
Seizure-onset zone (SOZ), 2, 156–157
Seizures. *See also* Epilepsy
block of inhibition, 18–19
brain excitability, 19
changes in energy metabolism, 212–216
chemical genetics, 287–288
closed-loop systems, 284, 285–287
convulsive and nonconvulsive, 10
direct electrical stimulation, 284
vs. earthquakes, analogy, 146–148, 147*f*
gene therapy, 284, 290–291
interneurons, 288–290
intrinsic and synaptic excitability of neurons, 17–18
neural mass model, 143, 144*f*
optogenetics, 284–287

Seizures (*Continued*)
outlasting activity, 19–20
partial cortical deafferentation, 20
prevention, 280–282
reduction of adverse effects, 283
spatial propagation of, 133–135, 145
through bifurcations, 141*f*
transition to, 162
trauma-induced epileptogenesis, 19–20
treatment and control, 291–294
Seizure termination
as critical transition
dynamic systems theory, 199–200, 201*f*
features of, 200
single-cell level, 202
status epilepticus, 201–202
graphs and functional networks, 197–199
metabolic mechanisms of
acidosis, 190–191
extracellular potassium concentration, 191–192
neuromodulators, 192
oxygen, glucose, and neurotransmitter depletion, 189–190
sustained neuronal activity, 188, 190*f*
synchronization, 193–197
Silent states, of slow-wave sleep, 14
duration of, 13
onset of, 14
Single-photon emission computed tomography (SPECT)
in epileptic seizures
ictal injection, 261–262
^{99m}Tc-ECD, 261–262
^{99m}Tc-HMPAO, 261–262
propagation pattern, 262–263
SISCOM, 263–265, 264*f*
in limbic epilepsy, 109, 110–111
SLC2A1 mutations, 281
Sleep
non-REM sleep, 11–15
REM sleep, 11
slow oscillation, 13, 22–25
slow-wave sleep, 12–15
Slow oscillation, 12–15
Slow-wave sleep, 12–15
Spike–wave discharges, 135–137, 136*f*
Statistical parametric mapping (SPM), 258
Stochastic models, 146
Subplate cells, 42–43
Sulfated neurosteroids, 78–80
Superoxide production site, 210–212
Surface-based morphometry (SBM), 251
Synchronization, 195–196
definition of, 193
of neuronal oscillations, 195

T

^{99m}Tc-ethyl cysteinate dimer (^{99m}Tc-ECD), 261–262
^{99m}Tc-hexamethylenepropylene amine (^{99m}Tc-HMPAO), 261–262
Temporal lobectomy, 73
Temporal lobe epilepsy (TLE), 4–5, 80–81. *See also* Mesial temporal lobe epilepsy (MTLE)
atrophy, 249–250
diffusion tensor imaging, 252–253
epileptic seizure in, 194*f*
MRI volumetry, 249
network topology, 130–131
TSPO, 260
VBM analysis, of mesial TLE, 250–251
Transient neurons and circuits, 42–43
Translocator protein (TSPO), 260

V

VBM. *See* Voxel-based morphometry (VBM)
Ventricular zone (VZ)
neuroepithelial progenitor cells, 38
pyramidal cell generation, 37–38
Very-fast oscillations (VFOs), 128
Voxel-based morphometry (VBM), 250–251

W

Waking state, 10–11

CONTENTS OF RECENT VOLUMES

Volume 37

Section I: Selectionist Ideas and Neurobiology

Selectionist and Instructionist Ideas in Neuroscience
Olaf Sporns

Population Thinking and Neuronal Selection: Metaphors or Concepts?
Ernst Mayr

Selection and the Origin of Information
Manfred Eigen

Section II: Development and Neuronal Populations

Morphoregulatory Molecules and Selectional Dynamics during Development
Kathryn L. Crossin

Exploration and Selection in the Early Acquisition of Skill
Esther Thelen and Daniela Corbetta

Population Activity in the Control of Movement
Apostolos P. Georgopoulos

Section III: Functional Segregation and Integration in the Brain

Reentry and the Problem of Cortical Integration
Giulio Tononi

Coherence as an Organizing Principle of Cortical Functions
Wolf Singerl

Temporal Mechanisms in Perception
Ernst Pöppel

Section IV: Memory and Models

Selection versus Instruction: Use of Computer Models to Compare Brain Theories
George N. Reeke, Jr.

Memory and Forgetting: Long-Term and Gradual Changes in Memory Storage
Larry R. Squire

Implicit Knowledge: New Perspectives on Unconscious Processes
Daniel L. Schacter

Section V: Psychophysics, Psychoanalysis, and Neuropsychology

Phantom Limbs, Neglect Syndromes, Repressed Memories, and Freudian Psychology
V. S. Ramachandran

Neural Darwinism and a Conceptual Crisis in Psychoanalysis
Arnold H. Modell

A New Vision of the Mind
Oliver Sacks

INDEX

Volume 38

Regulation of $GABA_A$ Receptor Function and Gene Expression in the Central Nervous System
A. Leslie Morrow

Genetics and the Organization of the Basal Ganglia
Robert Hitzemann, Yeang Olan, Stephen Kanes, Katherine Dains, and Barbara Hitzemann

Structure and Pharmacology of Vertebrate $GABA_A$ Receptor Subtypes
Paul J. Whiting, Ruth M. McKernan, and Keith A. Wafford

Neurotransmitter Transporters: Molecular Biology, Function, and Regulation
Beth Borowsky and Beth J. Hoffman

Presynaptic Excitability
Meyer B. Jackson

Monoamine Neurotransmitters in Invertebrates and Vertebrates: An Examination of the Diverse Enzymatic Pathways Utilized to Synthesize and Inactivate Biogenic Amines
B. D. Sloley and A. V. Juorio

Neurotransmitter Systems in Schizophrenia
Gavin P. Reynolds

Physiology of Bergmann Glial Cells
Thomas Müller and Helmut Kettenmann

INDEX

Volume 39

Modulation of Amino Acid-Gated Ion Channels by Protein Phosphorylation
Stephen J. Moss and Trevor G. Smart

Use-Dependent Regulation of $GABA_A$ Receptors
Eugene M. Barnes, Jr.

Synaptic Transmission and Modulation in the Neostriatum
David M. Lovinger and Elizabeth Tyler

The Cytoskeleton and Neurotransmitter Receptors
Valerie J. Whatley and R. Adron Harris

Endogenous Opioid Regulation of Hippocampal Function
Michele L. Simmons and Charles Chavkin

Molecular Neurobiology of the Cannabinoid Receptor
Mary E. Abood and Billy R. Martin

Genetic Models in the Study of Anesthetic Drug Action
Victoria J. Simpson and Thomas E. Johnson

Neurochemical Bases of Locomotion and Ethanol Stimulant Effects
Tamara J. Phillips and Elaine H. Shen

Effects of Ethanol on Ion Channels
Fulton T. Crews, A. Leslie Morrow, Hugh Criswell, and George Breese

INDEX

Volume 40

Mechanisms of Nerve Cell Death: Apoptosis or Necrosis after Cerebral Ischemia
R. M. E. Chalmers-Redman, A. D. Fraser, W. Y. H. Ju, J. Wadia, N. A. Tatton, and W. G. Tatton

Changes in Ionic Fluxes during Cerebral Ischemia
Tibor Kristian and Bo K. Siesjo

Techniques for Examining Neuroprotective Drugs *in Vitro*
A. Richard Green and Alan J. Cross

Techniques for Examining Neuroprotective Drugs *in Vivo*
Mark P. Goldberg, Uta Strasser, and Laura L. Dugan

Calcium Antagonists: Their Role in Neuroprotection
A. Jacqueline Hunter

Sodium and Potassium Channel Modulators: Their Role in Neuroprotection
Tihomir P. Obrenovich

NMDA Antagonists: Their Role in Neuroprotection
Danial L. Small

Development of the NMDA Ion-Channel Blocker, Aptiganel Hydrochloride, as a Neuroprotective Agent for Acute CNS Injury
Robert N. McBurney

The Pharmacology of AMPA Antagonists and Their Role in Neuroprotection
Rammy Gill and David Lodge

GABA and Neuroprotection
Patrick D. Lyden

Adenosine and Neuroprotection
Bertil B. Fredholm

Interleukins and Cerebral Ischemia
Nancy J. Rothwell, Sarah A. Loddick, and Paul Stroemer

Nitrone-Based Free Radical Traps as Neuroprotective Agents in Cerebral Ischemia and Other Pathologies
Kenneth Hensley, John M. Carney, Charles A. Stewart, Tahera Tabatabaie, Quentin Pye, and Robert A. Floyd

Neurotoxic and Neuroprotective Roles of Nitric Oxide in Cerebral Ischemia
Turgay Dalkara and Michael A. Moskowitz

A Review of Earlier Clinical Studies on Neuroprotective Agents and Current Approaches
Nils-Gunnar Wahlgren

INDEX

Volume 41

Section I: Historical Overview

Rediscovery of an Early Concept
Jeremy D. Schmahmann

Section II: Anatomic Substrates

The Cerebrocerebellar System
Jeremy D. Schmahmann and Deepak N. Pandya

Cerebellar Output Channels
Frank A. Middleton and Peter L. Strick

Cerebellar-Hypothalamic Axis: Basic Circuits and Clinical Observations
Duane E. Haines, Espen Dietrichs, Gregory A. Mihailoff, and E. Frank McDonald

Section III. Physiological Observations

Amelioration of Aggression: Response to Selective Cerebellar Lesions in the Rhesus Monkey
Aaron J. Berman

Autonomic and Vasomotor Regulation
Donald J. Reis and Eugene V. Golanov

Associative Learning
Richard F. Thompson, Shaowen Bao, Lu Chen, Benjamin D. Cipriano, Jeffrey S. Grethe, Jeansok J. Kim, Judith K. Thompson, Jo Anne Tracy, Martha S. Weninger, and David J. Krupa

Visuospatial Abilities
Robert Lalonde

Spatial Event Processing
Marco Molinari, Laura Petrosini, and Liliana G. Grammaldo

Section IV: Functional Neuroimaging Studies

Linguistic Processing
Julie A. Fiez and Marcus E. Raichle

Sensory and Cognitive Functions
Lawrence M. Parsons and Peter T. Fox

Skill Learning
Julien Doyon

Section V: Clinical and Neuropsychological Observations

Executive Function and Motor Skill Learning
Mark Hallett and Jordon Grafman

Verbal Fluency and Agrammatism
Marco Molinari, Maria G. Leggio, and Maria C. Silveri

Classical Conditioning
Diana S. Woodruff-Pak

Early Infantile Autism
Margaret L. Bauman, Pauline A. Filipek, and Thomas L. Kemper

Olivopontocerebellar Atrophy and Friedreich's Ataxia: Neuropsychological Consequences of Bilateral versus Unilateral Cerebellar Lesions
Thérèse Botez-Marquard and Mihai I. Botez

Posterior Fossa Syndrome
Ian F. Pollack

Cerebellar Cognitive Affective Syndrome
Jeremy D. Schmahmann and Janet C. Sherman

Inherited Cerebellar Diseases
Claus W. Wallesch and Claudius Bartels

Neuropsychological Abnormalities in Cerebellar Syndromes—Fact or Fiction?
Irene Daum and Hermann Ackermann

Section VI: Theoretical Considerations

Cerebellar Microcomplexes
Masao Ito

Control of Sensory Data Acquisition
James M. Bower

Neural Representations of Moving Systems
Michael Paulin

How Fibers Subserve Computing Capabilities: Similarities between Brains and Machines
Henrietta C. Leiner and Alan L. Leiner

Cerebellar Timing Systems
Richard Ivry

Attention Coordination and Anticipatory Control
Natacha A. Akshoomoff, Eric Courchesne, and Jeanne Townsend

Context-Response Linkage
W. Thomas Thach

Duality of Cerebellar Motor and Cognitive Functions
James R. Bloedel and Vlastislav Bracha

Section VII: Future Directions

Therapeutic and Research Implications
Jeremy D. Schmahmann

Volume 42

Alzheimer Disease
Mark A. Smith

Neurobiology of Stroke
W. Dalton Dietrich

Free Radicals, Calcium, and the Synaptic Plasticity-Cell Death Continuum: Emerging Roles of the Trascription Factor NFκB
Mark P. Mattson

AP-I Transcription Factors: Short- and Long-Term Modulators of Gene Expression in the Brain
Keith Pennypacker

Ion Channels in Epilepsy
Istvan Mody

Posttranslational Regulation of Ionotropic Glutamate Receptors and Synaptic Plasticity
Xiaoning Bi, Steve Standley, and Michel Baudry

Heritable Mutations in the Glycine, $GABA_A$, and Nicotinic Acetylcholine Receptors Provide New Insights into the Ligand-Gated Ion Channel Receptor Superfamily
Behnaz Vafa and Peter R. Schofield

INDEX

Volume 43

Early Development of the *Drosophila* Neuromuscular Junction: A Model for Studying Neuronal Networks in Development
Akira Chiba

Development of Larval Body Wall Muscles
Michael Bate, Matthias Landgraf, and Mar Ruiz Gómez Bate

Development of Electrical Properties and Synaptic Transmission at the Embryonic Neuromuscular Junction
Kendal S. Broadie

Ultrastructural Correlates of Neuromuscular Junction Development
Mary B. Rheuben, Motojiro Yoshihara, and Yoshiaki Kidokoro

Assembly and Maturation of the *Drosophila* Larval Neuromuscular Junction
L. Sian Gramates and Vivian Budnik

Second Messenger Systems Underlying Plasticity at the Neuromuscular Junction
Frances Hannan and Yi Zhong

Mechanisms of Neurotransmitter Release
J. Troy Littleton, Leo Pallanck, and Barry Ganetzky

Vesicle Recycling at the *Drosophila* Neuromuscular Junction
Daniel T. Stimson and Mani Ramaswami

Ionic Currents in Larval Muscles of *Drosophila*
Satpal Singh and Chun-Fang Wu

Development of the Adult Neuromuscular System
Joyce J. Fernandes and Haig Keshishian

Controlling the Motor Neuron
James R. Trimarchi, Ping Jin, and Rodney K. Murphey

Volume 44

Human Ego-Motion Perception
A. V. van den Berg

Optic Flow and Eye Movements
M. Lappe and K.-P. Hoffman

The Role of MST Neurons during Ocular Tracking in 3D Space
K. Kawano, U. Inoue, A. Takemura, Y. Kodaka, and F. A. Miles

Visual Navigation in Flying Insects
M. V. Srinivasan and S.-W. Zhang

Neuronal Matched Filters for Optic Flow Processing in Flying Insects
H. G. Krapp

A Common Frame of Reference for the Analysis of Optic Flow and Vestibular Information
B. J. Frost and D. R. W. Wylie

Optic Flow and the Visual Guidance of Locomotion in the Cat
H. Sherk and G. A. Fowler

Stages of Self-Motion Processing in Primate Posterior Parietal Cortex
F. Bremmer, J.-R. Duhamel, S. B. Hamed, and W. Graf

Optic Flow Analysis for Self-Movement Perception
C. J. Duffy

Neural Mechanisms for Self-Motion Perception in Area MST
R. A. Andersen, K. V. Shenoy, J. A. Crowell, and D. C. Bradley

Computational Mechanisms for Optic Flow Analysis in Primate Cortex
M. Lappe

Human Cortical Areas Underlying the Perception of Optic Flow: Brain Imaging Studies
M. W. Greenlee

What Neurological Patients Tell Us about the Use of Optic Flow
L. M. Vaina and S. K. Rushton

INDEX

Volume 45

Mechanisms of Brain Plasticity: From Normal Brain Function to Pathology
Philip. A. Schwartzkroin

Brain Development and Generation of Brain Pathologies
Gregory L. Holmes and Bridget McCabe

Maturation of Channels and Receptors: Consequences for Excitability
David F. Owens and Arnold R. Kriegstein

Neuronal Activity and the Establishment of Normal and Epileptic Circuits during Brain Development
John W. Swann, Karen L. Smith, and Chong L. Lee

The Effects of Seizures of the Hippocampus of the Immature Brain
Ellen F. Sperber and Solomon L. Moshe

Abnormal Development and Catastrophic Epilepsies: The Clinical Picture and Relation to Neuroimaging
Harry T. Chugani and Diane C. Chugani

Cortical Reorganization and Seizure Generation in Dysplastic Cortex
G. Avanzini, R. Preafico, S. Franceschetti, G. Sancini, G. Battaglia, and V. Scaioli

Rasmussen's Syndrome with Particular Reference to Cerebral Plasticity: A Tribute to Frank Morrell
Fredrick Andermann and Yuonne Hart

Structural Reorganization of Hippocampal Networks Caused by Seizure Activity
Daniel H. Lowenstein

Epilepsy-Associated Plasticity in gamma-Amniobutyric Acid Receptor Expression, Function and Inhibitory Synaptic Properties
Douglas A. Coulter

Synaptic Plasticity and Secondary Epileptogenesis
Timothy J. Teyler, Steven L. Morgan, Rebecca N. Russell, and Brian L. Woodside

Synaptic Plasticity in Epileptogenesis: Cellular Mechanisms Underlying Long-Lasting Synaptic Modifications that Require New Gene Expression
Oswald Steward, Christopher S. Wallace, and Paul F. Worley

Cellular Correlates of Behavior
Emma R. Wood, Paul A. Dudchenko, and Howard Eichenbaum

Mechanisms of Neuronal Conditioning
David A. T. King, David J. Krupa, Michael R. Foy, and Richard F. Thompson

Plasticity in the Aging Central Nervous System
C. A. Barnes

Secondary Epileptogenesis, Kindling, and Intractable Epilepsy: A Reappraisal from the Perspective of Neuronal Plasticity
Thomas P. Sutula

Kindling and the Mirror Focus
Dan C. McIntyre and Michael O. Poulter

Partial Kindling and Behavioral Pathologies
Robert E. Adamec

The Mirror Focus and Secondary Epileptogenesis
B. J. Wilder

Hippocampal Lesions in Epilepsy: A Historical Review
Robert Naquet

Clinical Evidence for Secondary Epileptogensis
Hans O. Luders

Epilepsy as a Progressive (or Nonprogressive "Benign") Disorder
John A. Wada

Pathophysiological Aspects of Landau-Kleffner Syndrome: From the Active Epileptic Phase to Recovery
Marie-Noelle Metz-Lutz, Pierre Maquet, Annd De Saint Martin, Gabrielle Rudolf, Norma Wioland, Edouard Hirsch, and Chriatian Marescaux

Local Pathways of Seizure Propagation in Neocortex
Barry W. Connors, David J. Pinto, and Albert E. Telefeian

Multiple Subpial Transection: A Clinical Assessment
C. E. Polkey

The Legacy of Frank Morrell
Jerome Engel, Jr.

Volume 46

Neurosteroids: Beginning of the Story
Etienne E. Baulieu, P. Robel, and M. Schumacher

Biosynthesis of Neurosteroids and Regulation of Their Synthesis
Synthia H. Mellon and Hubert Vaudry

Neurosteroid 7-Hydroxylation Products in the Brain
Robert Morfin and Luboslav Stárka

Neurosteroid Analysis
Ahmed A. Alomary, Robert L. Fitzgerald, and Robert H. Purdy

Role of the Peripheral-Type Benzodiazepine Receptor in Adrenal and Brain Steroidogenesis
Rachel C. Brown and Vassilios Papadopoulos

Formation and Effects of Neuroactive Steroids in the Central and Peripheral Nervous System
Roberto Cosimo Melcangi, Valerio Magnaghi, Mariarita Galbiati, and Luciano Martini

Neurosteroid Modulation of Recombinant and Synaptic $GABA_A$ Receptors
Jeremy J. Lambert, Sarah C. Harney, Delia Belelli, and John A. Peters

$GABA_A$-Receptor Plasticity during Long-Term Exposure to and Withdrawal from Progesterone
Giovanni Biggio, Paolo Follesa, Enrico Sanna, Robert H. Purdy, and Alessandra Concas

Stress and Neuroactive Steroids
Maria Luisa Barbaccia, Mariangela Serra, Robert H. Purdy, and Giovanni Biggio

Neurosteroids in Learning and Memory Processes
Monique Vallée, Willy Mayo, George F. Koob, and Michel Le Moal

Neurosteroids and Behavior
Sharon R. Engel and Kathleen A. Grant

Ethanol and Neurosteroid Interactions in the Brain
A. Leslie Morrow, Margaret J. VanDoren, Rebekah Fleming, and Shannon Penland

Preclinical Development of Neurosteroids as Neuroprotective Agents for the Treatment of Neurodegenerative Diseases
Paul A. Lapchak and Dalia M. Araujo

Clinical Implications of Circulating Neurosteroids
Andrea R. Genazzani, Patrizia Monteleone, Massimo Stomati, Francesca Bernardi, Luigi Cobellis, Elena Casarosa, Michele Luisi, Stefano Luisi, and Felice Petraglia

Neuroactive Steroids and Central Nervous System Disorders
Mingde Wang, Torbjörn Bäckström, Inger Sundström, Göran Wahlström, Tommy Olsson, Di Zhu, Inga-Maj Johansson, Inger Björn, and Marie Bixo

Neuroactive Steroids in Neuropsychopharmacology
Rainer Rupprecht and Florian Holsboer

Current Perspectives on the Role of Neurosteroids in PMS and Depression
Lisa D. Griffin, Susan C. Conrad, and Synthia H. Mellon

Index

Volume 47

Introduction: Studying Gene Expression in Neural Tissues by *in Situ* Hybridization
W. Wisden and B. J. Morris

Part I: *In Situ* Hybridization with Radiolabelled Oligonucleotides
In Situ Hybridization with Oligonucleotide Probes
Wl. Wisden and B. J. Morris

Cryostat Sectioning of Brains
Victoria Revilla and Alison Jones

Processing Rodent Embryonic and Early Postnatal Tissue for *in Situ* Hybridization with Radiolabelled Oligonucleotides
David J. Laurie, Petra C. U. Schrotz, Hannah Monyer, and Ulla Amtmann

Processing of Retinal Tissue for *in Situ* Hybridization
Frank Müller

Processing the Spinal Cord for *in Situ* Hybridization with Radiolabelled Oligonucleotides
A. Berthele and T. R. Tölle

Processing Human Brain Tissue for *in Situ* Hybridization with Radiolabelled Oligonucleotides
Louise F. B. Nicholson

In Situ Hybridization of Astrocytes and Neurons Cultured *in Vitro*
L. A. Arizza-McNaughton, C. De Felipe, and S. P. Hunt

In Situ Hybridization on Organotypic Slice Cultures
A. Gerfin-Moser and H. Monyer

Quantitative Analysis of *in Situ* Hybridization Histochemistry
Andrew L. Gundlach and Ross D. O'Shea

Part II: Nonradioactive *in Situ* hybridization

Nonradioactive *in Situ* Hybridization Using Alkaline Phosphatase-Labelled Oligonucleotides
S. J. Augood, E. M. McGowan, B. R. Finsen, B. Heppelmann, and P. C. Emson

Combining Nonradioactive *in Situ* Hybridization with Immunohistological and Anatomical Techniques
Petra Wahle

Nonradioactive *in Situ* Hybridization: Simplified Procedures for Use in Whole Mounts of Mouse and Chick Embryos
Linda Ariza-McNaughton and Robb Krumlauf

INDEX

Volume 48

Assembly and Intracellular Trafficking of $GABA_A$ Receptors Eugene
Barnes

Subcellular Localization and Regulation of $GABA_A$ Receptors and Associated Proteins Bernhard Lüscher and Jean-Marc Fritschy D_1 Dopamine Receptors
Richard Mailman

Molecular Modeling of Ligand-Gated Ion Channels: Progress and Challenges
Ed Bertaccini and James R. Trudel

Alzheimer's Disease: Its Diagnosis and Pathogenesis
Jillian J. Kril and Glenda M. Halliday

DNA Arrays and Functional Genomics in Neurobiology
Christelle Thibault, Long Wang, Li Zhang, and Michael F. Miles

INDEX

Volume 49

What Is West Syndrome?
Olivier Dulac, Christine Soufflet, Catherine Chiron, and Anna Kaminski

The Relationship between encephalopathy and Abnormal Neuronal Activity in the Developing Brain
Frances E. Jensen

Hypotheses from Functional Neuroimaging Studies
Csaba Juhász, Harry T. Chugani, Ouo Muzik, and Diane C. Chugani

Infantile Spasms: Unique Sydrome or General Age-Dependent Manifestation of a Diffuse Encephalopathy?
M. A. Koehn and M. Duchowny

Histopathology of Brain Tissue from Patients with Infantile Spasms
Harry V. Vinters

Generators of Ictal and Interictal Electroencephalograms Associated with Infantile Spasms: Intracellular Studies of Cortical and Thalamic Neurons
M. Steriade and I. Timofeev

Cortical and Subcortical Generators of Normal and Abnormal Rhythmicity
David A. McCormick

Role of Subcortical Structures in the Pathogenesis of Infantile Spasms: What Are Possible Subcortical Mediators?
F. A. Lado and S. L. Moshé

What Must We Know to Develop Better Therapies?
Jean Aicardi

The Treatment of Infantile Spasms: An Evidence-Based Approach
Mark Mackay, Shelly Weiss, and O. Carter Snead III

ACTH Treatment of Infantile Spasms: Mechanisms of Its Effects in Modulation of Neuronal Excitability
K. L. Brunson, S. Avishai-Eliner, and T. Z. Baram

Neurosteroids and Infantile Spasms: The Deoxycorticosterone Hypothesis
Michael A. Rogawski and Doodipala S. Reddy

Are there Specific Anatomical and/or Transmitter Systems (Cortical or Subcortical) That Should Be Targeted?
Phillip C. Jobe

Medical versus Surgical Treatment: Which Treatment When
W. Donald Shields

Developmental Outcome with and without Successful Intervention
Rochelle Caplan, Prabha Siddarth, Gary Mathern, Harry Vinters, Susan Curtiss, Jennifer Levitt, Robert Asarnow, and W. Donald Shields

Infantile Spasms versus Myoclonus: Is There a Connection?
Michael R. Pranzatelli

Tuberous Sclerosis as an Underlying Basis for Infantile Spasm
Raymond S. Yeung

Brain Malformation, Epilepsy, and Infantile Spasms
M. Elizabeth Ross

Brain Maturational Aspects Relevant to Pathophysiology of Infantile Spasms
G. Auanzini, F. Panzica, and S. Franceschetti

Gene Expression Analysis as a Strategy to Understand the Molecular Pathogenesis of Infantile Spasms
Peter B. Crino

Infantile Spasms: Criteria for an Animal Model
Carl E. Stafstrom and Gregory L. Holmes

INDEX

Volume 50

Part I: Primary Mechanisms

How Does Glucose Generate Oxidative Stress In Peripheral Nerve?
Irina G. Obrosova

Glycation in Diabetic Neuropathy: Characteristics, Consequences, Causes, and Therapeutic Options
Paul J. Thornalley

Part II: Secondary Changes

Protein Kinase C Changes in Diabetes: Is the Concept Relevant to Neuropathy?
Joseph Eichberg

Are Mitogen-Activated Protein Kinases Glucose Transducers for Diabetic Neuropathies?
Tertia D. Purves and David R. Tomlinson

Neurofilaments in Diabetic Neuropathy
Paul Fernyhough and Robert E. Schmidt

Apoptosis in Diabetic Neuropathy
Aviva Tolkovsky

Nerve and Ganglion Blood Flow in Diabetes: An Appraisal
Douglas W. Zochodne

Part III: Manifestations

Potential Mechanisms of Neuropathic Pain in Diabetes
Nigel A. Calcutt

Electrophysiologic Measures of Diabetic Neuropathy: Mechanism and Meaning
Joseph C. Arezzo and Elena Zotova

Neuropathology and Pathogenesis of Diabetic Autonomic Neuropathy
Robert E. Schmidt

Role of the Schwann Cell in Diabetic Neuropathy
Luke Eckersley

Part IV: Potential Treatment

Polyol Pathway and Diabetic Peripheral Neuropathy
Peter J. Oates

Nerve Growth Factor for the Treatment of Diabetic Neuropathy: What Went Wrong, What Went Right, and What Does the Future Hold?
Stuart C. Apfel

Angiotensin-Converting Enzyme Inhibitors: Are there Credible Mechanisms for Beneficial Effects in Diabetic Neuropathy?
Rayaz A. Malik and David R. Tomlinson

Clinical Trials for Drugs Against Diabetic Neuropathy: Can We Combine Scientific Needs With Clinical Practicalities?
Dan Ziegler and Dieter Luft

INDEX

Volume 51

Energy Metabolism in the Brain
Leif Hertz and Gerald A. Dienel

The Cerebral Glucose-Fatty Acid Cycle: Evolutionary Roots, Regulation, and (Patho) physiological Importance
Kurt Heininger

Expression, Regulation, and Functional Role of Glucose Transporters (GLUTs) in Brain
Donard S. Dwyer, Susan J. Vannucci, and Ian A. Simpson

Insulin-Like Growth Factor-1 Promotes Neuronal Glucose Utilization During Brain Development and Repair Processes
Carolyn A. Bondy and Clara M. Cheng

CNS Sensing and Regulation of Peripheral Glucose Levels
Barry E. Levin, Ambrose A. Dunn-Meynell, and Vanessa H. Routh

Glucose Transporter Protein Syndromes
Darryl C. De Vivo, Dong Wang, Juan M. Pascual, and Yuan Yuan Ho

Glucose, Stress, and Hippocampal Neuronal Vulnerability
Lawrence P. Reagan

Glucose/Mitochondria in Neurological Conditions
John P. Blass

Energy Utilization in the Ischemic/Reperfused Brain
John W. Phillis and Michael H. O'Regan

Diabetes Mellitus and the Central Nervous System
Anthony L. McCall

Diabetes, the Brain, and Behavior: Is There a Biological Mechanism Underlying the Association between Diabetes and Depression?
A. M. Jacobson, J. A. Samson, K. Weinger, and C. M. Ryan

Schizophrenia and Diabetes
David C. Henderson and Elissa R. Ettinger

Psychoactive Drugs Affect Glucose Transport and the Regulation of Glucose Metabolism
Donard S. Dwyer, Timothy D. Ardizzone, and Ronald J. Bradley

INDEX

Volume 52

Neuroimmune Relationships in Perspective
Frank Hucklebridge and Angela Clow

Sympathetic Nervous System Interaction with the Immune System
Virginia M. Sanders and Adam P. Kohm

Mechanisms by Which Cytokines Signal the Brain
Adrian J. Dunn

Neuropeptides: Modulators of Immune Responses in Health and Disease
David S. Jessop

Brain–Immune Interactions in Sleep
Lisa Marshall and Jan Born

Neuroendocrinology of Autoimmunity
Michael Harbuz

Systemic Stress-Induced Th2 Shift and Its Clinical Implications
Ibia J. Elenkov

Neural Control of Salivary S-IgA Secretion
Gordon B. Proctor and Guy H. Carpenter

Stress and Secretory Immunity
Jos A. Bosch, Christopher Ring, Eco J. C. de Geus, Enno C. I. Veerman, and Arie V. Nieuw Amerongen

Cytokines and Depression
Angela Clow

Immunity and Schizophrenia: Autoimmunity, Cytokines, and Immune Responses
Fiona Gaughran

Cerebral Lateralization and the Immune System
Pierre J. Neveu

Behavioral Conditioning of the Immune System
Frank Hucklebridge

Psychological and Neuroendocrine Correlates of Disease Progression
Julie M. Turner-Cobb

The Role of Psychological Intervention in Modulating Aspects of Immune Function in Relation to Health and Well-Being
J. H. Gruzelier

INDEX

Volume 53

Section I: Mitochondrial Structure and Function

Mitochondrial DNA Structure and Function
Carlos T. Moraes, Sarika Srivastava, Ilias Kirkinezos, Jose Oca-Cossio, Corina van Waveren, Markus Woischnick, and Francisca Diaz

Oxidative Phosphorylation: Structure, Function, and Intermediary Metabolism
Simon J. R. Heales, Matthew E. Gegg, and John B. Clark

Import of Mitochondrial Proteins
Matthias F. Bauer, Sabine Hofmann, and Walter Neupert

Section II: Primary Respiratory Chain Disorders

Mitochondrial Disorders of the Nervous System: Clinical, Biochemical, and Molecular Genetic Features
Dominic Thyagarajan and Edward Byrne

Section III: Secondary Respiratory Chain Disorders

Friedreich's Ataxia
J. M. Cooper and J. L. Bradley

Wilson Disease
C. A. Davie and A. H. V. Schapira

Hereditary Spastic Paraplegia
Christopher J. McDermott and Pamela J. Shaw

Cytochrome c Oxidase Deficiency
Giacomo P. Comi, Sandra Strazzer, Sara Galbiati, and Nereo Bresolin

Section IV: Toxin Induced Mitochondrial Dysfunction

Toxin-Induced Mitochondrial Dysfunction
Susan E. Browne and M. Flint Beal

Section V: Neurodegenerative Disorders

Parkinson's Disease
L. V. P. Korlipara and A. H. V. Schapira

Huntington's Disease: The Mystery Unfolds?
Åsa Petersén and Patrik Brundin

Mitochondria in Alzheimer's Disease
Russell H. Swerdlow and Stephen J. Kish

Contributions of Mitochondrial Alterations, Resulting from Bad Genes and a Hostile Environment, to the Pathogenesis of Alzheimer's Disease
Mark P. Mattson

Mitochondria and Amyotrophic Lateral Sclerosis
Richard W. Orrell and Anthony H. V. Schapira

Section VI: Models of Mitochondrial Disease

Models of Mitochondrial Disease
Danae Liolitsa and Michael G. Hanna

Section VII: Defects of β Oxidation Including Carnitine Deficiency

Defects of β Oxidation Including Carnitine Deficiency
K. Bartlett and M. Pourfarzam

Section VIII: Mitochondrial Involvement in Aging

The Mitochondrial Theory of Aging: Involvement of Mitochondrial DNA Damage and Repair
Nadja C. de Souza-Pinto and Vilhelm A. Bohr

INDEX

Volume 54

Unique General Anesthetic Binding Sites Within Distinct Conformational States of the Nicotinic Acetylcholine Receptor
Hugo R. Ariaas, William, R. Kem, James R. Truddell, and Michael P. Blanton

Signaling Molecules and Receptor Transduction Cascades That Regulate NMDA Receptor-Mediated Synaptic Transmission
Suhas. A. Kotecha and John F. MacDonald

Behavioral Measures of Alcohol Self-Administration and Intake Control: Rodent Models
Herman H. Samson and Cristine L. Czachowski

Dopaminergic Mouse Mutants: Investigating the Roles of the Different Dopamine Receptor Subtypes and the Dopamine Transporter
Shirlee Tan, Bettina Hermann, and Emiliana Borrelli

Drosophila melanogaster, A Genetic Model System for Alcohol Research
Douglas J. Guarnieri and Ulrike Heberlein

INDEX

Volume 55

Section I: Virsu Vectors For Use in the Nervous System

Non-Neurotropic Adenovirus: a Vector for Gene Transfer to the Brain and Gene Therapy of Neurological Disorders
P. R. Lowenstein, D. Suwelack, J. Hu, X. Yuan, M. Jimenez-Dalmaroni, S. Goverdhama, and M.G. Castro

Adeno-Associated Virus Vectors
E. Lehtonen and L. Tenenbaum

Problems in the Use of Herpes Simplex Virus as a Vector
L. T. Feldman

Lentiviral Vectors
J. Jakobsson, C. Ericson, N. Rosenquist, and C. Lundberg

Retroviral Vectors for Gene Delivery to Neural Precursor Cells
K. Kageyama, H. Hirata, and J. Hatakeyama

Section II: Gene Therapy with Virus Vectors for Specific Disease of the Nervous System

The Principles of Molecular Therapies for Glioblastoma
G. Karpati and J. Nalbatonglu

Oncolytic Herpes Simplex Virus
J. C. C. Hu and R. S. Coffin

Recombinant Retrovirus Vectors for Treatment of Brain Tumors
N. G. Rainov and C. M. Kramm

Adeno-Associated Viral Vectors for Parkinson's Disease
I. Muramatsu, L. Wang, K. Ikeguchi, K-i Fujimoto, T. Okada, H. Mizukami, Y. Hanazono, A. Kume, I. Nakano, and K. Ozawa

HSV Vectors for Parkinson's Disease
D. S. Latchman

Gene Therapy for Stroke
K. Abe and W. R. Zhang

Gene Therapy for Mucopolysaccharidosis
A. Bosch and J. M. Heard

INDEX

Volume 56

Behavioral Mechanisms and the Neurobiology of Conditioned Sexual Responding
Mark Krause

NMDA Receptors in Alcoholism
Paula L. Hoffman

Processing and Representation of Species-Specific Communication Calls in the Auditory System of Bats
George D. Pollak, Achim Klug, and Eric E. Bauer

Central Nervous System Control of Micturition
Gert Holstege and Leonora J. Mouton

The Structure and Physiology of the Rat Auditory System: An Overview
Manuel Malmierca

Neurobiology of Cat and Human Sexual Behavior
Gert Holstege and J. R. Georgiadis

INDEX

Volume 57

Cumulative Subject Index of Volumes 1–25

Volume 58

Cumulative Subject Index of Volumes 26–50

Volume 59

Loss of Spines and Neuropil
Liesl B. Jones

Schizophrenia as a Disorder of Neuroplasticity
Robert E. McCullumsmith, Sarah M. Clinton, and James H. Meador-Woodruff

The Synaptic Pathology of Schizophrenia: Is Aberrant Neurodevelopment and Plasticity to Blame?
Sharon L. Eastwood

Neurochemical Basis for an Epigenetic Vision of Synaptic Organization
E. Costa, D. R. Grayson, M. Veldic, and A. Guidotti

Muscarinic Receptors in Schizophrenia: Is There a Role for Synaptic Plasticity?
Thomas J. Raedler

Serotonin and Brain Development
Monsheel S. K. Sodhi and Elaine Sanders-Bush

Presynaptic Proteins and Schizophrenia
William G. Honer and Clint E. Young

Mitogen-Activated Protein Kinase Signaling
Svetlana V. Kyosseva

Postsynaptic Density Scaffolding Proteins at Excitatory Synapse and Disorders of Synaptic Plasticity: Implications for Human Behavior Pathologies
Andrea de Bartolomeis and Germano Fiore

Prostaglandin-Mediated Signaling in Schizophrenia
S. Smesny

Mitochondria, Synaptic Plasticity, and Schizophrenia
Dorit Ben-Shachar and Daphna Laifenfeld

Membrane Phospholipids and Cytokine Interaction in Schizophrenia
Jeffrey K. Yao and Daniel P. van Kammen

Neurotensin, Schizophrenia, and Antipsychotic Drug Action
Becky Kinkead and Charles B. Nemeroff

Schizophrenia, Vitamin D, and Brain Development
Alan Mackay-Sim, François FÉron, Darryl Eyles, Thomas Burne, and John McGrath

Possible Contributions of Myelin and Oligodendrocyte Dysfunction to Schizophrenia
Daniel G. Stewart and Kenneth L. Davis

Brain-Derived Neurotrophic Factor and the Plasticity of the Mesolimbic Dopamine Pathway
Oliver Guillin, Nathalie Griffon, Jorge Diaz, Bernard Le Foll, Erwan Bezard, Christian Gross, Chris Lammers, Holger Stark, Patrick Carroll, Jean-Charles Schwartz, and Pierre Sokoloff

S100B in Schizophrenic Psychosis
Matthias Rothermundt, Gerald Ponath, and Volker Arolt

Oct-6 Transcription Factor
Maria Ilia

NMDA Receptor Function, Neuroplasticity, and the Pathophysiology of Schizophrenia
Joseph T. Coyle and Guochuan Tsai

INDEX

Volume 60

Microarray Platforms: Introduction and Application to Neurobiology
Stanislav L. Karsten, Lili C. Kudo, and Daniel H. Geschwind

Experimental Design and Low-Level Analysis of Microarray Data
B. M. Bolstad, F. Collin, K. M. Simpson, R. A. Irizarry, and T. P. Speed

Brain Gene Expression: Genomics and Genetics
Elissa J. Chesler and Robert W. Williams

DNA Microarrays and Animal Models of Learning and Memory
Sebastiano Cavallaro

Microarray Analysis of Human Nervous System Gene Expression in Neurological Disease
Steven A. Greenberg

DNA Microarray Analysis of Postmortem Brain Tissue
Károly Mirnics, Pat Levitt, and David A. Lewis

INDEX

Volume 61

Section I: High-Throughput Technologies

Biomarker Discovery Using Molecular Profiling Approaches
Stephen J. Walker and Arron Xu

Proteomic Analysis of Mitochondrial Proteins
Mary F. Lopez, Simon Melov, Felicity Johnson, Nicole Nagulko, Eva Golenko, Scott Kuzdzal, Suzanne Ackloo, and Alvydas Mikulskis

Section II: Proteomic Applications

NMDA Receptors, Neural Pathways, and Protein Interaction Databases
Holger Husi

Dopamine Transporter Network and Pathways
Rajani Maiya and R. Dayne Mayfield

Proteomic Approaches in Drug Discovery and Development
Holly D. Soares, Stephen A. Williams, Peter J. Snyder, Feng Gao, Tom Stiger, Christian Rohlff, Athula Herath, Trey Sunderland, Karen Putnam, and W. Frost White

Section III: Informatics

Proteomic Informatics
Steven Russell, William Old, Katheryn Resing, and Lawrence Hunter

Section IV: Changes in the Proteome by Disease

Proteomics Analysis in Alzheimer's Disease: New Insights into Mechanisms of Neurodegeneration
D. Allan Butterfield and Debra Boyd-Kimball

Proteomics and Alcoholism
Frank A. Witzmann and Wendy N. Strother

Proteomics Studies of Traumatic Brain Injury
Kevin K. W. Wang, Andrew Ottens, William Haskins, Ming Cheng Liu, Firas Kobeissy, Nancy Denslow, SuShing Chen, and Ronald L. Hayes

Influence of Huntington's Disease on the Human and Mouse Proteome
Claus Zabel and Joachim Klose

Section V: Overview of the Neuroproteome

Proteomics—Application to the Brain
Katrin Marcus, Oliver Schmidt, Heike Schaefer, Michael Hamacher, AndrÅ van Hall, and Helmut E. Meyer

INDEX

Volume 62

$GABA_A$ Receptor Structure–Function Studies: A Reexamination in Light of New Acetylcholine Receptor Structures
Myles H. Akabas

Dopamine Mechanisms and Cocaine Reward
Aiko Ikegami and Christine L. Duvauchelle

Proteolytic Dysfunction in Neurodegenerative Disorders
Kevin St. P. McNaught

Neuroimaging Studies in Bipolar Children and Adolescents
Rene L. Olvera, David C. Glahn, Sheila C. Caetano, Steven R. Pliszka, and Jair C. Soares

Chemosensory G-Protein-Coupled Receptor Signaling in the Brain
Geoffrey E. Woodard

Disturbances of Emotion Regulation after Focal Brain Lesions
Antoine Bechara

The Use of *Caenorhabditis elegans* in Molecular Neuropharmacology
Jill C. Bettinger, Lucinda Carnell, Andrew G. Davies, and Steven L. McIntire

INDEX

Volume 63

Mapping Neuroreceptors at work: On the Definition and Interpretation of Binding Potentials after 20 years of Progress
Albert Gjedde, Dean F. Wong, Pedro Rosa-Neto, and Paul Cumming

Mitochondrial Dysfunction in Bipolar Disorder: From ^{31}P-Magnetic Resonance Spectroscopic Findings to Their Molecular Mechanisms
Tadafumi Kato

Large-Scale Microarray Studies of Gene Expression in Multiple Regions of the Brain in Schizophrenia and Alzeimer's Disease
Pavel L. Katsel, Kenneth L. Davis, and Vahram Haroutunian

Regulation of Serotonin 2C Receptor PRE-mRNA Editing By Serotonin
Claudia Schmauss

The Dopamine Hypothesis of Drug Addiction: Hypodopaminergic State
Miriam Melis, Saturnino Spiga, and Marco Diana

Human and Animal Spongiform Encephalopathies are Autoimmune Diseases: A Novel Theory and Its supporting Evidence
Bao Ting Zhu

Adenosine and Brain Function
Bertil B. Fredholm, Jiang-Fan Chen, Rodrigo A. Cunha, Per Svenningsson, and Jean-Marie Vaugeois

INDEX

Volume 64

Section I. The Cholinergic System
John Smythies

Section II. The Dopamine System
John Symythies

Section III. The Norepinephrine System
John Smythies

Section IV. The Adrenaline System
John Smythies

Section V. Serotonin System
John Smythies

INDEX

Volume 65

Insulin Resistance: Causes and Consequences
Zachary T. Bloomgarden

Antidepressant-Induced Manic Conversion: A Developmentally Informed Synthesis of the Literature
Christine J. Lim, James F. Leckman, Christopher Young, and AndrÉs Martin

Sites of Alcohol and Volatile Anesthetic Action on Glycine Receptors
Ingrid A. Lobo and R. Adron Harris

Role of the Orbitofrontal Cortex in Reinforcement Processing and Inhibitory Control: Evidence from Functional Magnetic Resonance Imaging Studies in Healthy Human Subjects
Rebecca Elliott and Bill Deakin

Common Substrates of Dysphoria in Stimulant Drug Abuse and Primary Depression: Therapeutic Targets
Kate Baicy, Carrie E. Bearden, John Monterosso, Arthur L. Brody, Andrew J. Isaacson, and Edythe D. London

The Role of cAMP Response Element–Binding Proteins in Mediating Stress-Induced Vulnerability to Drug Abuse
Arati Sadalge Kreibich and Julie A. Blendy

G-Protein–Coupled Receptor Deorphanizations
Yumiko Saito and Olivier Civelli

Mechanistic Connections Between Glucose/ Lipid Disturbances and Weight Gain Induced by Antipsychotic Drugs
Donard S. Dwyer, Dallas Donohoe, Xiao-Hong Lu, and Eric J. Aamodt

Serotonin Firing Activity as a Marker for Mood Disorders: Lessons from Knockout Mice
Gabriella Gobbi

INDEX

Volume 66

Brain Atlases of Normal and Diseased Populations
Arthur W. Toga and Paul M. Thompson

Neuroimaging Databases as a Resource for Scientific Discovery
John Darrell Van Horn, John Wolfe, Autumn Agnoli, Jeffrey Woodward, Michael Schmitt, James Dobson, Sarene Schumacher, and Bennet Vance

Modeling Brain Responses
Karl J. Friston, William Penny, and Olivier David

Voxel-Based Morphometric Analysis Using Shape Transformations
Christos Davatzikos

The Cutting Edge of fMRI and High-Field fMRI
Dae-Shik Kim

Quantification of White Matter Using Diffusion-Tensor Imaging
Hae-Jeong Park

Perfusion fMRI for Functional Neuroimaging
Geoffrey K. Aguirre, John A. Detre, and Jiongjiong Wang

Functional Near-Infrared Spectroscopy: Potential and Limitations in Neuroimaging Studies
Yoko Hoshi

Neural Modeling and Functional Brain Imaging: The Interplay Between the Data-Fitting and Simulation Approaches
Barry Horwitz and Michael F. Glabus

Combined EEG and fMRI Studies of Human Brain Function
V. Menon and S. Crottaz-Herbette

INDEX

Volume 67

Distinguishing Neural Substrates of Heterogeneity Among Anxiety Disorders
Jack B. Nitschke and Wendy Heller

Neuroimaging in Dementia
K. P. Ebmeier, C. Donaghey, and N. J. Dougall

Prefrontal and Anterior Cingulate Contributions to Volition in Depression
Jack B. Nitschke and Kristen L. Mackiewicz

Functional Imaging Research in Schizophrenia
H. Tost, G. Ende, M. Ruf, F. A. Henn, and A. Meyer-Lindenberg

Neuroimaging in Functional Somatic Syndromes
Patrick B. Wood

Neuroimaging in Multiple Sclerosis
Alireza Minagar, Eduardo Gonzalez-Toledo, James Pinkston, and Stephen L. Jaffe

Stroke
Roger E. Kelley and Eduardo Gonzalez-Toledo

Functional MRI in Pediatric Neurobehavioral Disorders
Michael Seyffert and F. Xavier Castellanos

Structural MRI and Brain Development
Paul M. Thompson, Elizabeth R. Sowell, Nitin Gogtay, Jay N. Giedd, Christine N. Vidal, Kiralee M. Hayashi, Alex Leow, Rob Nicolson, Judith L. Rapoport, and Arthur W. Toga

Neuroimaging and Human Genetics
Georg Winterer, Ahmad R. Hariri, David Goldman, and Daniel R. Weinberger

Neuroreceptor Imaging in Psychiatry: Theory and Applications
W. Gordon Frankle, Mark Slifstein, Peter S. Talbot, and Marc Laruelle

INDEX

Volume 68

Fetal Magnetoencephalography: Viewing the Developing Brain *In Utero*
Hubert Preissl, Curtis L. Lowery, and Hari Eswaran

Magnetoencephalography in Studies of Infants and Children
Minna Huotilainen

Let's Talk Together: Memory Traces Revealed by Cooperative Activation in the Cerebral Cortex
Jochen Kaiser, Susanne Leiberg, and Werner Lutzenberger

Human Communication Investigated With Magnetoencephalography: Speech, Music, and Gestures
Thomas R. Knösche, Burkhard Maess, Akinori Nakamura, and Angela D. Friederici

Combining Magnetoencephalography and Functional Magnetic Resonance Imaging
Klaus Mathiak and Andreas J. Fallgatter

Beamformer Analysis of MEG Data
Arjan Hillebrand and Gareth R. Barnes

Functional Connectivity Analysis in Magnetoencephalography
Alfons Schnitzler and Joachim Gross

Human Visual Processing as Revealed by Magnetoencephalographys
Yoshiki Kaneoke, Shoko Watanabe, and Ryusuke Kakigi

A Review of Clinical Applications of Magnetoencephalography
Andrew C. Papanicolaou, Eduardo M. Castillo, Rebecca Billingsley-Marshall, Ekaterina Pataraia, and Panagiotis G. Simos

INDEX

Volume 69

Nematode Neurons: Anatomy and Anatomical Methods in Caenorhabditis elegans
David H. Hall, Robyn Lints, and Zeynep Altun

Investigations of Learning and Memory in Caenorhabditis elegans
Andrew C. Giles, Jacqueline K. Rose, and Catharine H. Rankin

Neural Specification and Differentiation
Eric Aamodt and Stephanie Aamodt

Sexual Behavior of the Caenorhabditis elegans Male
Scott W. Emmons

The Motor Circuit
Stephen E. Von Stetina, Millet Treinin, and David M. Miller III

Mechanosensation in Caenorhabditis elegans
Robert O'Hagan and Martin Chalfie

Volume 70

Spectral Processing by the Peripheral Auditory System Facts and Models
Enrique A. Lopez-Poveda

Basic Psychophysics of Human Spectral Processing
Brian C. J. Moore

Across-Channel Spectral Processing
John H. Grose, Joseph W. Hall III, and Emily Buss

Speech and Music Have Different Requirements for Spectral Resolution
Robert V. Shannon

Non-Linearities and the Representation of Auditory Spectra
Eric D. Young, Jane J. Yu, and Lina A. J. Reiss

Spectral Processing in the Inferior Colliculus
Kevin A. Davis

Neural Mechanisms for Spectral Analysis in the Auditory Midbrain, Thalamus, and Cortex
Monty A. Escabí and Heather L. Read

Spectral Processing in the Auditory Cortex
Mitchell L. Sutter

Processing of Dynamic Spectral Properties of Sounds
Adrian Rees and Manuel S. Malmierca

Representations of Spectral Coding in the Human Brain
Deborah A. Hall, PhD

Spectral Processing and Sound Source Determination
Donal G. Sinex

Spectral Information in Sound Localization
Simon Carlile, Russell Martin, and Ken McAnally

Plasticity of Spectral Processing
Dexter R. F. Irvine and Beverly A. Wright

Spectral Processing In Cochlear Implants
Colette M. McKay

INDEX

Volume 71

Autism: Neuropathology, Alterations of the GABAergic System, and Animal Models
Christoph Schmitz, Imke A. J. van Kooten, Patrick R. Hof, Herman van Engeland, Paul H. Patterson, and Harry W. M. Steinbusch

The Role of GABA in the Early Neuronal Development
Marta Jelitai and Emı'lia Madarasz

GABAergic Signaling in the Developing Cerebellum
Chitoshi Takayama

Insights into GABA Functions in the Developing Cerebellum
Mo'nica L. Fiszman

Role of GABA in the Mechanism of the Onset of Puberty in Non-Human Primates
Ei Terasawa

Rett Syndrome: A Rosetta Stone for Understanding the Molecular Pathogenesis of Autism
Janine M. LaSalle, Amber Hogart, and Karen N. Thatcher

GABAergic Cerebellar System in Autism: A Neuropathological and Developmental Perspective
Gene J. Blatt

Reelin Glycoprotein in Autism and Schizophrenia
S. Hossein Fatemi

Is There A Connection Between Autism, Prader-Willi Syndrome, Catatonia, and GABA?
Dirk M. Dhossche, Yaru Song, and Yiming Liu

Alcohol, GABA Receptors, and Neurodevelopmental Disorders
Ujjwal K. Rout

Effects of Secretin on Extracellular GABA and Other Amino Acid Concentrations in the Rat Hippocampus
Hans-Willi Clement, Alexander Pschibul, and Eberhard Schulz

Predicted Role of Secretin and Oxytocin in the Treatment of Behavioral and Developmental Disorders: Implications for Autism
Martha G. Welch and David A. Ruggiero

Immunological Findings in Autism
Hari Har Parshad Cohly and Asit Panja

Correlates of Psychomotor Symptoms in Autism
Laura Stoppelbein, Sara Sytsma-Jordan, and Leilani Greening

GABRB3 Gene Deficient Mice: A Potential Model of Autism Spectrum Disorder
Timothy M. DeLorey

The Reeler Mouse: Anatomy of a Mutant
Gabriella D'Arcangelo

Shared Chromosomal Susceptibility Regions Between Autism and Other Mental Disorders
Yvon C. Chagnon index

INDEX

Volume 72

Classification Matters for Catatonia and Autism in Children
Klaus-Jürgen Neumärker

A Systematic Examination of Catatonia-Like Clinical Pictures in Autism Spectrum Disorders
Lorna Wing and Amitta Shah

Catatonia in Individuals with Autism Spectrum Disorders in Adolescence and Early Adulthood: A Long-Term Prospective Study
Masataka Ohta, Yukiko Kano, and Yoko Nagai

Are Autistic and Catatonic Regression Related? A Few Working Hypotheses Involving GABA, Purkinje Cell Survival, Neurogenesis, and ECT
Dirk Marcel Dhossche and Ujjwal Rout

Psychomotor Development and Psychopathology in Childhood
Dirk M. J. De Raeymaecker

The Importance of Catatonia and Stereotypies in Autistic Spectrum Disorders
Laura Stoppelbein, Leilani Greening, and Angelina Kakooza

Prader–Willi Syndrome: Atypical Psychoses and Motor Dysfunctions
Willem M. A. Verhoeven and Siegfried Tuinier

Towards a Valid Nosography and Psychopathology of Catatonia in Children and Adolescents
David Cohen

Is There a Common Neuronal Basis for Autism and Catatonia?
Dirk Marcel Dhossche, Brendan T. Carroll, and Tressa D. Carroll

Shared Susceptibility Region on Chromosome 15 Between Autism and Catatonia
Yvon C. Chagnon

Current Trends in Behavioral Interventions for Children with Autism
Dorothy Scattone and Kimberly R. Knight

Case Reports with a Child Psychiatric Exploration of Catatonia, Autism, and Delirium
Jan N. M. Schieveld

ECT and the Youth: Catatonia in Context
Frank K. M. Zaw

Catatonia in Autistic Spectrum Disorders: A Medical Treatment Algorithm
Max Fink, Michael A. Taylor, and Neera Ghaziuddin

Psychological Approaches to Chronic Catatonia-Like Deterioration in Autism Spectrum Disorders
Amitta Shah and Lorna Wing

Section V: Blueprints

Blueprints for the Assessment, Treatment, and Future Study of Catatonia in Autism Spectrum Disorders
Dirk Marcel, Dhossche, Amitta Shah, and Lorna Wing

INDEX

Volume 73

Chromosome 22 Deletion Syndrome and Schizophrenia
Nigel M. Williams, Michael C. O'Donovan, and Michael J. Owen

Characterization of Proteome of Human Cerebrospinal Fluid
Jing Xu, Jinzhi Chen, Elaine R. Peskind, Jinghua Jin, Jimmy Eng, Catherine Pan, Thomas J. Montine, David R. Goodlett, and Jing Zhang

Hormonal Pathways Regulating Intermale and Interfemale Aggression
Neal G. Simon, Qianxing Mo, Shan Hu, Carrie Garippa, and Shi-Fang Lu

Neuronal GAP Junctions: Expression, Function, and Implications for Behavior
Clinton B. McCracken and David C. S. Roberts

Effects of Genes and Stress on the Neurobiology of Depression
J. John Mann and Dianne Currier

Quantitative Imaging with the Micropet Small-Animal Pet Tomograph
Paul Vaska, Daniel J. Rubins, David L. Alexoff, and Wynne K. Schiffer

Understanding Myelination through Studying its Evolution
Rüdiger Schweigreiter, Betty I. Roots, Christine Bandtlow, and Robert M. Gould

INDEX

Volume 74

Evolutionary Neurobiology and Art
C. U. M. Smith

Section I: Visual Aspects

Perceptual Portraits
Nicholas Wade

The Neuropsychology of Visual Art: Conferring Capacity
Anjan Chatterjee

Vision, Illusions, and Reality
Christopher Kennard

Localization in the Visual Brain
George K. York

Section II: Episodic Disorders

Neurology, Synaesthesia, and Painting
Amy Ione

Fainting in Classical Art
Philip Smith

Migraine Art in the Internet: A Study of 450 Contemporary Artists
Klaus Podoll

Sarah Raphael's Migraine with Aura as Inspiration for the Foray of Her Work into Abstraction
Klaus Podoll and Debbie Ayles

The Visual Art of Contemporary Artists with Epilepsy
Steven C. Schachter

Section III: Brain Damage

Creativity in Painting and Style in Brain-Damaged Artists
Julien Bogousslavsky

Artistic Changes in Alzheimer's Disease
Sebastian J. Crutch and Martin N. Rossor

Section IV: Cerebrovascular Disease

Stroke in Painters
H. Bäzner and M. Hennerici

Visuospatial Neglect in Lovis Corinth's Self-Portraits
Olaf Blanke

Art, Constructional Apraxia, and the Brain
Louis Caplan

Section V: Genetic Diseases

Neurogenetics in Art
Alan E. H. Emery

A Naïve Artist of St Ives
F. Clifford Rose

Van Gogh's Madness
F. Clifford Rose

Absinthe, The Nervous System and Painting
Tiina Rekand

Section VI: Neurologists as Artists

Sir Charles Bell, KGH, FRS, FRSE (1774–1842)
Christopher Gardner-Thorpe

Section VII: Miscellaneous

Peg Leg Frieda
Espen Dietrichs

The Deafness of Goya (1746–1828)
F. Clifford Rose

INDEX

Volume 75

Introduction on the Use of the *Drosophila* Embryonic/Larval Neuromuscular Junction as a Model System to Study Synapse Development and Function, and a Brief Summary of Pathfinding and Target Recognition
Catalina Ruiz-Cañada and Vivian Budnik

Development and Structure of Motoneurons
Matthias Landgraf and Stefan Thor

The Development of the *Drosophila* Larval Body Wall Muscles
Karen Beckett and Mary K. Baylies

Organization of the Efferent System and Structure of Neuromuscular Junctions in *Drosophila*
Andreas Prokop

Development of Motoneuron Electrical Properties and Motor Output
Richard A. Baines

Transmitter Release at the Neuromuscular Junction
Thomas L. Schwarz

Vesicle Trafficking and Recycling at the Neuromuscular Junction: Two Pathways for Endocytosis
Yoshiaki Kidokoro

Glutamate Receptors at the *Drosophila* Neuromuscular Junction
Aaron DiAntonio

Scaffolding Proteins at the *Drosophila* Neuromuscular Junction
Bulent Ataman, Vivian Budnik, and Ulrich Thomas

Synaptic Cytoskeleton at the Neuromuscular Junction
Catalina Ruiz-Cañada and Vivian Budnik

Plasticity and Second Messengers During Synapse Development
Leslie C. Griffith and Vivian Budnik

Retrograde Signaling that Regulates Synaptic Development and Function at the *Drosophila* Neuromuscular Junction
Guillermo Marqués and Bing Zhang

Activity-Dependent Regulation of Transcription During Development of Synapses
Subhabrata Sanyal and Mani Ramaswami

Experience-Dependent Potentiation of Larval Neuromuscular Synapses
Christoph M. Schuster

Selected Methods for the Anatomical Study of *Drosophila* Embryonic and Larval Neuromuscular Junctions
Vivian Budnik, Michael Gorczyca, and Andreas Prokop

INDEX

Volume 76

Section I: Physiological Correlates of Freud's Theories

The ID, the Ego, and the Temporal Lobe
Shirley M. Ferguson and Mark Rayport

ID, Ego, and Temporal Lobe Revisited
Shirley M. Ferguson and Mark Rayport

Section II: Stereotaxic Studies

Olfactory Gustatory Responses Evoked by Electrical Stimulation of Amygdalar Region in Man Are Qualitatively Modifiable by Interview Content: Case Report and Review
Mark Rayport, Sepehr Sani, and Shirley M. Ferguson

Section III: Controversy in Definition of Behavioral Disturbance

Pathogenesis of Psychosis in Epilepsy. The "Seesaw" Theory: Myth or Reality?
Shirley M. Ferguson and Mark Rayport

Section IV: Outcome of Temporal Lobectomy

Memory Function After Temporal Lobectomy for Seizure Control: A Comparative Neuropsy chiatric and Neuropsychological Study
Shirley M. Ferguson, A. John McSweeny, and Mark Rayport

Life After Surgery for Temporolimbic Seizures
Shirley M. Ferguson, Mark Rayport, and Carolyn A. Schell

Appendix I
Mark Rayport

Appendix II: Conceptual Foundations of Studies of Patients Undergoing Temporal Lobe Surgery for Seizure Control
Mark Rayport

INDEX

Volume 77

Regenerating the Brain
David A. Greenberg and Kunlin Jin

Serotonin and Brain: Evolution, Neuroplasticity, and Homeostasis
Efrain C. Azmitia

Therapeutic Approaches to Promoting Axonal Regeneration in the Adult Mammalian Spinal Cord
Sari S. Hannila, Mustafa M. Siddiq, and Marie T. Filbin

Evidence for Neuroprotective Effects of Antipsychotic Drugs: Implications for the Pathophysiology and Treatment of Schizophrenia
Xin-Min Li and Haiyun Xu

Neurogenesis and Neuroenhancement in the Pathophysiology and Treatment of Bipolar Disorder
Robert J. Schloesser, Guang Chen, and Husseini K. Manji

Neuroreplacement, Growth Factor, and Small Molecule Neurotrophic Approaches for Treating Parkinson's Disease
Michael J. O'Neill, Marcus J. Messenger, Viktor Lakics, Tracey K. Murray, Eric H. Karran, Philip G. Szekeres, Eric S. Nisenbaum, and Kalpana M. Merchant

Using *Caenorhabditis elegans* Models of Neurodegenerative Disease to Identify Neuroprotective Strategies
Brian Kraemer and Gerard D. Schellenberg

Neuroprotection and Enhancement of Neurite Outgrowth With Small Molecular Weight Compounds From Screens of Chemical Libraries
Donard S. Dwyer and Addie Dickson

INDEX

Volume 78

Neurobiology of Dopamine in Schizophrenia
Olivier Guillin, Anissa Abi-Dargham, and Marc Laruelle

The Dopamine System and the Pathophysiology of Schizophrenia: A Basic Science Perspective
Yukiori Goto and Anthony A. Grace

Glutamate and Schizophrenia: Phencyclidine, N-methyl-D-aspartate Receptors, and Dopamine–Glutamate Interactions
Daniel C. Javitt

Deciphering the Disease Process of Schizophrenia: The Contribution of Cortical GABA Neurons
David A. Lewis and Takanori Hashimoto

Alterations of Serotonin Transmission in Schizophrenia
Anissa Abi-Dargham

Serotonin and Dopamine Interactions in Rodents and Primates: Implications for Psychosis and Antipsychotic Drug Development
Gerard J. Marek

Cholinergic Circuits and Signaling in the Pathophysiology of Schizophrenia
Joshua A. Berman, David A. Talmage, and Lorna W. Role

Schizophrenia and the α7 Nicotinic Acetylcholine Receptor
Laura F. Martin and Robert Freedman

Histamine and Schizophrenia
Jean-Michel Arrang

Cannabinoids and Psychosis
Deepak Cyril D'Souza

Involvement of Neuropeptide Systems in Schizophrenia: Human Studies
Ricardo Cáceda, Becky Kinkead, and Charles B. Nemeroff

Brain-Derived Neurotrophic Factor in Schizophrenia and Its Relation with Dopamine
Olivier Guillin, Caroline Demily, and Florence Thibaut

Schizophrenia Susceptibility Genes: In Search of a Molecular Logic and Novel Drug Targets for a Devastating Disorder
Joseph A. Gogos

INDEX

Volume 79

The Destructive Alliance: Interactions of Leukocytes, Cerebral Endothelial Cells, and the Immune Cascade in Pathogenesis of Multiple Sclerosis
Alireza Minagar, April Carpenter, and J. Steven Alexander

Role of B Cells in Pathogenesis of Multiple Sclerosis
Behrouz Nikbin, Mandana Mohyeddin Bonab, Farideh Khosravi, and Fatemeh Talebian

The Role of CD4 T Cells in the Pathogenesis of Multiple Sclerosis
Tanuja Chitnis

The CD8 T Cell in Multiple Sclerosis: Suppressor Cell or Mediator of Neuropathology?
Aaron J. Johnson, Georgette L. Suidan, Jeremiah McDole, and Istvan Pirko

Immunopathogenesis of Multiple Sclerosis
Smriti M. Agrawal and V. Wee Yong

Molecular Mimicry in Multiple Sclerosis
Jane E. Libbey, Lori L. McCoy, and Robert S. Fujinami

Molecular "Negativity" May Underlie Multiple Sclerosis: Role of the Myelin Basic Protein Family in the Pathogenesis of MS
Abdiwahab A. Musse and George Harauz

Microchimerism and Stem Cell Transplantation in Multiple Sclerosis
Behrouz Nikbin, Mandana Mohyeddin Bonab, and Fatemeh Talebian

The Insulin-Like Growth Factor System in Multiple Sclerosis
Daniel Chesik, Nadine Wilczak, and Jacques De Keyser

Cell-Derived Microparticles and Exosomes in Neuroinflammatory Disorders
Lawrence L. Horstman, Wenche Jy, Alireza Minagar, Carlos J. Bidot, Joaquin J. Jimenez, J. Steven Alexander, and Yeon S. Ahn

Multiple Sclerosis in Children: Clinical, Diagnostic, and Therapeutic Aspects
Kevin Rostásy

Migraine in Multiple Sclerosis
Debra G. Elliott

Multiple Sclerosis as a Painful Disease
Meghan Kenner, Uma Menon, and Debra Elliott

Multiple Sclerosis and Behavior
James B. Pinkston, Anita Kablinger, and Nadejda Alekseeva

Cerebrospinal Fluid Analysis in Multiple Sclerosis
Francisco A. Luque and Stephen L. Jaffe

Multiple Sclerosis in Isfahan, Iran
Mohammad Saadatnia, Masoud Etemadifar, and Amir Hadi Maghzi

Gender Issues in Multiple Sclerosis
Robert N. Schwendimann and Nadejda Alekseeva

Differential Diagnosis of Multiple Sclerosis
Halim Fadil, Roger E. Kelley, and Eduardo Gonzalez-Toledo

Prognostic Factors in Multiple Sclerosis
Roberto Bergamaschi

Neuroimaging in Multiple Sclerosis
Robert Zivadinov and Jennifer L. Cox

Detection of Cortical Lesions Is Dependent on Choice of Slice Thickness in Patients with Multiple Sclerosis
Ondrej Dolezal, Michael G. Dwyer, Dana Horakova, Eva Havrdova, Alireza Minagar, Srivats Balachandran, Niels Bergsland, Zdenek Seidl, Manuela Vaneckova, David Fritz, Jan Krasensky, and Robert Zivadinov

The Role of Quantitative Neuroimaging Indices in the Differentiation of Ischemia from Demyelination: An Analytical Study with Case Presentation
Romy Hoque, Christina Ledbetter, Eduardo Gonzalez-Toledo, Vivek Misra, Uma Menon, Meghan Kenner, Alejandro A. Rabinstein, Roger E. Kelley, Robert Zivadinov, and Alireza Minagar

HLA-DRB1*1501, -DQB1*0301, -DQB1*0302, -DQB1*0602, and -DQB1*0603 Alleles Are Associated with More Severe Disease Outcome on MRI in Patients with Multiple Sclerosis
Robert Zivadinov, Laura Uxa, Alessio Bratina, Antonio Bosco, Bhooma Srinivasaraghavan, Alireza Minagar, Maja Ukmar, Su yen Benedetto, and Marino Zorzon

Glatiramer Acetate: Mechanisms of Action in Multiple Sclerosis
Tjalf Ziemssen and Wiebke Schrempf

Evolving Therapies for Multiple Sclerosis
Elena Korniychuk, John M. Dempster, Eileen O'Connor, J. Steven Alexander, Roger E. Kelley, Meghan Kenner, Uma Menon, Vivek Misra, Romy Hoque, Eduardo C. Gonzalez-Toledo, Robert N. Schwendimann, Stacy Smith, and Alireza Minagar

Remyelination in Multiple Sclerosis
Divya M. Chari

Trigeminal Neuralgia: A Modern-Day Review
Kelly Hunt and Ravish Patwardhan

Optic Neuritis and the Neuro-Ophthalmology of Multiple Sclerosis
Paramjit Kaur and Jeffrey L. Bennett

Neuromyelitis Optica: New Findings on Pathogenesis
Dean M. Wingerchuk

INDEX

Volume 80

Epilepsy in the Elderly: Scope of the Problem
Ilo E. Leppik

Animal Models in Gerontology Research
Nancy L. Nadon

Animal Models of Geriatric Epilepsy
Lauren J. Murphree, Lynn M. Rundhaugen, and Kevin M. Kelly

Life and Death of Neurons in the Aging Cerebral Cortex
John H. Morrison and Patrick R. Hof

An In Vitro Model of Stroke-Induced Epilepsy: Elucidation of the Roles of Glutamate and Calcium in the Induction and Maintenance of Stroke-Induced Epileptogenesis
Robert J. DeLorenzo, David A. Sun, Robert E. Blair, and Sompong Sambati

Mechanisms of Action of Antiepileptic Drugs
H. Steve White, Misty D. Smith, and Karen S. Wilcox

Epidemiology and Outcomes of Status Epilepticus in the Elderly
Alan R. Towne

Diagnosing Epilepsy in the Elderly
R. Eugene Ramsay, Flavia M. Macias, and A. James Rowan

Pharmacoepidemiology in Community-Dwelling Elderly Taking Antiepileptic Drugs
Dan R. Berlowitz and Mary Jo V. Pugh

Use of Antiepileptic Medications in Nursing Homes
Judith Garrard, Susan L. Harms, Lynn E. Eberly, and Ilo E. Leppik

Differential Diagnosis of Multiple Sclerosis
Halim Fadil, Roger E. Kelley, and Eduardo Gonzalez-Toledo

Prognostic Factors in Multiple Sclerosis
Roberto Bergamaschi

Neuroimaging in Multiple Sclerosis
Robert Zivadinov and Jennifer L. Cox

Detection of Cortical Lesions Is Dependent on Choice of Slice Thickness in Patients with Multiple Sclerosis
Ondrej Dolezal, Michael G. Dwyer, Dana Horakova, Eva Havrdova, Alireza Minagar, Srivats Balachandran, Niels Bergsland, Zdenek Seidl, Manuela Vaneckova, David Fritz, Jan Krasensky, and Robert Zivadinov

The Role of Quantitative Neuroimaging Indices in the Differentiation of Ischemia from Demyelination: An Analytical Study with Case Presentation
Romy Hoque, Christina Ledbetter, Eduardo Gonzalez-Toledo, Vivek Misra, Uma Menon, Meghan Kenner, Alejandro A. Rabinstein, Roger E. Kelley, Robert Zivadinov, and Alireza Minagar

HLA-DRB1*1501, -DQB1*0301,-DQB1 *0302,-DQB1*0602, and -DQB1*0603 Alleles Are Associated with More Severe Disease Outcome on MRI in Patients with Multiple Sclerosis
Robert Zivadinov, Laura Uxa, Alessio Bratina, Antonio Bosco, Bhooma Srinivasaraghavan, Alireza Minagar, Maja Ukmar, Su yen Benedetto, and Marino Zorzon

Glatiramer Acetate: Mechanisms of Action in Multiple Sclerosis
Tjalf Ziemssen and Wiebke Schrempf

Evolving Therapies for Multiple Sclerosis
Elena Korniychuk, John M. Dempster, Eileen O'Connor, J. Steven Alexander, Roger E. Kelley, Meghan Kenner, Uma Menon, Vivek Misra, Romy Hoque, Eduardo C. Gonzalez-Toledo, Robert N. Schwendimann, Stacy Smith, and Alireza Minagar

Remyelination in Multiple Sclerosis
Divya M. Chari

Trigeminal Neuralgia: A Modern-Day Review
Kelly Hunt and Ravish Patwardhan

Optic Neuritis and the Neuro-Ophthalmology of Multiple Sclerosis
Paramjit Kaur and Jeffrey L. Bennett

Neuromyelitis Optica: New Findings on Pathogenesis
Dean M. Wingerchuk

INDEX

Volume 81

Epilepsy in the Elderly: Scope of the Problem
Ilo E. Leppik

Animal Models in Gerontology Research
Nancy L. Nadon

Animal Models of Geriatric Epilepsy
Lauren J. Murphree, Lynn M. Rundhaugen, and Kevin M. Kelly

Life and Death of Neurons in the Aging Cerebral Cortex
John H. Morrison and Patrick R. Hof

An In Vitro Model of Stroke-Induced Epilepsy: Elucidation of the Roles of Glutamate and Calcium in the Induction and Maintenance of Stroke-Induced Epileptogenesis
Robert J. DeLorenzo, David A. Sun, Robert E. Blair, and Sompong Sambati

Mechanisms of Action of Antiepileptic Drugs
H. Steve White, Misty D. Smith, and Karen S. Wilcox

Epidemiology and Outcomes of Status Epilepticus in the Elderly
Alan R. Towne

Diagnosing Epilepsy in the Elderly
R. Eugene Ramsay, Flavia M. Macias, and A. James Rowan

Pharmacoepidemiology in Community-Dwelling Elderly Taking Antiepileptic Drugs
Dan R. Berlowitz and Mary Jo V. Pugh

Use of Antiepileptic Medications in Nursing Homes
Judith Garrard, Susan L. Harms, Lynn E. Eberly, and Ilo E. Leppik

Age-Related Changes in Pharmacokinetics: Predictability and Assessment Methods
Emilio Perucca

Factors Affecting Antiepileptic Drug Pharmacokinetics in Community-Dwelling Elderly
James C. Cloyd, Susan Marino, and Angela K. Birnbaum

Pharmacokinetics of Antiepileptic Drugs in Elderly Nursing Home Residents
Angela K. Birnbaum

The Impact of Epilepsy on Older Veterans
Mary Jo V. Pugh, Dan R. Berlowitz, and Lewis Kazis

Risk and Predictability of Drug Interactions in the Elderly
René H. Levy and Carol Collins

Outcomes in Elderly Patients With Newly Diagnosed and Treated Epilepsy
Martin J. Brodie and Linda J. Stephen

Recruitment and Retention in Clinical Trials of the Elderly
Flavia M. Macias, R. Eugene Ramsay, and A. James Rowan

Treatment of Convulsive Status Epilepticus
David M. Treiman

Treatment of Nonconvulsive Status Epilepticus
Matthew C. Walker

Antiepileptic Drug Formulation and Treatment in the Elderly: Biopharmaceutical Considerations
Barry E. Gidal

INDEX

Volume 82

Inflammatory Mediators Leading to Protein Misfolding and Uncompetitive/Fast Off-Rate Drug Therapy for Neurodegenerative Disorders
Stuart A. Lipton, Zezong Gu, and Tomohiro Nakamura

Innate Immunity and Protective Neuroinflammation: New Emphasis on the Role of Neuroimmune Regulatory Proteins
M. Griffiths, J. W. Neal, and P. Gasque

Glutamate Release from Astrocytes in Physiological Conditions and in Neurodegenerative Disorders Characterized by Neuroinflammation
Sabino Vesce, Daniela Rossi, Liliana Brambilla, and Andrea Volterra

The High-Mobility Group Box 1 Cytokine Induces Transporter-Mediated Release of Glutamate from Glial Subcellular Particles (Gliosomes) Prepared from *In Situ*-Matured Astrocytes
Giambattista Bonanno, Luca Raiteri, Marco Milanese, Simona Zappettini, Edon Melloni, Marco Pedrazzi, Mario Passalacqua, Carlo Tacchetti, Cesare Usai, and Bianca Sparatore

The Role of Astrocytes and Complement System in Neural Plasticity
Milos Pekny, Ulrika Wilhelmsson, Yalda Rahpeymai Bogestål, and Marcela Pekna

New Insights into the Roles of Metalloproteinases in Neurodegeneration and Neuroprotection
A. J. Turner and N. N. Nalivaeva

Relevance of High-Mobility Group Protein Box 1 to Neurodegeneration
Silvia Fossati and Alberto Chiarugi

Early Upregulation of Matrix Metalloproteinases Following Reperfusion Triggers Neuro-inflammatory Mediators in Brain Ischemia in Rat
Diana Amantea, Rossella Russo, Micaela Gliozzi, Vincenza Fratto, Laura Berliocchi, G. Bagetta, G. Bernardi, and M. Tiziana Corasaniti

The (Endo)Cannabinoid System in Multiple Sclerosis and Amyotrophic Lateral Sclerosis
Diego Centonze, Silvia Rossi, Alessandro Finazzi-Agrò, Giorgio Bernardi, and Mauro Maccarrone

Chemokines and Chemokine Receptors: Multi-purpose Players in Neuroinflammation
Richard M. Ransohoff, LiPing Liu, and Astrid E. Cardona

Systemic and Acquired Immune Responses in Alzheimer's Disease
Markus Britschgi and Tony Wyss-Coray

Neuroinflammation in Alzheimer's Disease and Parkinson's Disease: Are Microglia Pathogenic in Either Disorder?
Joseph Rogers, Diego Mastroeni, Brian Leonard, Jeffrey Joyce, and Andrew Grover

Cytokines and Neuronal Ion Channels in Health and Disease
Barbara Viviani, Fabrizio Gardoni, and Marina Marinovich

Cyclooxygenase-2, Prostaglandin E_2, and Micro-glial Activation in Prion Diseases
Luisa Minghetti and Maurizio Pocchiari

Glia Proinflammatory Cytokine Upregulation as a Therapeutic Target for Neurodegenerative Diseases: Function-Based and Target-Based Discovery Approaches
Linda J. Van Eldik, Wendy L. Thompson, Hantamalala Ralay Ranaivo, Heather A. Behanna, and D. Martin Watterson

Oxidative Stress and the Pathogenesis of Neuro-degenerative Disorders
Ashley Reynolds, Chad Laurie, R. Lee Mosley, and Howard E. Gendelman

Differential Modulation of Type 1 and Type 2 Cannabinoid Receptors Along the Neuroimmune Axis
Sergio Oddi, Paola Spagnuolo, Monica Bari, Antonella D'Agostino, and Mauro Maccarrone

Effects of the HIV-1 Viral Protein Tat on Central Neurotransmission: Role of Group I Meta-botropic Glutamate Receptors
Elisa Neri, Veronica Musante, and Anna Pittaluga

Evidence to Implicate Early Modulation of Inter-leukin-1β Expression in the Neuroprotection Afforded by 17β-Estradiol in Male Rats Under-gone Transient Middle Cerebral Artery Occlusion
Olga Chiappetta, Micaela Gliozzi, Elisa Siviglia, Diana Amantea, Luigi A. Morrone, Laura Berliocchi, G. Bagetta, and M. Tiziana Corasaniti

A Role for Brain Cyclooxygenase-2 and Prosta-glandin-E_2 in Migraine: Effects of Nitroglycerin
Cristina Tassorelli, Rosaria Greco, Marie Therèse Armentero, Fabio Blandini, Giorgio Sandrini, and Giuseppe Nappi

The Blockade of K+-ATP Channels has Neuro-protective Effects in an *In Vitro* Model of Brain Ischemia
Robert Nisticò, Silvia Piccirilli, L. Sebastianelli, Giuseppe Nisticò, G. Bernardi, and N. B. Mercuri

Retinal Damage Caused by High Intraocular Pressure-Induced Transient Ischemia is Prevented by Coenzyme Q10 in Rat
Carlo Nucci, Rosanna Tartaglione, Angelica Cerulli, R. Mancino, A. Spanò, Federica Cavaliere, Laura Rombolà, G. Bagetta, M. Tiziana Corasaniti, and Luigi A. Morrone

Evidence Implicating Matrix Metalloproteinases in the Mechanism Underlying Accumulation of IL-1β and Neuronal Apoptosis in the Neocortex of HIV/gp120-Exposed Rats
Rossella Russo, Elisa Siviglia, Micaela Gliozzi, Diana Amantea, Annamaria Paoletti, Laura Berliocchi, G. Bagetta, and M. Tiziana Corasaniti

Neuroprotective Effect of Nitroglycerin in a Rodent Model of Ischemic Stroke: Evaluation of Bcl-2 Expression
Rosaria Greco, Diana Amantea, Fabio Blandini, Giuseppe Nappi, Giacinto Bagetta, M. Tiziana Corasaniti, and Cristina Tassorelli

INDEX

Volume 83

Gender Differences in Pharmacological Response
Gail D. Anderson

Epidemiology and Classification of Epilepsy: Gender Comparisons
John C. McHugh and Norman Delanty

Hormonal Influences on Seizures: Basic Neurobiology
Cheryl A. Frye

Catamenial Epilepsy
Patricia E. Penovich and Sandra Helmers

Epilepsy in Women: Special Considerations for Adolescents
Mary L. Zupanc and Sheryl Haut

Contraception in Women with Epilepsy: Pharmacokinetic Interactions, Contraceptive Options, and Management
Caryn Dutton and Nancy Foldvary-Schaefer

Reproductive Dysfunction in Women with Epilepsy: Menstrual Cycle Abnormalities, Fertility, and Polycystic Ovary Syndrome
Jürgen Bauer and Déirdre Cooper-Mahkorn

Sexual Dysfunction in Women with Epilepsy: Role of Antiepileptic Drugs and Psychotropic Medications
Mary A. Gutierrez, Romila Mushtaq, and Glen Stimmel

Pregnancy in Epilepsy: Issues of Concern
John DeToledo

Teratogenicity and Antiepileptic Drugs: Potential Mechanisms
Mark S. Yerby

Antiepileptic Drug Teratogenesis: What are the Risks for Congenital Malformations and Adverse Cognitive Outcomes?
Cynthia L. Harden

Teratogenicity of Antiepileptic Drugs: Role of Pharmacogenomics
Raman Sankar and Jason T. Lerner

Antiepileptic Drug Therapy in Pregnancy I: Gestation-Induced Effects on AED Pharmacokinetics
Page B. Pennell and Collin A. Hovinga

Antiepileptic Drug Therapy in Pregnancy II: Fetal and Neonatal Exposure
Collin A. Hovinga and Page B. Pennell

Seizures in Pregnancy: Diagnosis and Management
Robert L. Beach and Peter W. Kaplan

Management of Epilepsy and Pregnancy: An Obstetrical Perspective
Julian N. Robinson and Jane Cleary-Goldman

Pregnancy Registries: Strengths, Weaknesses, and Bias Interpretation of Pregnancy Registry Data
Marianne Cunnington and John Messenheimer

Bone Health in Women With Epilepsy: Clinical Features and Potential Mechanisms
Alison M. Pack and Thaddeus S. Walczak

Metabolic Effects of AEDs: Impact on Body Weight, Lipids and Glucose Metabolism
Raj D. Sheth and Georgia Montouris

Psychiatric Comorbidities in Epilepsy
W. Curt Lafrance, Jr., Andres M. Kanner, and Bruce Hermann

Issues for Mature Women with Epilepsy
Cynthia L. Harden

Pharmacodynamic and Pharmacokinetic Interactions of Psychotropic Drugs with Antiepileptic Drugs
Andres M. Kanner and Barry E. Gidal

Health Disparities in Epilepsy: How Patient-Oriented Outcomes in Women Differ from Men
Frank Gilliam

INDEX

Volume 84

Normal Brain Aging: Clinical, Immunological, Neuropsychological, and Neuroimaging Features
Maria T. Caserta, Yvonne Bannon, Francisco Fernandez, Brian Giunta, Mike R. Schoenberg, and Jun Tan

Subcortical Ischemic Cerebrovascular Dementia
Uma Menon and Roger E. Kelley

Cerebrovascular and Cardiovascular Pathology in Alzheimer's Disease
Jack C. de la Torre

Neuroimaging of Cognitive Impairments in Vascular Disease
Carol Di Perri, Turi O. Dalaker, Mona K. Beyer, and Robert Zivadinov

Contributions of Neuropsychology and Neuroimaging to Understanding Clinical Subtypes of Mild Cognitive Impairment
Amy J. Jak, Katherine J. Bangen, Christina E. Wierenga, Lisa Delano-Wood, Jody Corey-Bloom, and Mark W. Bondi

Proton Magnetic Resonance Spectroscopy in Dementias and Mild Cognitive Impairment
H. Randall Griffith, Christopher C. Stewart, and Jan A. den Hollander

Application of PET Imaging to Diagnosis of Alzheimer's Disease and Mild Cognitive Impairment
James M. Noble and Nikolaos Scarmeas

The Molecular and Cellular Pathogenesis of Dementia of the Alzheimer's Type: An Overview
Francisco A. Luque and Stephen L. Jaffe

Alzheimer's Disease Genetics: Current Status and Future Perspectives
Lars Bertram

Frontotemporal Lobar Degeneration: Insights from Neuropsychology and Neuroimaging
Andrea C. Bozoki and Muhammad U. Farooq

Lewy Body Dementia
Jennifer C. Hanson and Carol F. Lippa

Dementia in Parkinson's Disease
Bradley J. Robottom and William J. Weiner

Early Onset Dementia
Halim Fadil, Aimee Borazanci, Elhachmia Ait Ben Haddou, Mohamed Yahyaoui, Elena Korniychuk, Stephen L. Jaffe, and Alireza Minagar

Normal Pressure Hydrocephalus
Glen R. Finney

Reversible Dementias
Anahid Kabasakalian and Glen R. Finney

INDEX

Volume 85

Involvement of the Prefrontal Cortex in Problem Solving
Hajime Mushiake, Kazuhiro Sakamoto, Naohiro Saito, Toshiro Inui, Kazuyuki Aihara, and Jun Tanji

GluK1 Receptor Antagonists and Hippocampal Mossy Fiber Function
Robert Nisticò, Sheila Dargan, Stephen M. Fitzjohn, David Lodge, David E. Jane, Graham L. Collingridge, and Zuner A. Bortolotto

Monoamine Transporter as a Target Molecule for Psychostimulants
Ichiro Sora, BingJin Li, Setsu Fumushima, Asami Fukui, Yosefu Arime, Yoshiyuki Kasahara, Hiroaki Tomita, and Kazutaka Ikeda

Targeted Lipidomics as a Tool to Investigate Endocannabinoid Function
Giuseppe Astarita, Jennifer Geaga, Faizy Ahmed, and Daniele Piomelli

The Endocannabinoid System as a Target for Novel Anxiolytic and Antidepressant Drugs
Silvana Gaetani, Pasqua Dipasquale, Adele Romano, Laura Righetti, Tommaso Cassano, Daniele Piomelli, and Vincenzo Cuomo

$GABA_A$ Receptor Function and Gene Expression During Pregnancy and Postpartum
Giovanni Biggio, Maria Cristina Mostallino, Paolo Follesa, Alessandra Concas, and Enrico Sanna

Early Postnatal Stress and Neural Circuit Underlying Emotional Regulation
Machiko Matsumoto, Mitsuhiro Yoshioka, and Hiroko Togashi

Roles of the Histaminergic Neurotransmission on Methamphetamine-Induced Locomotor Sensitization and Reward: A Study of Receptors Gene Knockout Mice
Naoko Takino, Eiko Sakurai, Atsuo Kuramasu, Nobuyuki Okamura, and Kazuhiko Yanai

Developmental Exposure to Cannabinoids Causes Subtle and Enduring Neurofunctional Alterations
Patrizia Campolongo, Viviana Trezza, Maura Palmery, Luigia Trabace, and Vincenzo Cuomo

Neuronal Mechanisms for Pain-Induced Aversion: Behavioral Studies Using a Conditioned Place Aversion Test
Masabumi Minami

Bv8/Prokineticins and their Receptors: A New Pronociceptive System
Lucia Negri, Roberta Lattanzi, Elisa Giannini, Michela Canestrelli, Annalisa Nicotra, and Pietro Melchiorri

$P2Y_6$-Evoked Microglial Phagocytosis
Kazuhide Inoue, Schuichi Koizumi, Ayako Kataoka, Hidetoshi Tozaki-Saitoh, and Makoto Tsuda

PPAR and Pain
Takehiko Maeda and Shiroh Kishioka

Involvement of Inflammatory Mediators in Neuropathic Pain Caused by Vincristine
Norikazu Kiguchi, Takehiko Maeda, Yuka Kobayashi, Fumihiro Saika, and Shiroh Kishioka

Nociceptive Behavior Induced by the Endogenous Opioid Peptides Dynorphins in Uninjured Mice: Evidence with Intrathecal *N*-ethylmaleimide Inhibiting Dynorphin Degradation
Koichi Tan-No, Hiroaki Takahashi, Osamu Nakagawasai, Fukie Niijima, Shinobu Sakurada, Georgy Bakalkin, Lars Terenius, and Takeshi Tadano

Mechanism of Allodynia Evoked by Intrathecal Morphine-3-Glucuronide in Mice
Takaaki Komatsu, Shinobu Sakurada, Sou Katsuyama, Kengo Sanai, and Tsukasa Sakurada

(–)-Linalool Attenuates Allodynia in Neuropathic Pain Induced by Spinal Nerve Ligation in C57/Bl6 Mice
Laura Berliocchi, Rossella Russo, Alessandra Levato, Vincenza Fratto, Giacinto Bagetta, Shinobu Sakurada, Tsukasa Sakurada, Nicola Biagio Mercuri, and Maria Tiziana Corasaniti

Intraplantar Injection of Bergamot Essential Oil into the Mouse Hindpaw: Effects on Capsaicin-Induced Nociceptive Behaviors
Tsukasa Sakurada, Hikari Kuwahata, Soh Katsuyama, Takaaki Komatsu, Luigi A. Morrone, M. Tiziana Corasaniti, Giacinto Bagetta, and Shinobu Sakurada

New Therapy for Neuropathic Pain
Hirokazu Mizoguchi, Chizuko Watanabe, Akihiko Yonezawa, and Shinobu Sakurada

Regulated Exocytosis from Astrocytes: Physiological and Pathological Related Aspects
Corrado Calì, Julie Marchaland, Paola Spagnuolo, Julien Gremion, and Paola Bezzi

Glutamate Release from Astrocytic Gliosomes Under Physiological and Pathological Conditions
Marco Milanese, Tiziana Bonifacino, Simona Zappettini, Cesare Usai, Carlo Tacchetti, Mario Nobile, and Giambattista Bonanno

Neurotrophic and Neuroprotective Actions of an Enhancer of Ganglioside Biosynthesis
Jin-ichi Inokuchi

Involvement of Endocannabinoid Signaling in the Neuroprotective Effects of Subtype 1 Metabotropic Glutamate Receptor Antagonists in Models of Cerebral Ischemia
Elisa Landucci, Francesca Boscia, Elisabetta Gerace, Tania Scartabelli, Andrea Cozzi, Flavio Moroni, Guido Mannaioni, and Domenico E. Pellegrini-Giampietro

NF-kappaB Dimers in the Regulation of Neuronal Survival
Ilenia Sarnico, Annamaria Lanzillotta, Marina Benarese, Manuela Alghisi, Cristina Baiguera, Leontino Battistin, PierFranco Spano, and Marina Pizzi

Oxidative Stress in Stroke Pathophysiology: Validation of Hydrogen Peroxide Metabolism as a Pharmacological Target to Afford Neuroprotection
Diana Amantea, Maria Cristina Marrone, Robert Nisticò, Mauro Federici, Giacinto Bagetta, Giorgio Bernardi, and Nicola Biagio Mercuri

Role of Akt and ERK Signaling in the Neurogenesis following Brain Ischemia
Norifumi Shioda, Feng Han, and Kohji Fukunaga

Prevention of Glutamate Accumulation and Upregulation of Phospho-Akt may Account for Neuroprotection Afforded by Bergamot Essential Oil against Brain Injury Induced by Focal Cerebral Ischemia in Rat
Diana Amantea, Vincenza Fratto, Simona Maida, Domenicantonio Rotiroti, Salvatore Ragusa, Giuseppe Nappi, Giacinto Bagetta, and Maria Tiziana Corasaniti

Identification of Novel Pharmacological Targets to Minimize Excitotoxic Retinal Damage
Rossella Russo, Domenicantonio Rotiroti, Cristina Tassorelli, Carlo Nucci, Giacinto Bagetta, Massimo Gilberto Bucci, Maria Tiziana Corasaniti, and Luigi Antonio Morrone

INDEX

Volume 86

Section One: Hybrid Bionic Systems
EMG-Based and Gaze-Tracking-Based Man–Machine Interfaces
Federico Carpi and Danilo De Rossi

Bidirectional Interfaces with the Peripheral Nervous System
Silvestro Micera and Xavier Navarro

Interfacing Insect Brain for Space Applications
Giovanni Di Pino, Tobias Seidl, Antonella Benvenuto, Fabrizio Sergi, Domenico Campolo, Dino Accoto, Paolo Maria Rossini, and Eugenio Guglielmelli

Section Two: Meet the Brain
Meet the Brain: Neurophysiology
John Rothwell

Fundamentals of Electroencefalography, Magnetoencefalography, and Functional Magnetic Resonance Imaging
Claudio Babiloni, Vittorio Pizzella, Cosimo Del Gratta, Antonio Ferretti, and Gian Luca Romani

Implications of Brain Plasticity to Brain–Machine Interfaces Operation: A Potential Paradox?
Paolo Maria Rossini

Section Three: Brain Machine Interfaces, A New Brain-to-Environment Communication Channel
An Overview of BMIs
Francisco Sepulveda

Neurofeedback and Brain–Computer Interface: Clinical Applications
Niels Birbaumer, Ander Ramos Murguialday, Cornelia Weber, and Pedro Montoya

Flexibility and Practicality: Graz Brain–Computer Interface Approach
Reinhold Scherer, Gernot R. Müller-Putz, and Gert Pfurtscheller

On the Use of Brain–Computer Interfaces Outside Scientific Laboratories: Toward an Application in Domotic Environments
F. Babiloni, F. Cincotti, M. Marciani, S. Salinari, L. Astolfi, F. Aloise, F. De Vico Fallani, and D. Mattia

Brain–Computer Interface Research at the Wadsworth Center: Developments in Noninvasive Communication and Control
Dean J. Krusienski and Jonathan R. Wolpaw

Watching Brain TV and Playing Brain Ball: Exploring Novel BCL Strategies Using Real–Time Analysis of Human Intercranial Data
Karim Jerbi, Samson Freyermuth, Lorella Minotti, Philippe Kahane, Alain Berthoz, and Jean-Philippe Lachaux

Section Four: Brain-Machine Interfaces and Space
Adaptive Changes of Rhythmic EEG Oscillations in Space: Implications for Brain–Machine Interface Applications
G. Cheron, A. M. Cebolla, M. Petieau, A. Bengoetxea, E. Palmero-Soler, A. Leroy, and B. Dan

Validation of Brain–Machine Interfaces During Parabolic Flight
José del R. Millán, Pierre W. Ferrez, and Tobias Seidl

Matching Brain–Machine Interface Performance to Space Applications
Luca Citi, Oliver Tonet, and Martina Marinelli

Brain–Machine Interfaces for Space Applications—Research, Technological Development, and Opportunities
Leopold Summerer, Dario Izzo, and Luca Rossini

INDEX

Volume 87

Peripheral Nerve Repair and Regeneration Research: A Historical Note
Bruno Battiston, Igor Papalia, Pierluigi Tos, and Stefano Geuna

Development of the Peripheral Nerve
Suleyman Kaplan, Ersan Odaci, Bunyami Unal, Bunyamin Sahin, and Michele Fornaro

Histology of the Peripheral Nerve and Changes Occurring During Nerve Regeneration
Stefano Geuna, Stefania Raimondo, Giulia Ronchi, Federica Di Scipio, Pierluigi Tos, Krzysztof Czaja, and Michele Fornaro

Methods and Protocols in Peripheral Nerve Regeneration Experimental Research: Part I—Experimental Models
Pierluigi Tos, Giulia Ronchi, Igor Papalia, Vera Sallen, Josette Legagneux, Stefano Geuna, and Maria G. Giacobini-Robecchi

Methods and Protocols in Peripheral Nerve Regeneration Experimental Research: Part II—Morphological Techniques
Stefania Raimondo, Michele Fornaro, Federica Di Scipio, Giulia Ronchi, Maria G. Giacobini-Robecchi, and Stefano Geuna

Methods and Protocols in Peripheral Nerve Regeneration Experimental Research: Part III—Electrophysiological Evaluation
Xavier Navarro and Esther Udina

Methods and Protocols in Peripheral Nerve Regeneration Experimental Research: Part IV—Kinematic Gait Analysis to Quantify Peripheral Nerve Regeneration in the Rat
Luís M. Costa, Maria J. Simões, Ana C. Maurício and Artur S.P. Varejão

Current Techniques and Concepts in Peripheral Nerve Repair
Maria Siemionow and Grzegorz Brzezicki

Artificial Scaffolds for Peripheral Nerve Reconstruction
Valeria Chiono, Chiara Tonda-Turo, and Gianluca Ciardelli

Conduit Luminal Additives for Peripheral Nerve Repair
Hede Yan, Feng Zhang, Michael B. Chen, and William C. Lineaweaver

Tissue Engineering of Peripheral Nerves
Bruno Battiston, Stefania Raimondo, Pierluigi Tos, Valentina Gaidano, Chiara Audisio, Anna Scevola, Isabelle Perroteau, and Stefano Geuna

Mechanisms Underlying The End-to-Side Nerve Regeneration
Eleana Bontioti and Lars B. Dahlin

Experimental Results in End-To-Side Neurorrhaphy
Alexandros E. Beris and Marios G. Lykissas

End-to-Side Nerve Regeneration: From the Laboratory Bench to Clinical Applications
Pierluigi Tos, Stefano Artiaco, Igor Papalia, Ignazio Marcoccio, Stefano Geuna, and Bruno Battiston

Novel Pharmacological Approaches to Schwann Cells as Neuroprotective Agents for Peripheral Nerve Regeneration
Valerio Magnaghi, Patrizia Procacci, and Ada Maria Tata

Melatonin and Nerve Regeneration
Ersan Odaci and Suleyman Kaplan

Transthyretin: An Enhancer of Nerve Regeneration
Carolina E. Fleming, Fernando Milhazes Mar, Filipa Franquinho, and Mónica M. Sousa

Enhancement of Nerve Regeneration and Recovery by Immunosuppressive Agents
Damien P. Kuffler

The Role of Collagen in Peripheral Nerve Repair
Guido Koopmans, Birgit Hasse, and Nektarios Sinis

Gene Therapy Perspectives for Nerve Repair
Serena Zacchigna and Mauro Giacca

Use of Stem Cells for Improving Nerve Regeneration
Giorgio Terenghi, Mikael Wiberg, and Paul J. Kingham

Transplantation of Olfactory Ensheathing Cells for Peripheral Nerve Regeneration
Christine Radtke, Jeffery D. Kocsis, and Peter M. Vogt

Manual Stimulation of Target Muscles has Different Impact on Functional Recovery after Injury of Pure Motor or Mixed Nerves
Nektarios Sinis, Thodora Manoli, Frank Werdin, Armin Kraus, Hans E. Schaller, Orlando Guntinas-Lichius, Maria Grosheva, Andrey Irintchev, Emanouil Skouras, Sarah Dunlop, and Doychin N. Angelov

Electrical Stimulation for Improving Nerve Regeneration: Where do we Stand?
Tessa Gordon, Olewale A. R. Sulaiman, and Adil Ladak

Phototherapy in Peripheral Nerve Injury: Effects on Muscle Preservation and Nerve Regeneration
Shimon Rochkind, Stefano Geuna, and Asher Shainberg

Age-Related Differences in the Reinnervation after Peripheral Nerve Injury
Uroš Kovačič, Janez Sketelj, and Fajko F. Bajrović

Neural Plasticity After Nerve Injury and Regeneration
Xavier Navarro

Future Perspective in Peripheral Nerve Reconstruction
Lars Dahlin, Fredrik Johansson, Charlotta Lindwall, and Martin Kanje

INDEX

Volume 88

Effects Of Psychostimulants On Neurotrophins: Implications For Psychostimulant-Induced Neurotoxicity
Francesco Angelucci, Valerio Ricci, Gianfranco Spalletta, Carlo Caltagirone, Aleksander A. Mathé, and Pietro Bria

Dosing Time-Dependent Actions of Psychostimulants
H. Manev and T. Uz

Dopamine-Induced Behavioral Changes and Oxidative Stress in Methamphetamine-Induced Neurotoxicity
Taizo Kita, Ikuko Miyazaki, Masato Asanuma, Mika Takeshima, and George C. Wagner

Acute Methamphetamine Intoxication: Brain Hyperthermia, Blood–Brain Barrier, Brain Edema, and morphological cell abnormalities
Eugene A. Kiyatkin and Hari S. Sharma

Molecular Bases of Methamphetamine-Induced Neurodegeneration
Jean Lud Cadet and Irina N. Krasnova

Involvement of Nicotinic Receptors in Methamphetamine- and MDMA-Induced Neurotoxicity: Pharmacological Implications
E. Escubedo, J. Camarasa, C. Chipana, S. García-Ratés, and D.Pubill

Ethanol Alters the Physiology of Neuron–Glia Communication
Antonio González and Ginés M. Salido

Therapeutic Targeting of "DARPP-32": A Key Signaling Molecule in the Dopiminergic Pathway for the Treatment of Opiate Addiction
Supriya D. Mahajan, Ravikumar Aalinkeel, Jessica L. Reynolds, Bindukumar B. Nair, Donald E. Sykes, Zihua Hu, Adela Bonoiu, Hong Ding, Paras N. Prasad, and Stanley A. Schwartz

Pharmacological and Neurotoxicological Actions Mediated By Bupropion and Diethylpropion
Hugo R. Arias, Abel Santamaría, and Syed F. Ali

Neural and Cardiac Toxicities Associated With 3,4-Methylenedioxymethamphetamine (MDMA)
Michael H. Baumann and Richard B. Rothman

Cocaine-Induced Breakdown of the Blood–Brain Barrier and Neurotoxicity
Hari S. Sharma, Dafin Muresanu, Aruna Sharma, and Ranjana Patnaik

Cannabinoid Receptors in Brain: Pharmacogenetics, Neuropharmacology, Neurotoxicology, and Potential Therapeutic Applications
Emmanuel S. Onaivi

Intermittent Dopaminergic Stimulation causes Behavioral Sensitization in the Addicted Brain and Parkinsonism
Francesco Fornai, Francesca Biagioni, Federica Fulceri, Luigi Murri, Stefano Ruggieri, Antonio Paparelli

The Role of the Somatotrophic Axis in Neuroprotection and Neuroregeneration of the Addictive Brain
Fred Nyberg

INDEX

Volume 89

Molecular Profiling of Striatonigral and Striatopallidal Medium Spiny Neurons: Past, Present, and Future
Mary Kay Lobo

BAC to Degeneration: Bacterial Artificial Chromosome (Bac)-Mediated Transgenesis for Modeling Basal Ganglia Neurodegenerative Disorders
Xiao-Hong Lu

Behavioral Outcome Measures for the Assessment of Sensorimotor Function in Animal Models of Movement Disorders
Sheila M. Fleming

The Role of DNA Methylation in the Central Nervous System and Neuropsychiatric Disorders
Jian Feng and Guoping Fan

Heritability of Structural Brain Traits: An Endo-phenotype Approach to Deconstruct Schizophrenia
Nil Kaymaz and J. Van Os

The Role of Striatal NMDA Receptors in Drug Addiction
Yao-Ying Ma, Carlos Cepeda, and Cai-Lian Cui

Deciphering Rett Syndrome With Mouse Genetics, Epigenomics, and Human Neurons
Jifang Tao, Hao Wu, and Yi Eve Sun

INDEX

Volume 90

Part I: Introduction

Introductory Remarks on the History and Current Applications of TCS
Matthew B. Stern

Method and Validity of Transcranial Sonography in Movement Disorders
David Školoudík and Uwe Walter

Transcranial Sonography—Anatomy
Heiko Huber

Part II: Transcranial Sonography in Parkinsons Disease

Transcranial Sonography in Relation to SPECT and MIBG
Yoshinori Kajimoto, Hideto Miwa and Tomoyoshi Kondo

Diagnosis of Parkinson's Disease—Transcranial Sonography in Relation to MRI
Ludwig Niehaus and Kai Boelmans

Early Diagnosis of Parkinson's Disease
Alexandra Gaenslen and Daniela Berg

Transcranial Sonography in the Premotor Diagnosis of Parkinson's Disease
Stefanie Behnke, Ute Schroder and Daniela Berg

Pathophysiology of Transcranial Sonography Signal Changes in the Human Substantia Nigra
K. L. Double, G. Todd and S. R. Duma

Transcranial Sonography for the Discrimination of Idiopathic Parkinson's Disease from the Atypical Parkinsonian Syndromes
A. E. P. Bouwmans, A. M. M. Vlaar, K. Srulijes, W. H. Mess AND W. E. J. Weber

Transcranial Sonography in the Discrimination of Parkinson's Disease Versus Vascular Parkinsonism
Pablo Venegas-Francke

TCS in Monogenic Forms of Parkinson's Disease
Kathrin Brockmann and Johann Hagenah

Part III—Transcranial Sonography in other Movement Disorders and Depression

Transcranial Sonography in Brain Disorders with Trace Metal Accumulation
Uwe Walter

Transcranial Sonography in Dystonia
Alexandra Gaenslen

Transcranial Sonography in Essential Tremor
Heike Stockner and Isabel Wurster

VII—Transcranial Sonography in Restless Legs Syndrome
Jana Godau and Martin Sojer

Transcranial Sonography in Ataxia
Christos Krogias, Thomas Postert and Jens Eyding

Transcranial Sonography in Huntington's Disease
Christos Krogias, Jens Eyding and Thomas Postert

Transcranial Sonography in Depression
Milija D. Mijajlovic

Part IV: Future Applications and Conclusion

Transcranial Sonography-Assisted Stereotaxy and Follow-Up of Deep Brain Implants in Patients with Movement Disorders
Uwe Walter

Conclusions
Daniela Berg

INDEX

Volume 91

The Role of microRNAs in Drug Addiction: A Big Lesson from Tiny Molecules
Andrzej Zbigniew Pietrzykowski

The Genetics of Behavioral Alcohol Responses in *Drosophila*
Aylin R. Rodan and Adrian Rothenfluh

Neural Plasticity, Human Genetics, and Risk for Alcohol Dependence
Shirley Y. Hill

Using Expression Genetics to Study the Neurobiology of Ethanol and Alcoholism
Sean P. Farris, Aaron R. Wolen and Michael F. Miles

Genetic Variation and Brain Gene Expression in Rodent Models of Alcoholism: Implications for Medication Development
Karl Björk, Anita C. Hansson and W. olfgang H. Sommer

Identifying Quantitative Trait Loci (QTLs) and Genes (QTGs) for Alcohol-Related Phenotypes in Mice
Lauren C. Milner and Kari J. Buck

Glutamate Plasticity in the Drunken Amygdala: The Making of an Anxious Synapse
Brian A. Mccool, Daniel T. Christian, Marvin R. Diaz and Anna K. Läck

Ethanol Action on Dopaminergic Neurons in the Ventral Tegmental Area: Interaction with Intrinsic Ion Channels and Neurotransmitter Inputs
Hitoshi Morikawa and Richard A. Morrisett

Alcohol and the Prefrontal Cortex
Kenneth Abernathy, L. Judson Chandler and John J. Woodward

BK Channel and Alcohol, A Complicated Affair
Gilles Erwan Martin

A Review of Synaptic Plasticity at Purkinje Neurons with a Focus on Ethanol-Induced Cerebellar Dysfunction
C. Fernando Valenzuela, Britta Lindquist and Paula A. Zflmudio-Bulcock

INDEX

Volume 92

The Development of the Science of Dreaming
Claude Gottesmann

Dreaming as Inspiration: Evidence from Religion, Philosophy, Literature, and Film
Kelly Bulkeley

Developmental Perspective: Dreaming Across the Lifespan and What This Tells Us
Melissa M. Burnham and Christian Conte

REM and NREM Sleep Mentation
Patrick Mcnamara, Patricia Johnson, Deirdre McLaren, Erica Harris,Catherine Beauharnais and Sanford Auerbach

Neuroimaging of Dreaming: State of the Art and Limitations
Caroline Kussé, Vincenzo Muto, Laura Mascetti, Luca Matarazzo, Ariane Foret, Anahita Shaffii-Le Bourdiec and Pierre Maquet

Memory Consolidation, The Diurnal Rhythm of Cortisol, and The Nature of Dreams: A New Hypothesis
Jessica D. Payne

Characteristics and Contents of Dreams
Michael Schredl

Trait and Neurobiological Correlates of Individual Differences in Dream Recall and Dream Content
Mark Blagrove and Edward F. Pace-Schott

Consciousness in Dreams
David Kahn and Tzivia Gover

The Underlying Emotion and the Dream: Relating Dream Imagery to the Dreamer's Underlying Emotion can Help Elucidate the Nature of Dreaming
Ernest Hartmann

Dreaming, Handedness, and Sleep Architecture: Interhemispheric Mechanisms
Stephen D. Christman and Ruth E. Propper

To What Extent Do Neurobiological Sleep-Waking Processes Support Psychoanalysis?
Claude Gottesmann

The Use of Dreams in Modern Psychotherapy
Clara E. Hill and Sarah Knox

INDEX

Volume 93

Underlying Brain Mechanisms that Regulate Sleep-Wakefulness Cycles
Irma Gvilia

What Keeps Us Awake?—the Role of Clocks and Hourglasses, Light, and Melatonin
Christian Cajochen, Sarah Chellappa and Christina Schmidt

Suprachiasmatic Nucleus and Autonomic Nervous System Influences on Awakening From Sleep
Andries Kalsbeek, Chun-xia Yi, Susanne E. la Fleur, Ruud m. Buijs, and Eric Fliers

Preparation for Awakening: Self-Awakening Vs. Forced Awakening: Preparatory Changes in the Pre-Awakening Period
Mitsuo Hayashi, Noriko Matsuura and Hiroki Ikeda

Circadian and Sleep Episode Duration Influences on Cognitive Performance Following the Process of Awakening
Robert L. Matchock

The Cortisol Awakening Response in Context
Angela Clow, Frank Hucklebridge and Lisa Thorn

Causes and Correlates of Frequent Night Awakenings in Early Childhood
Amy Jo Schwichtenberg and Beth Goodlin-Jones

Pathologies of Awakenings: The Clinical Problem of Insomnia Considered From Multiple Theory Levels
Douglas E. Moul

The Neurochemistry of Awakening: Findings from Sleep Disorder Narcolepsy
Seiji Nishino and Yohei Sagawa

INDEX

Volume 94

5-HT_6 Medicinal Chemistry
Kevin G. Liu and Albert J. Robichaud

Patents
Nicolas Vincent Ruiz and Gloria Oranias

5-HT_6 Receptor Charactertization
Teresa Riccioni

5-HT_6 Receptor Signal Transduction: Second Messenger Systems
Xavier Codony, Javier Burgueño, Maria Javier Ramírez and José Miguel Vela

Electrophysiology of 5-HT_6 Receptors
Annalisa Tassone, Graziella Madeo, Giuseppe Sciamanna, Antonio Pisani and Paola Bonsi

Genetic Variations and Association
Massimo Gennarelli and Annamaria Cattaneo

Pharmacokinetics of 5-HT_6 Receptor Ligands
Angelo Mancinelli

INDEX

Volume 95

Introductory Remarks: Catechol-O-Methyltransferase Inhibition–An Innovative Approach to Enhance L-dopa Therapy in Parkinson's Disease with Dual Enzyme Inhibition
Erkki Nissinen

The Catechol-O-Methyltransferase Gene: its Regulation and Polymorphisms
Elizabeth M. Tunbridge

Distribution and Functions of Catechol-O-Methyltransferase Proteins: Do Recent Findings Change the Picture?
Timo T. Myöhänen and Pekka T. Männistö

Catechol-O-Methyltransferase Enzyme: Cofactor S-Adenosyl-L-Methionine and Related Mechanisms
Thomas Müller

Biochemistry and Pharmacology of Catechol-O-Methyltransferase Inhibitors
Erkki nissinen and Pekka T. Männisto

The Chemistry of Catechol-O-Methyltransferase Inhibitors
David A. Learmonth, László E. Kiss, and Patrício Soares-da-Silva

Toxicology and Safety of COMT Inhibitors
Kristiina Haasio

Catechol-O-Methyltransferase Inhibitors in Preclinical Models as Adjuncts of L-dopa Treatment
Concepció Marin and J. A. Obeso

Problems with the Present Inhibitors and a Relevance of New and Improved COMT Inhibitors in Parkinson's Disease
Seppo Kaakkola

Catechol-O-Methyltransferase and Pain
Oleg Kambur and Pekka T. Männistö

INDEX

Volume 96

The Central Role of 5-HT_6 Receptors in Modulating Brain Neurochemistry
Lee A. Dawson

5-HT_6 Receptor Memory and Amnesia: Behavioral Pharmacology – Learning and Memory Processes
Alfredo Meneses, G. Pérez-García, R. Tellez, T. Ponce-Lopez and C. Castillo

Behavioral Pharmacology: Potential Antidepressant and Anxiolytic Properties
Anna Wesołowska and Magdalena Jastrzbska-Wisek

The 5-HT_6 Receptor as a Target for Developing Novel Antiobesity Drugs
David Heal, Jane Gosden and Sharon Smith

Behavioral and Neurochemical Pharmacology of 5-HT_6 Receptors Related to Reward and Reinforcement
Gaetano Di Chiara, Valentina Valentini and Sandro Fenu

5-HT_6 Receptor Ligands and their Antipsychotic Potential
Jørn Arnt and Christina Kurre Olsen

5-HT_6 Receptor Ligands as Antidementia Drugs
Ellen Siobhan Mitchell

Other 5-HT_6 Receptor-Mediated Effects
Franco Borsini

INDEX

Volume 97

Behavioral Pharmacology of Orofacial Movement Disorders
Noriaki Koshikawa, Satoshi Fujita and Kazunori Adachi

Regulation of Orofacial Movement: Dopamine Receptor Mechanisms and Mutant Models
John L. Waddington, Gerard J. O'Sullivan and Katsunori Tomiyama

Regulation of Orofacial Movement: Amino Acid Mechanisms and Mutant Models
Katsunori Tomiyama, Colm M.P. O'Tuathaigh, and John L. Waddington

The Trigeminal Circuits Responsible for Chewing
Karl-Gunnar Westberg and Arlette Kolta

Ultrastructural Basis for Craniofacial Sensory Processing in the Brainstem
Yong Chul Bae and Atsushi Yoshida

Mechanisms of Nociceptive Transduction and Transmission: A Machinery for Pain Sensation and Tools for Selective Analgesia
Alexander M. Binshtok

Peripheral and Central Mechanisms of Orofacial Inflammatory Pain
Barry J. Sessle

The Role of Trigeminal Interpolaris-Caudalis Transition Zone in Persistent Orofacial Pain
Ke Ren and Ronald Dubner

Physiological Mechanisms of Neuropathic Pain: The Orofacial Region
Koichi Iwata, Yoshiki Imamura, Kuniya Honda and Masamichi Shinoda

Neurobiology of Estrogen Status in Deep Craniofacial Pain
David A Bereiter and Keiichiro Okamoto

Macroscopic Connection of Rat Insular Cortex: Anatomical Bases Underlying its Physiological Functions
Masayuki Kobayashi

The Balance Between Excitation And Inhibition And Functional Sensory Processing in the Somatosensory Cortex
Zhi Zhang and Qian-Quan Sun

INDEX

Volume 98

An Introduction to Dyskinesia—the Clinical Spectrum
Ainhi Ha and Joseph Jankovic

L-dopa-induced Dyskinesia—Clinical Presentation, Genetics, And Treatment
L.K. Prashanth, Susan Fox and Wassilios G. Meissner

Experimental Models of L-DOPA-induced Dyskinesia
Tom H. Johnston and Emma L. Lane

Molecular Mechanisms of L-DOPA-induced Dyskinesia
Gilberto Fisone and Erwan Bezard

New Approaches to Therapy
Jonathan Brotchie and Peter Jenner

Surgical Approach to L-DOPA-induced Dyskinesias
Tejas Sankar and Andres M. Lozano

Clinical and Experimental Experiences of Graft-induced Dyskinesia
Emma L. Lane

Tardive Dyskinesia: Clinical Presentation and Treatment
P.N. van Harten and D.E. Tenback

Epidemiology and Risk Factors for (Tardive) Dyskinesia
D.E. Tenback and P.N. van Harten

Genetics of Tardive Dyskinesia
Heon-Jeong Lee and Seung-Gul Kang

Animal Models of Tardive Dyskinesia
S.K. Kulkarni and Ashish Dhir

Surgery for Tardive Dyskinesia
Stephane Thobois, Alice Poisson and Philippe Damier

Huntington's Disease: Clinical Presentation and Treatment
M.J.U. Novak and S.J. Tabrizi

Genetics and Neuropathology of Huntington's Disease: Huntington's Disease
Anton Reiner, Ioannis Dragatsis and Paula Dietrich

Pathogenic Mechanisms in Huntington's Disease
Lesley Jones and Alis Hughes

Experimental Models of HD And Reflection on Therapeutic Strategies
Olivia L. Bordiuk, Jinho Kim and Robert J. Ferrante

Cell-based Treatments for Huntington's Disease
Stephen B. Dunnett and Anne E. Rosser

Clinical Phenomenology of Dystonia
Carlo Colosimo and Alfredo Berardelli

Genetics and Pharmacological Treatment of Dystonia
Susan Bressman and Matthew James

Experimental Models of Dystonia
A. Tassone, G. Sciamanna, P. Bonsi, G. Martella and A. Pisani

Surgical Treatment of Dystonia
John Yianni, Alexander L. Green and Tipu Z. Aziz

INDEX

Volume 99

Seizure and Epilepsy: Studies of Seizure-disorders in *Drosophila*
Louise Parker, Iris C. Howlett, Zeid M. Rusan and Mark A. Tanouye

Homeostatic Control of Neural Activity: A *Drosophila* Model for Drug Tolerance and Dependence
Alfredo Ghezzi and Nigel S. Atkinson

Attention in *Drosophila*
Bruno van Swinderen

The roles of *Fruitless* and *Doublesex* in the Control of Male Courtship
Brigitte Dauwalder

Circadian Plasticity: from Structure to Behavior
Lia Frenkel and María Fernanda Ceriani

Learning and Memory in *Drosophila*: Behavior, Genetics, and Neural Systems
Lily Kahsai and Troy Zars

Studying Sensorimotor Processing with Physiology in Behaving *Drosophila*
Johannes D. Seelig and Vivek Jayaraman

Modeling Human Trinucleotide Repeat Diseases in *Drosophila*
Zhenming Yu and Nancy M. Bonini

From Genetics to Structure to Function: Exploring Sleep in *Drosophila*
Daniel Bushey and Chiara Cirelli

INDEX

Volume 100

Structural Properties of Human Monoamine Oxidases A and B
Claudia Binda, Andrea Mattevi and Dale E. Edmondson

Behavioral Outcomes of Monoamine Oxidase Deficiency: Preclinical and Clinical Evidence
Marco Bortolato and Jean C. Shih

Kinetic Behavior and Reversible Inhibition of Monoamine Oxidases—Enzymes that Many Want Dead
Keith F. Tipton, Gavin P. Davey and Andrew G. McDonald

The Pharmacology of Selegiline
Kálmán Magyar

Type A Monoamine Oxidase Regulates Life and Death of Neurons in Neurodegeneration and Neuroprotection
Makoto Naoi, Wakako Maruyama, Keiko Inaba-Hasegawa and Yukihiro Akao

Multimodal Drugs and their Future for Alzheimer's and Parkinson's Disease
Cornelis J. Van der Schyf and Werner J. Geldenhuys

Neuroprotective Profile of the Multitarget Drug Rasagiline in Parkinson's Disease
Orly Weinreb, Tamar Amit, Peter Riederer, Moussa B.H. Youdim and Silvia A. Mandel

Rasagiline in Parkinson's Disease
L.M. Chahine and M.B. Stern

Selective Inhibitors of Monoamine Oxidase Type B and the "Cheese Effect"
John P.M. Finberg and Ken Gillman

A Novel Anti-Alzheimer's Disease Drug, Ladostigil: Neuroprotective, Multimodal Brain-Selective Monoamine Oxidase and Cholinesterase Inhibitor
Orly Weinreb, Tamar Amit, Orit Bar-Am and Moussa B.H. Youdim

Novel MAO-B Inhibitors: Potential Therapeutic Use of the Selective MAO-B Inhibitor PF9601N in Parkinson's Disease
Mercedes Unzeta and Elisenda Sanz

INDEX

Volume 101

General Overview: Biomarkers in Neuroscience Research
Michaela D. Filiou and Christoph W. Turck

Imaging Brain Microglial Activation Using Positron Emission Tomography and Translocator Protein-Specific Radioligands
David R.J. Owen and Paul M. Matthews

The Utility of Gene Expression in Blood Cells for Diagnosing Neuropsychiatric Disorders
Christopher H. Woelk, Akul Singhania, Josué Pérez-Santiago, Stephen J. Glatt and Ming T. Tsuang

Proteomic Technologies for Biomarker Studies in Psychiatry: Advances and Needs
Daniel Martins-de-Souza, Paul C. Guest, Natacha Vanattou-Saifoudine, Laura W. Harris and Sabine Bahn

Converging Evidence of Blood-Based Biomarkers for Schizophrenia: An update
Man K. Chan, Paul C. Guest, Yishai Levin, Yagnesh Umrania, Emanuel Schwarz, Sabine Bahn and Hassan Rahmoune

Abnormalities in Metabolism and Hypothalamic–Pituitary–Adrenal Axis Function in Schizophrenia
Paul C. Guest, Daniel Martins-de-Souza, Natacha Vanattou-Saifoudine, Laura W. Harris and Sabine Bahn

Immune and Neuroimmune Alterations in Mood Disorders and Schizophrenia
Roosmarijn C. Drexhage, Karin Weigelt, Nico van Beveren, Dan Cohen, Marjan A. Versnel, Willem A. Nolen and Hemmo A. Drexhage

Behavioral and Molecular Biomarkers in Translational Animal Models for Neuropsychiatric Disorders
Zoltán Sarnyai, Murtada Alsaif, Sabine Bahn, Agnes Ernst, Paul C. Guest, Eva Hradetzky, Wolfgang Kluge, Viktoria Stelzhammer and Hendrik Wesseling

Stem Cell Models for Biomarker Discovery in Brain Disease
Alan Mackay-Sim, George Mellick and Stephen Wood

The Application of Multiplexed Assay Systems for Molecular Diagnostics
Emanuel Schwarz, Nico J.M. VanBeveren, Paul C. Guest, Rauf Izmailov and Sabine Bahn

Algorithm Development for Diagnostic Biomarker Assays
Rauf Izmailov, Paul C. Guest, Sabine Bahn and Emanuel Schwarz

Challenges of Introducing New Biomarker Products for Neuropsychiatric Disorders into the Market
Sabine Bahn, Richard Noll, Anthony Barnes, Emanuel Schwarz and Paul C. Guest

Toward Personalized Medicine in the Neuropsychiatric Field
Erik H.F. Wong, Jayne C. Fox, Mandy Y.M. Ng and Chi-Ming Lee

Clinical Utility of Serum Biomarkers for Major Psychiatric Disorders
Nico J.M. van Beveren and Witte J.G. Hoogendijk

The Future: Biomarkers, Biosensors, Neuroinformatics, and E-Neuropsychiatry
Christopher R. Lowe

SUBJECT INDEX

Volume 102

The Function and Mechanisms of Nurr1 Action in Midbrain Dopaminergic Neurons, from Development and Maintenance to Survival
Yu Luo

Monoclonal Antibodies as Novel Neurotherapeutic Agents in CNS Injury and Repair
Aruna Sharma and Hari Shanker Sharma

The Blood–Brain Barrier in Alzheimer's Disease: Novel Therapeutic Targets and Nanodrug delivery
Hari Shanker Sharma, Rudy J. Castellani, Mark A. Smith and Aruna Sharma

Neurovascular Aspects of Amyotrophic Lateral Sclerosis
Maria Carolina O. Rodrigues, Diana G. Hernandez-Ontiveros, Michael K. Louis, Alison E. Willing, Cesario V. Borlongan, Paul R. Sanberg, Júlio C. Voltarelli and Svitlana Garbuzova-Davis

Quercetin in Hypoxia-Induced Oxidative Stress: Novel Target for Neuroprotection
Anand Kumar Pandey, Ranjana Patnaik, Dafin F. Muresanu, Aruna Sharma and Hari Shanker Sharma

Environmental Conditions Modulate Neurotoxic Effects of Psychomotor Stimulant Drugs of Abuse
Eugene A. Kiyatkin and Hari Shanker Sharma

Central Nervous Tissue Damage after Hypoxia and Reperfusion in Conjunction with Cardiac Arrest and Cardiopulmonary Resuscitation: Mechanisms of Action and Possibilities for Mitigation
Lars Wiklund, Cecile Martijn, Adriana Miclescu, Egidijus Semenas, Sten Rubertsson and Hari Shanker Sharma

Interactions Between Opioids and Anabolic Androgenic Steroids: Implications for the Development of Addictive Behavior
Fred Nyberg and Mathias Hallberg

Neurotrophic Factors and Neurodegenerative Diseases: A Delivery Issue
Barbara Ruozi, Daniela Belletti, Lucia Bondioli, Alessandro De Vita, Flavio Forni, Maria Angela Vandelli and Giovanni Tosi

Neuroprotective Effects of Cerebrolysin, a Combination of Different Active Fragments of Neurotrophic Factors and Peptides on the Whole Body Hyperthermia-Induced Neurotoxicity: Modulatory Roles of Co-morbidity Factors and Nanoparticle Intoxication
Hari Shanker Sharma, Aruna Sharma, Herbert Mössler and Dafin Fior Muresanu

Alzheimer's Disease and Amyloid: Culprit or Coincidence?
Stephen D. Skaper

Vascular Endothelial Growth Factor and Other Angioglioneurins: Key Molecules in Brain Development and Restoration
José Vicente Lafuente, Naiara Ortuzar, Harkaitz Bengoetxea, Susana Bulnes and Enrike G. Argandoña

INDEX

Volume 103

Lost and Found in Behavioral Informatics
Melissa A. Haendel and Elissa J. Chesler

Biological Databases for Behavioral Neurobiology
Erich J. Baker

A Survey of the Neuroscience Resource Landscape: Perspectives from the Neuroscience Information Framework
Jonathan Cachat, Anita Bandrowski, Jeffery S. Grethe, Amarnath Gupta, Vadim Astakhov, Fahim Imam, Stephen D. Larson, and Maryann E. Martone

The Neurobehavior Ontology: An Ontology for Annotation and Integration of Behavior and Behavioral Phenotypes
Georgios V. Gkoutos, Paul N. Schofield, and Robert Hoehndorf

Ontologies for Human Behavior Analysis and Their Application to Clinical Data
Janna Hastings and Stefan Schulz

Text-Mining and Neuroscience
Kyle H. Ambert and Aaron M. Cohen

Applying *In Silico* Integrative Genomics to Genetic Studies of Human Disease: A Review
Scott F. Saccone

SUBJECT INDEX

Volume 104

Cross Species Integration of Functional Genomics Experiments
Jeremy J. Jay

Model Organism Databases in Behavioral Neuroscience
Mary Shimoyama, Jennifer R. Smith, G. Thomas Hayman, Victoria Petri, and Rajni Nigam

Accessing and Mining Data from Large-Scale Mouse Phenotyping Projects
Hugh Morgan, Michelle Simon, and Ann-Marie Mallon

Bioinformatics Resources for Behavior Studies in the Laboratory Mouse
Carol J. Bult

Using Genome-Wide Expression Profiling to Define Gene Networks Relevant to the Study of Complex Traits: From RNA Integrity to Network Topology
M.A. O'Brien, B.N. Costin, and M.F. Miles

Genetic and Molecular Network Analysis of Behavior
Robert W. Williams and Megan K. Mulligan

Large-Scale Neuroinformatics for *In Situ* Hybridization Data in the Mouse Brain
Lydia L. Ng, Susan M. Sunkin, David Feng, Chris Lau, Chinh Dang, and Michael J. Hawrylycz

Opportunities for Bioinformatics in the Classification of Behavior and Psychiatric Disorders
Elissa J. Chesler and Ryan W. Logan

Subject Index

Volume 105

Optic Nerve Disease and Axon Pathophysiology
Alireza Ghaffarieh and Leonard A. Levin

Role of Electrical Activity of Neurons for Neuroprotection
Takeshi Morimoto

Molecular Control of Axon Growth: Insights from Comparative Gene Profiling and High-Throughput Screening
Murray G. Blackmore

Gatekeeper Between Quiescence and Differentiation: p53 in Axonal Outgrowth and Neurogenesis
Giorgia Quadrato and Simone Di Giovanni

Cyclin-Dependent Kinase 5 in Axon Growth and Regeneration
Tao Ye, Amy K. Y. Fu, and Nancy Y. Ip

Rho Signaling and Axon Regeneration
L. McKerracher, Gino B. Ferraro, and Alyson E. Fournier

Neuron-Intrinsic Inhibitors of Axon Regeneration: PTEN and SOCS3
Xueting Luo and Kevin K. Park

Index

Volume 106

Neurotrophic Factors and the Regeneration of Adult Retinal Ganglion Cell Axons
Alan R. Harvey, Jacob Wei Wei Ooi, and Jennifer Rodger

MBS: Signaling Endosomes and Growth Cone Motility in Axon Regeneration
Michael B. Steketee and Jeffrey L. Goldberg

Intrinsic Mechanisms Regulating Axon Regeneration: An Integrin Perspective
Richard Eva, Melissa R. Andrews, Elske H.P. Franssen, and James W. Fawcett

The Role of Serotonin in Axon and Dendrite Growth
Ephraim F. Trakhtenberg and Jeffrey L. Goldberg

Inflammatory Pathways in Spinal Cord Injury
Samuel David, Juan Guillermo Zarruk, and Nader Ghasemlou

Combinatorial Therapy Stimulates Long-Distance Regeneration, Target Reinnervation, and Partial Recovery of Vision After Optic Nerve Injury in Mice
Silmara de Lima, Ghaith Habboub, and Larry I. Benowitz

From Bench to Beside to Cure Spinal Cord Injury: Lost in Translation?
Andreas Hug and Norbert Weidner

Subject Index

Volume 107

Neuromodulation: A More Comprehensive Concept Beyond Deep Brain Stimulation
Clement Hamani and Elena Moro

Computational Models of Neuromodulation
Christopher R. Butson

Neurophysiology of Deep Brain Stimulation
Manuela Rosa, Gaia Giannicola, Sara Marceglia, Manuela Fumagalli, Sergio Barbieri, and Alberto Priori

Neurophysiology of Cortical Stimulation
Jean-Pascal Lefaucheur

Neural Mechanisms of Spinal Cord Stimulation
Robert D. Foreman and Bengt Linderoth

Magnetoencephalography and Neuromodulation
Alfons Schnitzler and Jan Hirschmann

Current Challenges to the Clinical Translation of Brain Machine Interface Technology
Charles W. Lu, Parag G. Patil, and Cynthia A. Chestek

Nanotechnology in Neuromodulation
Russell J. Andrews

Optogenetic Neuromodulation
Paul S. A. Kalanithi and Jaimie M. Henderson

Diffusion Tensor Imaging and Neuromodulation: DTI as Key Technology for Deep Brain Stimulation
Volker Arnd Coenen, Thomas E. Schlaepfer, Niels Allert, and Burkhard Mädler

DBS and Electrical Neuro-Network Modulation to Treat Neurological Disorders
Amanda Thompson, Takashi Morishita, and Michael S. Okun

Neuromodulation in Psychiatric Disorders
Yasin Temel, Sarah A. Hescham, Ali Jahanshahi, Marcus L. F. Janssen, Sonny K. H. Tan, Jacobus J. van Overbeeke, Linda Ackermans, Mayke Oosterloo, Annelien Duits, Albert F. G. Leentjens, and LeeWei Lim

Ethical Aspects of Neuromodulation
Christiane Woopen

SUBJECT INDEX

Volume 108

Tissue Engineering and Regenerative Medicine: Past, Present, and Future
António J. Salgado, Joaquim M. Oliveira, Albino Martins, Fábio G. Teixeira, Nuno A. Silva, Nuno M. Neves, Nuno Sousa, and Rui L. Reis

Tissue Engineering and Peripheral Nerve Reconstruction: An Overview
Stefano Geuna, S. Gnavi, I. Perroteau, Pierluigi Tos, and B. Battiston

Bone Marrow Mesenchymal Stem Cell Transplantation for Improving Nerve Regeneration
Júlia Teixeira Oliveira, Klauss Mostacada, Silmara de Lima, and Ana Maria Blanco Martinez

Perspectives of Employing Mesenchymal Stem Cells from the Wharton's Jelly of the Umbilical Cord for Peripheral Nerve Repair
Jorge Ribeiro, Andrea Gartner, Tiago Pereira, Raquel Gomes, Maria Ascensão Lopes, Carolina Gonçalves, Artur Varejão, Ana Lúcia Luís, and Ana Colette Maurício

Adipose-Derived Stem Cells and Nerve Regeneration: Promises and Pitfalls
Alessandro Faroni, Giorgio Terenghi, and Adam J. Reid

The Pros and Cons of Growth Factors and Cytokines in Peripheral Axon Regeneration
Lars Klimaschewski, Barbara Hausott, and Doychin N. Angelov

Role of Inflammation and Cytokines in Peripheral Nerve Regeneration
P. Dubový, R. Jančálek, and T. Kubek

Ghrelin: A Novel Neuromuscular Recovery Promoting Factor?
Raimondo Stefania, Ronchi Giulia, Geuna Stefano, Pascal Davide, Reano Simone, Filigheddu Nicoletta, and Graziani Andrea

Neuregulin 1 Role in Schwann Cell Regulation and Potential Applications to Promote Peripheral Nerve Regeneration
Giovanna Gambarotta, Federica Fregnan, Sara Gnavi, and Isabelle Perroteau

Extracellular Matrix Components in Peripheral Nerve Regeneration
Francisco Gonzalez-Perez, Esther Udina, and Xavier Navarro

SUBJECT INDEX

Volume 109

The Use of Chitosan-Based Scaffold to Enhance Regeneration in the Nervous System
Sara Gnavi, Christina Barwig, Thomas Freier, Kirsten Haarstert-Talini, Claudia Grothe, and Stefano Geuna

Interfaces with the Peripheral Nerve for the Control of Neuroprostheses
Jaume del Valle and Xavier Navarro

The Use of Shock Waves in Peripheral Nerve Regeneration: New Perspectives?
Thomas Hausner and Antal Nógrádi

Phototherapy and Nerve Injury: Focus on Muscle Response
Shimon Rochkind, Stefano Geuna, and Asher Shainberg

Electrical Stimulation for Promoting Peripheral Nerve Regeneration
Kirsten Haastert-Talini and Claudia Grothe

Role of Physical Exercise for Improving Posttraumatic Nerve Regeneration
Paulo A.S. Armada-da-Silva, Cátia Pereira, SandraAmado, and António P. Veloso

The Role of Timing in Nerve Reconstruction
Lars B. Dahlin

Future Perspectives in Nerve Repair and Regeneration
Pierluigi Tos, Giulia Ronchi, Stefano Geuna, and Bruno Battiston

INDEX

Volume 110

The Relevance of Metals in the Pathophysiology of Neurodegeneration, Pathological Considerations
Kurt A. Jellinger

Pantothenate Kinase-Associated Neurodegeneration (PKAN) and *PLA2G6*-Associated Neurodegeneration (PLAN): Review of Two Major Neurodegeneration with Brain Iron Accumulation (NBIA) Phenotypes
Manju A. Kurian and Susan J. Hayflick

Mitochondrial Membrane Protein-Associated Neurodegeneration (MPAN)
Monika Hartig, Holger Prokisch, Thomas Meitinger, and Thomas Klopstock

BPAN: The Only X-Linked Dominant NBIA Disorder
T.B. Haack, P. Hogarth, A. Gregory, P. Prokisch, and S.J. Hayflick

Neuroferritinopathy
M.J. Keogh, C.M. Morris, and P.F. Chinnery

Aceruloplasminemia: An Update
Satoshi Kono

Therapeutic Advances in Neurodegeneration with Brain Iron Accumulation
Giovanna Zorzi and Nardo Nardocci

The Neuropathology of Neurodegeneration with Brain Iron Accumulation
Michael C. Kruer

Imaging of Iron
Petr Dusek, Monika Dezortova, and Jens Wuerfel

The Role of Iron Imaging in Huntington's Disease
S.J.A. van den Bogaard, E.M. Dumas, and R.A.C. Roos

Lysosomal Storage Disorders and Iron
Jose Miguel Bras

Manganese and the Brain
Karin Tuschl, Philippa B. Mills, and Peter T. Clayton

Update on Wilson Disease
Aggarwal Annu and Bhatt Mohit

An Update on Primary Familial Brain Calcification
R.R. Lemos, J.B.M.M. Ferreira, M.P. Keasey, and J.R.M. Oliveira

INDEX

Volume 111

History of Acupuncture Research
Yi Zhuang, Jing-jing Xing, Juan Li, Bai-Yun Zeng, and Fan-rong Liang

Effects of Acupuncture Needling with Specific Sensation on Cerebral Hemodynamics and Autonomic Nervous Activity in Humans
Kouich Takamoto, Susumu Urakawa, Kazushige Sakai, Taketoshi Ono, and Hisao Nishijo

Acupuncture Point Specificity
Jing-jing Xing, Bai-Yun Zeng, Juan Li, Yi Zhuang, and Fan-rong Liang

Acupuncture Stimulation Induces Neurogenesis in Adult Brain
Min-Ho Nam, Kwang Seok Ahn, and Seung-Hoon Choi

Acupuncture and Neurotrophin Modulation
Marzia Soligo, Stefania Lucia Nori, Virginia Protto, Fulvio Florenzano, and Luigi Manni

Acupuncture Stimulation and Neuroendocrine Regulation
Jung-Sheng Yu, Bai-Yun Zeng, and Ching-Liang Hsieh

Current Development of Acupuncture Research in Parkinson's Disease
Bai-Yun Zeng, Sarah Salvage, and Peter Jenner

Acupuncture Therapy for Stroke Patients
Xin Li and Qiang Wang

Effects of Acupuncture Therapy on Alzheimer's Disease
Bai-Yun Zeng, Sarah Salvage, and Peter Jenner

Acupuncture Therapy for Psychiatric Illness
Karen Pilkington

Acupuncture for the Treatment of Insomnia
Kaicun Zhao

Acupuncture for the Treatment of Drug Addiction
Cai-Lian Cui, Liu-Zhen Wu, and Yi-jing Li

Acupuncture Regulation of Blood Pressure: Two Decades of Research
John C. Longhurst and Stephanie Tjen-A-Looi

Effect and Mechanism of Acupuncture on Gastrointestinal Diseases
Toku Takahashi

INDEX

Volume 112

An Introduction to the Clinical Phenomenology of Tourette Syndrome
Davide Martino, Namrata Madhusudan, Panagiotis Zis, and Andrea E. Cavanna

Functional Neuroanatomy of Tics
Irene Neuner, Frank Schneider, and N. Jon Shah

Functional Imaging of Dopaminergic Neurotransmission in Tourette Syndrome
Bàrbara Segura and Antonio P. Strafella

Nondopaminergic Neurotransmission in the Pathophysiology of Tourette Syndrome
Patrick T. Udvardi, Ester Nespoli, Francesca Rizzo, Bastian Hengerer, and Andrea G. Ludolph

Reinforcement Learning and Tourette Syndrome
Stefano Palminteri and Mathias Pessiglione

Genetic Susceptibility and Neurotransmitters in Tourette Syndrome
Peristera Paschou, Thomas V. Fernandez, Frank Sharp, Gary A. Heiman, and Pieter J. Hoekstra

Pharmacological Animal Models of Tic Disorders
Kevin W. McCairn and Masaki Isoda

Animal Models Recapitulating the Multifactorial Origin of Tourette Syndrome
Simone Macrì, Martina Proietti Onori, Veit Roessner, and Giovanni Laviola

Neuroendocrine Aspects of Tourette Syndrome
Davide Martino, Antonella Macerollo, and James F. Leckman

Clinical Pharmacology of Dopamine-Modulating Agents in Tourette's Syndrome
Sabine Mogwitz, Judith Buse, Stefan Ehrlich, and Veit Roessner

Clinical Pharmacology of Nondopaminergic Drugs in Tourette Syndrome
Andreas Hartmann

Antiepileptic Drugs and Tourette Syndrome
Andrea E. Cavanna and Andrea Nani

Clinical Pharmacology of Comorbid Obsessive–Compulsive Disorder in Tourette Syndrome
Valeria Neri and Francesco Cardona

Clinical Pharmacology of Comorbid Attention Deficit Hyperactivity Disorder in Tourette Syndrome
Renata Rizzo and Mariangela Gulisano

Emerging Treatment Strategies in Tourette Syndrome: What's in the Pipeline?
C. Termine, C. Selvini, G. Rossi, and U. Balottin

Tics and Other Stereotyped Movements as Side Effects of Pharmacological Treatment
Marcos Madruga-Garrido and Pablo Mir

INDEX

Volume 113

Autism Spectrum Disorder and the Cerebellum
Esther B.E. Becker and Catherine J. Stoodley

Contribution of Long Noncoding RNAs to Autism Spectrum Disorder Risk
Brent Wilkinson and Daniel B. Campbell

Identifying Essential Cell Types and Circuits in Autism Spectrum Disorders
Susan E. Maloney, Michael A. Rieger, and Joseph D. Dougherty

Connecting Signaling Pathways Underlying Communication to ASD Vulnerability
Stephanie Lepp, Ashley Anderson, and Genevieve Konopka

MET Receptor Tyrosine Kinase as an Autism Genetic Risk Factor
Yun Peng, Matthew Huentelman, Christopher Smith, and Shenfeng Qiu

Transcriptional Dysregulation of Neocortical Circuit Assembly in ASD
Kenneth Y. Kwan

Motor Skill in Autism Spectrum Disorders: A Subcortical View
Leanne Chukoskie, Jeanne Townsend, and Marissa Westerfield

Orchestration of Neurodevelopmental Programs by RBFOX1: Implications for Autism Spectrum Disorder
Brent R. Bill, Jennifer K. Lowe, Christina T. DyBuncio, and Brent L. Fogel

Immune Dysregulation in Autism Spectrum Disorder
Elaine Y. Hsiao

Autism Susceptibility Genes and the Transcriptional Landscape of the Human Brain
Shingo Miyauchi and Irina Voineagu

Index